European Yearbook of International Economic Law

EYIEL Monographs - Studies in European and International Economic Law

Volume 41

EYIEL Monographs is a subseries of the European Yearbook of International Economic Law (EYIEL). It contains scholarly works in the fields of European and international economic law, in particular WTO law, international investment law, international monetary law, law of regional economic integration, external trade law of the EU and EU internal market law. The series does not include edited volumes. EYIEL Monographs are peer-reviewed by the series editors and external reviewers.

Xinyan Zhao

Integrating the UN SDGs into WTO Law

Springer Open

Xinyan Zhao
School of Law
Durham University
Durham, UK

ISSN 2364-8392 ISSN 2364-8406 (electronic)
European Yearbook of International Economic Law
ISSN 2524-6658 ISSN 2524-6666 (electronic)
EYIEL Monographs - Studies in European and International Economic Law
ISBN 978-3-031-73875-3 ISBN 978-3-031-73876-0 (eBook)
https://doi.org/10.1007/978-3-031-73876-0

This work was supported by Schweizerischer Nationalfonds zur Förderung der Wissenschaftlichen Forschung 10BP12_229815 / 1

This Springer imprint is published by the registered company Springer Nature Switzerland AG
The registered company address is: Gewerbestrasse 11, 6330 Cham, Switzerland

Dedicated to my parents, with gratitude for their constant love and support, to my supervisor for professional supervision, and to the University of Lausanne, especially the Faculty of Law, as well as the Swiss Institute of Comparative Law and the University of Montreal, for providing an enabling environment for academic research.

—Xinyan Zhao

Acknowledgements

This monograph explores the integration of the United Nations (UN) Sustainable Development Goals (SDGs) into World Trade Organisation (WTO) law, aiming to safeguard environmental and social interests from the negative externalities of international trade. It is based on research conducted during my doctoral studies at the University of Lausanne from 2018 to 2022. My initial motivation for this research stemmed from my concern about environmental pollution in my hometown of Beijing. My supervisor, Professor Andreas R. Ziegler, is interested in interdisciplinary research on trade and environmental and social issues. He encouraged and supported me throughout this journey, introducing me to the theory of global constitutionalism, which provided valuable insights into the nexus of trade and sustainable development.

The inclusion of sustainability values in WTO rules is often perceived as an unattainable utopian goal. Throughout my doctoral research, Professor Ziegler offered unwavering moral support and encouragement, enabling me to pursue my academic ambitions so that I could walk the long road without being alone. I am deeply grateful to him for allowing me to explore my ideas and for his professional guidance.

I extend my heartfelt thanks to Professor Andreas R. Ziegler, Professor Ernst-Ulrich Petersmann, Professor Belén Olmos Giupponi, Professor Damiano Canapa, and Professor Francesco Maiani for their invaluable comments on this work. It has been an honour to receive guidance from such esteemed scholars. Their constructive feedback and critiques have been instrumental in completing this monograph. In particular, I would like to thank Professors Petersmann, Giupponi and Ziegler. Their pioneering work on global constitutionalism, international trade, and sustainable development profoundly inspired me.

The endless support of my family in my life was also essential for me to complete this monograph. I am deeply grateful to my mother, Mrs. Meizhen Liu, and my father, Mr. Hongbin Zhao, whose constant love and support have sustained me through four years of research. I sincerely appreciate their understanding and support of my personal choices. Moreover, I am indebted to many friends and colleagues for their understanding and moral support.

Durham, Winter 2024

Xinyan Zhao

Contents

List of Abbreviations

AAAA	Addis Ababa Action Agenda
AB	Appellate Body
ACHPR	African Charter on Human and Peoples' Rights
ACtHPR	African Court on Human and Peoples' Rights
ADA	Antidumping Agreement
ALOP	Appropriate Level of Protection
AOA	Agreement on Agriculture
ASCM	Agreement on Subsidies and Countervailing Measures
ATC	Agreement on Textiles and Clothing
AUKUS	Australia, the United Kingdom and the United States Pact
CBAM	Carbon Border Adjustment Mechanism
CETA	EU-Canada Comprehensive Economic and Trade Agreement
CJEU	Court of Justice of the European Union
CO_{2e}	Greenhouse Gas Embedded in Their Importing Goods
CPTPP	Comprehensive and Progressive Agreement for Trans-Pacific Partnership
CSOs	Civil Society Organisations
CSR	Corporate Social Responsibility
DSB	Dispute Settlement Body
DSU	Understanding on Rules and Procedures Governing the Settlement of Disputes
EC	European Communities
ECJ	European Court of Justice
ECtHR	European Court of Human Rights
EEC	European Economic Community
EKC	Environmental Kuznets Curve

EMIT	GATT Group on Environmental Measures and International Trade
EU	European Union
EUGS	European Union's Global Strategy
FAO	Food and Agriculture Organisation
FTA	Free Trade Agreement
G20	Group of Twenty
GATS	General Agreement on Trade in Services
GATT	General Agreement on Tariffs and Trade
GC	General Council
GHG	Greenhouse Gas
Goal 1 'No Poverty'	Goal 1 End Poverty in All Its Forms Everywhere
Goal 2 'Zero Hunger'	Goal 2 End Hunger, Achieve Food Security and Improved Nutrition and Promote Sustainable Agriculture
Goal 3 'Good Health and Well-being'	Goal 3 Ensure Healthy Lives and Promote Well-Being for All at All Ages
Goal 4 'Quality Education'	Goal 4 Ensure Inclusive and Equitable Quality Education and Promote Life-Long Learning Opportunities for All
Goal 5 'Gender Equality'	Goal 5 Achieve Gender Equality and Empower All Women and Girls
Goal 6 'Clean Water and Sanitation'	Goal 6 Ensure Availability and Sustainable Management of Water and Sanitation for All
Goal 7 'Affordable and Clean Energy'	Goal 7 Ensure Access to Affordable, Reliable, Sustainable and Modern Energy for All
Goal 8 'Decent Work and Economic Growth'	Goal 8 Promote Sustained, Inclusive and Sustainable Economic Growth, Full and Productive Employment and Decent Work for All
Goal 9 'Industry, Innovation and Infrastructure'	Goal 9 Build Resilient Infrastructure, Promote Inclusive and Sustainable Industrialisation and Foster Innovation
Goal 10 'Reduced Inequalities'	Goal 10 Reduce Inequality within and among Countries
Goal 11 'Sustainable Cities and Communities'	Goal 11 Make Cities and Human Settlements Inclusive, Safe, Resilient and Sustainable
Goal 12 'Responsible Consumption and Production'	Goal 12 Ensure Sustainable Consumption and Production Patterns
Goal 13 'Climate Action'	Goal 13 Take Urgent Action to Combat Climate Change and Its Impacts
Goal 14 'Life below Water'	Goal 14 Conserve and Sustainably Use the Oceans, Seas and Marine Resources for Sustainable Development

Goal 15 'Life on Land'	Goal 15 Protect, Restore and Promote Sustainable Use of Terrestrial Ecosystems, Sustainably Manage Forests, Combat Desertification, and Halt and Reverse Land Degradation and Halt Biodiversity Loss
Goal 16 'Peace, Justice and Strong Institutions'	Goal 16 Achieve Peaceful and Inclusive Societies, Rule of Law, Effective and Capable Institutions
Goal 17 'Partnerships for the Goals'	Goal 17 Strengthen Means of Implementation and Revitalise the Global Partnership for Sustainable Development
GSP	Generalised Scheme of Preferences
IACtHR	Inter-American Court of Human Rights
ICC	International Criminal Court
ICCPR	International Covenant on Civil and Political Rights
ICESCR	International Covenant on Economic, Social and Cultural Rights
ICJ	International Court of Justice
ICTs	International Criminal Tribunals
ICTY	International Criminal Tribunal for the Former Yugoslavia
IEL	International Economic Law
ILO	International Labour Organisation
IPCC	Intergovernmental Panel on Climate Change
IPRs	Intellectual Property Rights
ISMPs	International Standards for Phytosanitary Measures
LGBTQ+	Lesbian, Gay, Bisexual and Transgender+
LON	League of Nations
MC	Ministerial Conference
MDGs	Millennium Development Goals
MEAs	Multilateral Environmental Agreements
MFN	Most-Favoured-Nation
MSMEs	Micro Small and Medium Enterprises
NAFTA	North American Free Trade Agreement
NATO	North Atlantic Treaty Organisation
NT	National Treatment
OECD	Organisation for Economic Cooperation and Development
PEDP	Principles of the European Development Policy
RCEP	Regional Comprehensive Economic Partnership
SCM Agreement	Agreement on Subsidies and Countervailing Measures
SDGs	Sustainable Development Goals
SDIs	Sustainable Development Issues
SDT	Special and Differential Treatment
SGA	The Agreement on Safeguards
SPC	Supplementary Protection Certificates

SPS	Sanitary and Phytosanitary
TBT	Technical Barriers to Trade
TESSD	Trade and Environmental Sustainability Structured Discussions
TEU	Treaty on European Union
TPP	Trans-Pacific Partnership
TRIMs Agreement	Agreement on Trade-Related Investment Measures
TRIPs Agreement	Agreement on Trade-Related Aspects of Intellectual Property Rights
TSDA	Trade and Sustainable Development Agreement on Establishing the Sustainable Development Club
TSDCs	Trade-Related Sustainable Development Commitments
UDHR	Universal Declaration of Human Rights
UK	United Kingdom
UN	United Nations
UNFCCC	United Nations Framework Convention on Climate Change
US	United States
USMCA	United States-Mexico-Canada Agreement on Trade
VCLT	Vienna Convention on Law of Treaties
WHO	World Health Organisation
WIPO	World Intellectual Property Organisation
WTO	World Trade Organisation
WWII	Second World War

Chapter 1
Introduction

Abstract The concept of development has evolved to address a wide range of human needs, including poverty eradication, human rights protection, and environmental conservation. Following WWII, the UN and scientific progress shaped the concept of sustainable development, culminating in the 2015 Sustainable Development Goals. However, the implementation of the 17 UN SDGs is currently hindered by the non-binding nature of international law and the lack of an international judicial system for enforcement. Although the WTO's role in influencing domestic legislation could potentially advance SDGs, current WTO rules and practices are inadequate for ensuring comprehensive SDG implementation. Efforts to integrate sustainability standards into WTO law, such as the US proposal for actionable subsidies, have not gained widespread support. While WTO provisions like GATT Article XX provide some flexibility for sustainable practices, the overall framework needs reform. This monograph argues for applying a 'sustainability test' by WTO panels to align trade practices with the SDGs and examines potential reforms to integrate these changes into WTO rules through a sustainable development club.

1.1 Setting the Scene

Development has long been a central concern for humanity, though its meaning has evolved across different historical contexts. The term 'development' carries diverse connotations depending on the challenges and aspirations of individuals and societies. For those grappling with hunger and poverty, development signifies the eradication of these fundamental issues. For communities affected by natural disasters, it involves enhancing environmental conditions and resilience. For individuals deprived of freedom and dignity, development equates to the fulfilment of human rights and the promotion of personal freedoms. Thus, development encompasses a broad spectrum of needs and goals.

Therefore, it is insufficient to achieve progress in just one area; people must simultaneously have access to wealth, food, a healthy environment, and the right to pursue these necessities. In an era of globalisation, the interactions between people

X. Zhao, *Integrating the UN SDGs into WTO Law*, EYIEL Monographs - Studies in European and International Economic Law 41,
https://doi.org/10.1007/978-3-031-73876-0_1

from different regions have made development issues extremely complex. Moreover, many development matters have evolved into global governance challenges due to the close interactions between regulators. For instance, the economic policies of one country can significantly impact the economic success or failure of others, and the environmental policies of one state can influence global climate governance. Additionally, human rights issues have transcended domestic boundaries, affecting international relations profoundly. Consequently, development issues must be addressed comprehensively and globally.

Development issues have garnered increasing attention from academics since the end of the Second World War (WWII). Post-WWII developments significantly contributed to the evolution of the concept of sustainable development. Key factors included the establishment of the United Nations (UN) and its emphasis on human rights, the decolonisation of Asia and Africa, and advancements in scientific research. Academics began to consider various development issues holistically, gradually formulating theories to guide human development.

The establishment of the UN and the decolonisation of Asia and Africa significantly advanced the right to development. Following the First World War, the League of Nations (LON) was created to maintain global peace. However, the outbreak of the Second World War revealed the LON's ineffectiveness in regulating the behaviour of nations and preventing conflicts. Instead, global superpowers dominated international relations and engaged in invasions. To enhance the international rule of law and prevent future wars, the victorious nations of WWII founded the UN in 1945. Unlike the LON, the UN quickly grew to represent a far broader array of states. Many new member states were those that had gained independence from colonial powers after WWII, and a significant number were developing or least-developed countries.

According to the UN Charter, the United Nations consists of a General Assembly, a Security Council, and several specialised agencies.[1] Through these bodies, UN members can actively engage in the decision-making process. The creation of the UN enabled both developed and developing countries to participate in international rulemaking and protect their national interests. For many developing nations, eradicating hunger and poverty is of paramount importance. Consequently, agriculture and economic growth in developing and least-developed countries became central topics in international negotiations. Over time, the international community has gradually reached a consensus on the right to development.[2]

Second, advances in scientific research in the 1950s led to a growing consensus on the importance of environmental protection. During this era, scientists began to recognise the existence of climate change and warned about its potential to cause numerous crises for humanity.[3] Research indicated that economic activities were

[1] UN Charter (26 June 1945), Article 7.

[2] Declaration on the Right to Development, UNGA Res 41/128 (4 December 1986).

[3] Landsberg (1958), p. 749.

increasing greenhouse gas (GHG) emissions,[4] leading to catastrophic consequences such as altered climate patterns and devastating weather events.[5] In response to these findings, UN members have been working since the late 1980s to develop international rules for climate change adaptation and mitigation. This effort has resulted in significant agreements, including the United Nations Framework Convention on Climate Change (UNFCCC), the Kyoto Protocol, the Paris Agreement on Climate Change, and fifteen other global multilateral environmental agreements (MEAs). These agreements represent a collective effort by the international community to address and mitigate the impacts of climate change.[6]

Third, the shifts in international politics following WWII significantly advanced the pursuit of human rights, transforming them into universal values.[7] The aftermath of the war saw substantial developments in human rights regulations. Not only was the protection of human rights enshrined in the preamble of the UN Charter (1945), but it also became one of the UN's core missions.[8] Between 1947 and 1991, UN members adopted seven of the nine core international human rights instruments.[9] Among these, the Universal Declaration of Human Rights (UDHR), the International Covenant on Economic, Social and Cultural Rights (ICESCR), and the International Covenant on Civil and Political Rights (ICCPR) established fundamental principles and protections for human rights.

On December 10, 1948, the Universal Declaration of Human Rights (UDHR) was adopted by the UN General Assembly with 48 votes in favour, none against, and eight abstentions.[10] The International Covenant on Economic, Social and Cultural Rights (ICESCR) and the International Covenant on Civil and Political Rights (ICCPR) were subsequently adopted by UN members in 1966.[11] Nearly all UN member states have signed and ratified these three key human rights instruments. Among the five permanent members of the UN Security Council, only China and the United States (US) have not fully committed to all three instruments. Specifically, China has signed but not ratified the ICCPR, while the United States has signed but not ratified the ICESCR.[12]

---

[4] Plass Gilbert (1955), p. 140; Wiseman (1954), p. 296; See also Pachauri and Meyer (2014), p. 151.

[5] Eggleton (2013).

[6] UNEP, Global Multilateral Environmental Agreements (MEAs). https://www.unep.org/about-un-environment/why-does-un-environment-matter/secretariats-and-conventions. Accessed 21 July 2024.

[7] See Navanethem Pillay, 'Are Human Rights Universal?' (UN Chronicle). https://www.un.org/en/chronicle/article/are-human-rights-universal. Accessed 21 July 2024.

[8] UN Charter (26 June 1945), Article 1; See also Freedman (2013), p. 14.

[9] See UN, The Core International Human Rights Instruments and Their Monitoring Bodies. https://www.ohchr.org/en/professionalinterest/pages/coreinstruments.aspx. Accessed 21 July 2024.

[10] See Drafting of the Universal Declaration of Human Rights (UN). https://research.un.org/en/undhr/ga/plenary. Accessed 21 July 2024.

[11] See UN, The Core International Human Rights Instruments and Their Monitoring Bodies. https://www.ohchr.org/en/professionalinterest/pages/coreinstruments.aspx. Accessed 21 July 2024.

[12] See UN Treaty Body Database (2023b); See UN Treaty Body Database (2023a).

Thus, when UN members began addressing sustainable development in the late 1980s, their primary focus was economic growth, environmental protection, and social rights. In 1983, the UN established the Brundtland Commission to prepare for international negotiations on sustainable development.[13] The Commission, in its 1987 report, known as the Brundtland Report, defined sustainable development as encompassing economic, environmental, and social dimensions.[14] This definition underscores that each dimension holds equal importance, advocating for a balanced approach to sustainable development.[15] In other words, achieving sustainable development requires a country to simultaneously foster economic growth, protect the environment, and uphold the social rights of its citizens. This foundational definition has remained consistent despite ongoing negotiations among UN members since 1987.

In 2000, UN members adopted the United Nations Millennium Declaration, a pivotal initiative that set detailed goals for sustainable development.[16] This declaration marked the first time the UN established specific targets with deadlines for achieving sustainable development objectives. It outlined eight Millennium Development Goals (MDGs) to be accomplished by 2015: MDG 1, 'Eradicate Extreme Poverty and Hunger', MDG 2, 'Achieve Universal Primary Education', MDG 3, 'Promote Gender Equality and Empower Women', MDG 4, 'Reduce Child Mortality', MDG 5, 'Improve Maternal Health', MDG 6, 'Combat HIV/AIDS, Malaria and Other Diseases', MDG 7, 'Ensure Environmental Sustainability', and the MDG 8, 'Global Partnership for Development'.[17]

Following the adoption of the Millennium Declaration, UN members adopted a report from the Secretary-General titled 'Road Map towards the Implementation of the United Nations Millennium Declaration.' This report aimed to guide the implementation of the eight MDGs.[18] However, while the report provided valuable guidelines, it did not establish enforceable measures for implementation. Consequently, the MDGs remained voluntary standards rather than legally binding commitments. As a result, by the 2015 deadline, UN members fell short of fully achieving these goals.

In response to the shortcomings in achieving the MDGs, UN members adopted the UN 2030 Agenda in 2015. This agenda sets out to realise 17 Sustainable Development Goals (SDGs) by 2030. These goals encompass a broad range of objectives: SDG 1, 'No Poverty', SDG 2, 'Zero Hunger', SDG 3, 'Good Health and Well-being', SDG 4, 'Quality Education', SDG 5, 'Gender Equality', SDG

---

[13] UNGA (1987).

[14] Ibid., p. 6.

[15] Ibid., p. 37.

[16] UNGA (2000).

[17] See UN Documentation: Development (UN) https://research.un.org/en/docs/dev/2000-2015. Accessed 21 July 2024.

[18] UNGA (2001).

6, 'Clean Water and Sanitation', SDG 7, 'Affordable and Clean Energy', SDG 8, 'Decent Work and Economic Growth', SDG 9, 'Industry, Innovation and Infrastructure', SDG 10, 'Reduced Inequalities', SDG 11, 'Sustainable Cities and Communities', SDG 12, 'Responsible Consumption and Production', SDG 13, 'Climate Action', SDG 14, 'Life Below Water', SDG 15, 'Life on Land', SDG16, 'Peace, Justice and Strong Institutions', and SDG 17, 'Partnerships for the Goals'.[19]

The 2030 Agenda represents a significant evolution from the Millennium Declaration, functioning as the UN's second 15-year plan for advancing sustainable development. On the one hand, the 2030 Agenda incorporates and refines the eight MDGs. For instance, MDG 1, which aimed to 'Eradicate Extreme Poverty and Hunger,' was split into SDG 1, 'No Poverty,' and SDG 2, 'Zero Hunger.' MDGs 4, 5, and 6, which focused on 'Reduce Child Mortality,' 'Improve Maternal Health,' and 'Combat HIV/AIDS, Malaria, and Other Diseases,' respectively, were consolidated into SDG 3, 'Good Health and Well-Being.' Similarly, MDGs 2, 3, and 8—concerning 'Achieve Universal Primary Education,' 'Promote Gender Equality and Empower Women,' and 'Global Partnership for Development'—were transformed into SDG 4, 'Quality Education,' SDG 5, 'Gender Equality,' and SDG 17, 'Partnerships for the Goals.' Furthermore, MDG 7, which aimed for 'Environmental Sustainability,' was replaced by three distinct SDGs addressing various aspects of environmental protection: SDG 6, 'Clean Water and Sanitation,' SDG 7, 'Affordable and Clean Energy,' and SDG 13, 'Climate Action.'

On the other hand, the 2030 Agenda has expanded both in number and scope, introducing eight new sustainable development goals. These include SDG 8, 'Decent Work and Economic Growth'; SDG 9, 'Industry, Innovation, and Infrastructure'; SDG 10, 'Reduced Inequalities'; SDG 11, 'Sustainable Cities and Communities'; SDG 12, 'Responsible Consumption and Production'; SDG 14, 'Life Below Water'; SDG 15, 'Life on Land'; and SDG 16, 'Peace, Justice, and Strong Institutions.' This expansion means that the 2030 Agenda now addresses a broader range of environmental and social dimensions than its predecessor.

Recognising that the lack of implementation led to the failure of the Millennium Declaration,[20] UN members emphasised the importance of effective implementation when drafting the UN 2030 Agenda. To support the realisation of the 17 SDGs, they adopted the Addis Ababa Action Agenda on Financing for Development (AAAA) in 2015.[21] This agenda focuses on facilitating investment in eight key areas essential for achieving the SDGs: (1) domestic public resources[22]; (2) domestic and international private business and finance[23]; (3) international

[19] UNGA (2015b).

[20] Fehling et al. (2013), p. 1109.

[21] UNGA (2015a).

[22] Ibid., 10.

[23] Ibid., 17.

development cooperation[24]; (4) international trade[25]; (5) debt sustainability[26]; (6) systemic issues[27]; (7) science, technology, innovation, and capacity-building[28]; and (8) data, monitoring, and follow-up.[29] The AAAA outlines over 100 specific measures designed to finance sustainable development, transform the global economy, and achieve the 17 SDGs. It also establishes a new framework to ensure stable and sustainable financial support for countries implementing the SDGs.[30]

Compared to the Millennium Declaration, the UN 2030 Agenda offers a broader policy framework for implementing sustainability standards. It not only encourages UN members to pursue a more comprehensive set of SDGs but also provides increased financial support to aid their achievement. Despite these advancements, a significant limitation of the UN 2030 Agenda is the absence of legally binding commitments. This lack of enforceability means that UN members are not obligated to implement the 17 SDGs, potentially allowing them to opt out of fulfilling these crucial sustainability targets.

1.2 The World Trade Organisation and the SDGs

It is imperative to move the 17 UN SDGs from soft law commitments to enforceable standards. However, international law in this field is characterised by non-binding rules.[31] Unlike domestic judicial systems, there is no international judicial system capable of ensuring the enforcement of the 17 UN SDGs. For example, the enforcement of MEAs relies on a voluntary arbitration process, which can only be invoked if both disputing parties notify the MEAs Secretariat of their acceptance of this dispute resolution method.[32] Consequently, even if a country refuses to implement environmental standards, it will not face sanctions under international law. Given this context, how can the implementation of the 17 UN SDGs be ensured?

It is argued that international trade rules significantly influence the domestic legislation of countries.[33] By modifying these rules, a coalition of countries can prompt others to enact and implement domestic laws or regulations promoting sustainable development, thereby ensuring the implementation of the SDGs through

[24] Ibid., 26.

[25] Ibid., 37.

[26] Ibid., 43.

[27] Ibid., 46–47.

[28] Ibid., 51.

[29] Ibid., 63–68.

[30] UN Department of Economic and Social Affairs (2015).

[31] Janoušková et al. (2018), p. 1540.

[32] Rose (2011), p. 9.

[33] Wang (2011).

their domestic judicial systems. For example, if international trade rules mandate that a state must phase out polluting industries to participate in the global market, the state will need to establish and enforce domestic laws and regulations, such as environmental laws, to ensure that local businesses produce goods and provide services in an environmentally friendly manner. Given that the World Trade Organisation (WTO) is the primary international body where states negotiate trade rules and regulate trade practices, it has a crucial role in enhancing the implementation of SDGs. I contend that leveraging WTO law to improve the enforcement of these SDGs can be a viable solution for three reasons.

The first reason is that the WTO is committed to supporting the UN in its initiatives. Although the WTO is not a specialised UN agency, it maintains a close working relationship with the UN. This collaboration began with the UN Conference on the Human Environment in Stockholm in 1972. During this conference, the Secretariat of the General Agreement on Tariffs and Trade (GATT), the WTO's predecessor, presented its first environment-related report, which explained the implications of environmental protection policies on international trade. That same year, the GATT Council of Representatives established a group focused on environmental measures and international trade (the EMIT group). Although the EMIT group was inactive for 20 years, its creation signalled a commitment to incorporating environmental considerations into GATT rules. This commitment was further reflected in the development of GATT rules on sanitary and phytosanitary (SPS) measures and technical regulation measures in the late 1980s and early 1990s. These developments built sufficient momentum for a more significant evolution in integrating environmental issues into international trade regulations.

Upon its establishment, the WTO included an independent committee on trade and environment, as well as the SPS and TBT Agreements. This new multilateral trading system marked significant progress in regulating environment-related trade issues at both institutional and normative levels. Immediately following the foundation of the WTO, its members adopted a communication titled 'Arrangements for Effective Cooperation with Other Intergovernmental Organisations' to govern the WTO-UN relationship.[34] This communication stipulates that the WTO and the UN should collaborate to ensure the efficient functioning of both organisations.[35] Given that sustainable development is one of the UN's core missions, the WTO should support the UN in enhancing the implementation of the 17 UN SDGs. In fact, ongoing WTO negotiations include many sustainable development issues (SDIs), including economic growth, gender equality, the protection of sea animals, solid waste management, and climate change adaptation and mitigation.[36]

The second reason is that the WTO has a Dispute Settlement Body (DSB) with jurisdiction over disputes related to SDIs. According to the Marrakesh Agreement

---

[34] Most WTO members are also UN members.

[35] WTO General Council (1995).

[36] See WTO, The WTO and the Sustainable Development Goals. https://www.wto.org/english/thewto_e/coher_e/sdgs_e/sdgs_e.htm. Accessed 21 July 2024.

establishing the WTO, sustainable development is one of the organisation's fundamental objectives.[37] Therefore, WTO panels and the Appellate Body (AB) are obligated to protect interests related to sustainable development through dispute settlement proceedings. Indeed, WTO panels and the AB have resolved numerous disputes concerning SDIs, including those related to public health, animal life, and environmental protection.[38]

The third reason is that the rulings and recommendations of WTO panels and the AB are legally enforceable. Once these bodies determine that a disputed measure is inconsistent with WTO law, they recommend that the DSB request the responding party to bring its measure into conformity with WTO obligations.[39] The final report of the AB must be adopted by the DSB and accepted by the disputing parties unless all WTO member representatives present at the DSB meeting, including the representative of the member that won the dispute, unanimously reject the AB report.[40] If the responding party ultimately fails to implement the rulings and recommendations of the WTO panels and the AB, the complaining party has the right to retaliate.[41] This means the complaining party can impose countermeasures to enforce the rulings and recommendations, such as suspending trade concessions or imposing trade barriers against the responding party. While such countermeasures may not compel the responding party to amend its domestic laws or regulations, they help the complaining party mitigate trade losses. In this way, the rulings and recommendations of WTO panels and the AB are effectively legally enforceable.

The three reasons outlined above suggest that WTO law can assist the UN in strengthening the implementation of the SDGs. Ideally, the WTO Agreements should include provisions on sustainable development commitments that establish clear sustainability standards. For example, take SDG 8, 'Decent Work and Economic Growth.' Relevant sustainability standards could specify exact working hours and guidelines for calculating minimum wage rates for workers. WTO members should commit to ensuring that their domestic laws or regulations do not set sustainability standards lower than those established by the WTO. Otherwise, domestic regulatory measures (e.g., labour laws) could undermine the level playing field of trade markets, as imported goods would bear higher costs than those manufactured domestically.

Violations of potential WTO sustainability standards could be considered actionable subsidies. If WTO panels determine that a member's domestic regulatory measure constitutes such a subsidy, they will recommend that the DSB request the member to bring its measure into conformity with WTO commitments. Should the member fail to implement the rulings and recommendations, it will face the possibility of retaliation. Other WTO members can impose countermeasures on imports

---

[37] The Marrakesh Agreement Establishing the World Trade Organization (15 April 1994), Preamble.

[38] Sampson (2008); Reid (2015).

[39] The Understanding on Rules and Procedures Governing the Settlement of Disputes Article 19.

[40] Ibid., Article 17.

[41] Ibid., Article 22.

from non-compliant members. In this way, WTO law plays a crucial role in reinforcing the implementation of the 17 UN SDGs.

There is already a proposal regarding the enforcement of sustainability standards. On December 17, 2020, the US proposed that the failure of a government to adopt, maintain, implement, and effectively enforce laws and regulations ensuring environmental protection at or above a threshold of fundamental standards should be considered an actionable subsidy under the Agreement on Subsidies and Countervailing Measures (ASCM). According to this proposal, if an industry disproportionately benefits from pollution controls or other environmental measures that fall below these fundamental standards, a WTO member could impose a countervailing duty equivalent to the benefit received by the industry when its goods enter the importing member's customs territory.[42] However, this proposal has not yet garnered support from other WTO members. Indeed, sustainability standards have not been incorporated into WTO members' commitments, making it premature for them to agree on such enforcement measures.

To date, the implementation of sustainability standards largely depends on the domestic laws and regulations of WTO members. The WTO Agreements include provisions that provide policy space for such domestic regulatory measures. For instance, GATT Articles II and III acknowledge the sovereign rights of WTO members to enact and enforce national regulations and laws. Among these provisions, GATT Article XX is particularly significant as it serves as the general exception clause of the WTO Agreements.

Article XX contains ten subparagraphs that allow WTO members to deviate from their trade commitments under certain conditions. Subparagraph (b) permits members to adopt domestic regulations aimed at protecting human, animal, or plant life or health. Subparagraph (g) allows for measures to conserve exhaustible natural resources. These provisions enable WTO members to implement measures related to SDG 3, 'Good Health and Well-Being,' SDG 7, 'Affordable and Clean Energy,' and SDG 13, 'Climate Action.'

Furthermore, WTO members can also draw on general protections under subparagraphs (a) and (d), which address the protection of public morals and the enforcement of domestic laws and regulations. Given the broad scope of public morals and domestic laws, these subparagraphs can be interpreted expansively, allowing members to enact and enforce laws or regulations related to the implementation of trade-related SDGs.

The only caveat is that the chapeau of GATT Article XX imposes two requirements on the domestic regulatory measures of WTO members. First, these measures should not constitute arbitrary or unjustifiable discrimination between countries where the same conditions prevail. Second, they should not constitute a disguised restriction on international trade.[43] WTO members can enact and enforce sustainability measures, provided they comply with these two requirements. Hence, GATT

[42] WTO General Council (2020).

[43] Marceau and Wyatt (2009), p. 228.

Article XX provides a legal basis for defending the legality of WTO members' domestic regulations on sustainable development.[44]

Moreover, major WTO Agreements include the General Agreement on Trade in Services (GATS) and several agreements that elaborate rules for applying GATT provisions, such as the Agreement on the Application of Sanitary and Phytosanitary Measures (the SPS Agreement) and the Agreement on Technical Barriers to Trade (the TBT Agreement). WTO panels *mutatis mutandis* apply GATT Article XX's analytical framework when assessing the legality of domestic regulatory measures covered under these Agreements.

Given the role of WTO provisions—particularly GATT Article XX—in advancing the UN SDGs, some scholars argue that WTO law has allowed the international trading system to evolve into a more balanced framework. This framework acknowledges both contemporary priorities, such as achieving the 17 UN SDGs, and its trade-focused objectives.[45] Does this suggest that WTO law has enabled the WTO dispute settlement system to support the UN in enhancing the implementation of SDGs? This monograph addresses this question. It concludes that, currently, WTO law hinders the effective implementation of the UN SDGs. The analysis suggests that WTO panels should incorporate a sustainability test to ensure that international trade practices align with sustainable development goals. Furthermore, this monograph explores how WTO members can reform WTO rules to provide a legal basis for such a test.

1.3 The Structure of the Book

This book is structured as follows. Part I, consisting of Chaps. 2 and 3, explains why WTO law impedes the implementation of trade-related SDGs. Chapter 2 analyses the trade-related sustainable development commitments (TSDCs) contained in the WTO Agreements. This Chapter helps understand WTO rules' limitations in promoting sustainable trade. It also explains why WTO members rely on domestic laws and regulations to implement SDGs. Chapter 3 analyses how WTO panels assess the legitimacy of WTO members' domestic sustainability measures. This Chapter systematically examines WTO panels' standard of review and analytical frameworks. It describes how WTO rules have limited the policy space for WTO members' domestic regulation.

Part II consists of Chaps. 4 and 5, in which I propose the legal approach that WTO panels should use to review the legitimacy of WTO members' domestic sustainability measures. I argue that panels should use the sustainability test. Chapter 4 defines this test. Chapter 5 specifies how panels should apply the test to address disputes related to sustainable development.

---

[44] Hartwick and Peet (2003), p. 188.

[45] Marceau and Wyatt (2009), p. 235.

Part III comprises Chaps. 6 through 10, which examine how WTO members can integrate the legal basis of the sustainability test into WTO rules. Chapter 6 introduces the theory of the WTO's constitutionalisation. It contends that the WTO's constitutionalisation is a means of reforming WTO rules and integrating the UN SDGs into WTO law. The Chapter also discusses the theoretical challenges to the WTO's constitutionalisation. Chapter 7 analyses the ongoing WTO plurilateral negotiations on sustainability. It reveals the difficulties facing the WTO's constitutionalisation in reality. Chapter 8 discusses a pragmatic solution to constitutionalise the WTO, namely, establishing a sustainable development club to reform WTO rules. This Chapter describes the idea of the sustainable development club in detail, discusses its legal issues, and looks at its prospects. Chapter 9 sets out the potential trade rules and policies of the sustainable development club. It helps understand the rules based on which WTO members should establish the sustainable development club and advance the subsequent reform of international trade rules. Chapter 10 suggests the next steps for WTO members. It explains specifically how WTO members implement the solutions proposed in this monograph. Finally, the book ends with a short conclusion.

References

Bibliography

Eggleton RA (2013) A short introduction to climate change. CUP, New York

Fehling M et al (2013) Limitations of the millennium development goals: a literature review. Glob Public Health 8(10):1109–1122

Freedman R (2013) The United Nations human rights council. Routledge, New York

Hartwick E, Peet R (2003) Rethinking sustainable development: neoliberalism and nature: the case of the WTO. Ann Am Acad Pol Soc Sci 590(3):188–211

Janoušková S et al (2018) Global SDGs assessments: helping or confusing indicators? Sustain For 10(5):1540–1554

Landsberg HE (1958) Trends in climatology. Science 128(3327):749–758

Marceau GZ, Wyatt JG (2009) Trade and the environment: the WTO's efforts to balance economic and sustainable development. In: Trindade RT et al (eds) Économie, environment, éthique: de la responsabilité sociale et sociétale: Liber amicorum Anne Petitpierre-Sauvain. Schulthess, Zürich, pp 225–235

Pachauri RK, Meyer LA (eds) (2014) Climate Change 2014. In: Synthesis Report. Contribution of Working Groups I, II and III to the Fifth Assessment Report of the Intergovernmental Panel on Climate Change. Available via https://www.ipcc.ch/report/ar5/syr/. Accessed 21 July 2024

Plass Gilbert N (1955) The carbon dioxide theory of climate change. Tellus 8(2):140–154

Reid E (2015) Balancing human rights, environmental protection and international trade: lessons from the EU experience. Hart, Oxford

Rose G (2011) Interlinkages between multi-lateral environmental agreements: international compliance cooperation. In: Paddock LC et al. (eds) Compliance and enforcement in environmental law: toward more effective implementation. Edward Elgar, Cheltenham, pp 3–34

Sampson GP (2008) The WTO and sustainable development. United Nations University Press, Tokyo
Wang G (2011) Radiating impact of WTO on its members' legal system: the Chinese perspective. Hague Academy of International Law, The Hague
Wiseman JDH (1954) The determination and significance of past temperature changes in the upper layer of the equatorial Atlantic Ocean. Proc R Soc Lond A Math Phys Sci 222(1150):296–323

Documents

UN Department of Economic and Social Affairs (2015) A DESA Briefing Note on the Addis Ababa Action Agenda. Available via https://www.un.org/esa/ffd/ffd3/wp-content/uploads/sites/2/2015/07/DESA-Briefing-Note-Addis-Action-Agenda.pdf. Accessed 21 July 2024
UN Treaty Body Database (2023a) Ratification Status for China https://tbinternet.ohchr.org/_layouts/15/TreatyBodyExternal/Treaty.aspx?CountryID=36&Lang=EN. Accessed 21 July 2024
UN Treaty Body Database (2023b) Ratification Status for United States of America. https://tbinternet.ohchr.org/_layouts/15/TreatyBodyExternal/Treaty.aspx?CountryID=187&Lang=EN. Accessed 21 July 2024
UNGA (1986) Declaration on the Right to Development. UNGA Res 41/128. 4 December 1986
UNGA (1987) Report of the World Commission on Environment and Development: Our Common Future. UNGA, UN Doc A/42/427
UNGA (2000) United Nations Millennium Declaration. UNGA RES/55/2. 18 September 2000
UNGA (2001) Road Map Towards the Implementation of the United Nations Millennium Declaration. UN Doc A/56/326
UNGA (2015a) Addis Ababa Action Agenda of the Third International Conference on Financing for Development. UNGA Res 69/313. 27 July 2015
UNGA (2015b) Transforming Our World: the 2030 Agenda for Sustainable Development. UNGA Res/70/1. 21 October 2015
WTO General Council (1995) Arrangements for Effective Cooperation with Other Intergovernmental Organizations-Relations Between the WTO and the United Nations. WT/GC/W/10. 15 November 1995
WTO General Council (2020) Advancing Sustainability Goals Through Trade Rules to Level the Playing Field. WT/GC/W/814. 17 December 2020

Part I
WTO Law: An Impediment to the Implementation of the UN SDGs

Chapter 2
The Lack of Trade-Related Sustainable Development Commitments in WTO Law

Abstract Most of the 17 SDGs outlined in the UN 2030 Agenda for Sustainable Development are intricately linked to trade. To achieve these goals, WTO members must ensure sustainability in international trade. Some international trade rules provide the legal norms for sustainable development, forming trade-related sustainable development commitments (TSDCs) that encompass economic development, social rights, and environmental protection. These TSDCs are crucial for achieving the UN SDGs. Trade can significantly contribute to sustainable development by spurring economic growth, reducing poverty and hunger, and facilitating the exchange of goods, commodities, and technologies essential for sustainable development. Economic prosperity also enables countries to invest more resources in protecting the environment and upholding social rights. However, while international trade can drive economic prosperity, it may also harm social rights and the environment. TSDCs are vital in mitigating these negative externalities. What exactly are TSDCs? What are their implications for sustainable development? Which TSDCs are included in the WTO Agreements, and are they sufficient to ensure that WTO members implement trade-related SDGs? This Chapter explores these questions in detail.

2.1 Definition of Trade-Related Sustainable Development Commitments

In this book, TSDCs are defined as commitments obligating countries to achieve the UN SDGs related to international trade. Their enforcement is guaranteed by judicial institutions at various levels, including the WTO DSB, FTA dispute settlement panels, and international commercial courts.[1]

[1] International commercial courts are a new type of judicial institution. To know more information, please read Dimitropoulos (2021), pp. 361–379.

X. Zhao, *Integrating the UN SDGs into WTO Law*, EYIEL Monographs - Studies in European and International Economic Law 41,
https://doi.org/10.1007/978-3-031-73876-0_2

Therefore, the definition of TSDCs is twofold. First, TSDCs can be advanced through international trade. Second, violations of these commitments will be sanctioned under international trade law.

TSDCs have been written into various legal texts, including national laws, regulations, and FTAs. While WTO Agreements do not explicitly contain TSDCs, they implicitly provide these commitments through many provisions related to the UN SDGs. Given the interactions between trade and sustainable development, TSDCs include three groups of commitments:

1. Commitments to ensure inalienable individual economic rights, such as property rights and the freedom of trade.[2]
2. Commitments to fulfil social rights necessary for a decent life, including the right to health, work, education, food and water, development, gender equality, and civil and political rights.[3]
3. Commitments to promote green economic activities, such as reducing GHG emissions in accordance with nationally determined contributions under the Paris Agreement,[4] preventing environmental pollution,[5] and protecting biodiversity.[6]

2.2 Implications of Trade-Related Sustainable Development Commitments for Achieving the UN SDGs

2.2.1 Contributing to Sustainable Development

TSDCs directly contribute to the three dimensions of sustainable development. Concerning the economic dimension, TSDCs contain free trade commitments, which aim to reduce barriers, in any form, to trade in goods and services and ensure a level playing field of domestic markets. Carrying out these commitments helps countries, including developing countries, to join the global supply chains of goods and services. Although imports may significantly influence the income of agricultural sectors of non-industrialised countries, WTO rules have reduced these side-effects. Compared to this potential risk, foreign exchange earnings from international trade undoubtedly helped to reduce poverty.[7] The stories of the Four Asian

---

[2] See Petersmann (2005), pp. 30–94.

[3] See Bunn (2012), pp. 15–20.

[4] The Paris Agreement, Article 4.

[5] The UN SDGs target five dimensions of environmental pollution, including chemicals and waste pollution, air pollution, water pollution, soil pollution, and marine and coastal pollution. See The United Nations Environment Assembly (2018).

[6] The Convention on Biological Diversity requires the conservation of the biodiversity of the ecosystem. See United Nations (1992) Convention on Biological Diversity. 22 May 1992.

[7] Petersmann (2005), p. 59.

Tigers during the 1960s–1990s[8] and the contemporary BRICS countries have shown the contribution of international trade to the economic rise of developing countries in the recent past.[9]

Moreover, free trade commitments are crucial for the fulfilment of social rights. One of the objectives of sustainable development is to ensure decent living standards.[10] To this end, countries shall provide sufficient supplies of goods pertinent to public interests (hereinafter, public goods), including food, water, medicines, energy, infrastructure facilities, education, legal services, and other commodities necessary for daily life.[11] Liberalised international trade facilitates the circulation of these public goods around the world. For example, the global agricultural market enables the supply of affordable agricultural products in the least developed countries and those dependent on agricultural imports. These cheap imported goods are crucial for eradicating hunger in these regions.

Furthermore, liberalised international trade will play an increasingly important role in climate change adaptation and mitigation. According to a special report of the Intergovernmental Panel on Climate Change (IPCC), climate change has a far-reaching impact on food security. The report states that:

> [...] climate change affecting the amount of food, both from direct impacts on yields and indirect effects through climate change's impacts on water availability and quality, pests and disease, and pollination services. Another route is via changing CO_2 in the atmosphere, affecting biomass and nutritional quality. Food safety risks during transport and storage can also be exacerbated by changing climate. Further, the direct impacts of changing weather can affect human health through the agricultural workforce's exposure to extreme temperatures. Through changing metabolic demands and physiological stress for people exposed to extreme temperatures, there is also the potential for interactions with food availability; people may require more food to cope, whilst at the same time being impaired from producing it.[12]

As a result of climate change's impact on food production and availability, many countries will struggle to produce enough agricultural products to meet the demands of their domestic markets. International agricultural trade will be crucial in ensuring

---

[8] The Four Asian Tigers (also known as the Four Asian Little Dragons) are the economies of South Korea, Taiwan, Singapore and Hong Kong. Between the early 1960s and 1990s, they underwent rapid industrialisation and maintained exceptionally high growth rates of more than 7 per cent a year.

[9] BRICS is the abbreviation of five major emerging economies: Brazil, Russia, India, China, and South Africa.

[10] See UNGA (2015). The preamble to the agenda states that 'we resolve also to create conditions for sustainable, inclusive and sustained economic growth, shared prosperity and decent work for all [...].' In addition, paragraph 50 of the preamble claims that 'we resolve to build a better future for all people, including the millions who have been denied the chance to lead decent, dignified and rewarding lives and to achieve their full human potential.' See also Goal 4 target 4 'Decent Jobs and Entrepreneurship' and Goal 8 'Decent Work'.

[11] See Bunn (2012), pp. 15–20; Breining-Kaufmann (2005), p. 342–381; See also Temmerman (2017), p. 64.

[12] See Mbow et al. (2019), p. 450.

food security by allowing these countries to address food shortage through the importation of agricultural products.

2.2.2 *Mitigating the Negative Externalities of International Trade*

Although international trade can significantly reduce poverty and hunger, fulfil social rights, and help adapt and mitigate the impact of climate change, it can also adversely affect sustainable development.[13] Currently, multilateral economic institutions tend to focus predominantly on the role of economic growth in advancing sustainability,[14] often overlooking the negative consequences of international trade. Indeed, a thriving global market can exacerbate economic and gender inequalities, undermine labour rights, damage ecosystems, and exacerbate climate change.[15]

To contribute to sustainable development and maximise the contribution of international trade, trade policies at various levels must address and mitigate the negative externalities of international trade. Two types of TSDCs are particularly effective in this regard: (1) commitments to adhere to international standards and (2) commitments to uphold high levels of protection. Many international trade agreements, especially FTAs, include such commitments.

A commitment to international standards requires countries to regulate their international trade according to established global norms. These standards include, for example, international food standards set by the Codex Alimentarius,[16] international standards for phytosanitary measures (ISMPs),[17] technical regulations,[18] international labour standards established by the International Labour Organisation (ILO),[19] and environmental standards outlined in multilateral environmental

[13] See Baumann-Pauly and Posner (2016), pp. 11–21.

[14] See Castellino and Bradshaw (2015), p. 464.

[15] Ibid., 464.

[16] The Codex Alimentarius is a collection of standards, guidelines and codes of practice adopted by the Codex Alimentarius Commission. The Commission was established by FAO and WHO to protect consumer health and promote fair practices in food trade.

[17] International Standards for Phytosanitary Measures are standards adopted by the Commission on Phytosanitary Measures, which is the governing body of the International Plant Protection Convention.

[18] International standards relating to technical regulations are adopted by international standardization organisations. The most important international standardization organisations include the International Organisation for Standardization (ISO), International Electrotechnical Commission (IEC), and the International Telecommunication Union (ITU).

[19] International labour standards included in eight fundamental conventions: freedom of association and protection of the Right to Organise Convention (1948); Right to Organise and Collective Bargaining Convention (1949); Forced Labour Convention (1930); Abolition of Forced Labour Convention (1957); Minimum Age Convention (1973), Worst Forms of Child Labour Convention (1999); Equal Remuneration Convention (1951); and Discrimination (Employment and

agreements.[20] Many international trade agreements recognise these international standards, including FTAs such as CETA, USMCA, RCEP,[21] and CPTPP.[22] Additionally, WTO Agreements explicitly incorporate some of these international standards. For example, the SPS Agreement and the TBT Agreement provide that WTO members shall design and apply their SPS measures and technical regulations based on relevant international standards.[23] This commitment integrates social rights and environmental norms into international trade rules, effectively updating the trade rulebook and helping reconcile the objective of a thriving global market with sustainable development.

International standards play a crucial role in mitigating the negative externalities of international trade. International food and technical standards safeguard consumer health. ISPMs harmonise quarantine measures, protecting both consumer health and the ecosystem of importing countries. The ILO Declaration on Fundamental Principles and Rights at Work provides fundamental principles and rights at work, including freedom of association, collective bargaining, eliminating forced and child labour, equal remuneration for equal work, and non-discrimination in employment.[24] These rights are crucial for empowering workers, including women. Particularly, freedom of association and the right to collective bargaining enable workers to form labour unions, strengthening their negotiating power and helping reduce economic and gender inequalities over time. Additionally, multilateral environmental agreements (MEAs) impose a range of obligations, such as prohibiting the use of ozone-depleting substances,[25] reducing GHG emissions,[26] conserving

Occupation) Convention (1958). See ILO, Fundamental Conventions. https://www.ilo.org/global/standards/introduction-to-international-labour-standards/conventions-and-recommendations/lang%2D%2Den/index.htm. Accessed 21 July 2024.

[20]There are eleven Multilateral Environmental Agreements and six Regional Seas Conventions. See https://www.unep. org/gef/multilateral-environmental-agreements. Accessed 21 July 2024.

[21]The RCEP has no environmental and labour chapters. However, it recognises the WTO's sustainability rules, such as the rules of SPS and TBT measures.

[22]In these FTAs, there are many provisions on international standards. They recognise the WTO's SPS and TBT rules that recognise the relevant international standards. See Chapsters 4 and 5 of the CETA, Chapters 9 and 11 of the USMCA, and Chapters 7 and 8 of the CPTPP. These FTAs also contain provisions on environmental and labour standards. Labour standards are included in Chapter 23 of the CETA, Chapter 23 of the USMCA, and Chapter 19 of the CPTPP. Environmental standards are included in Chapter 24 of the CETA, Chapter 24 of the USMCA, and Chapter 20 of the CPTPP. Regarding academic research, see Harrison (2019), pp. 635–657; See also Bronckers and Gruni (2021), pp. 25–51.

[23]Article 3 of the SPS Agreement and Article 2.4 of the TBT Agreements.

[24]See International Labour Organisation (1998) ILO Declaration on Fundamental Principles and Rights at Work, para. 2.

[25]See The Vienna Convention for the Protection of the Ozone Layer and the Montreal Protocol on Substances that Deplete the Ozone Layer. Available via https://ozone.unep. org/sites/default/files/2019-12/The%20Ozone%20Treaties%20EN%20-%20WEB_final.pdf. Accessed 21 July 2024.

[26]See The Paris Agreement.

biodiversity,[27] managing hazardous wastes,[28] protecting marine environments and coastal regions,[29] and safeguarding endangered species of flora and fauna.[30]

Regrettably, as shown below, the WTO Agreements are insufficient for ensuring the effective implementation of international standards, particularly those related to environmental and social rights. FTAs appear to be the primary mechanism through which countries implement these standards. For example, FTAs such as the USMCA and CPTPP include provisions on labour rights, biodiversity, air quality, and marine and fisheries management.[31] These provisions explicitly incorporate sustainable development commitments into trade agreements, providing clear legal bases for labour and environmental protection. However, FTAs also have significant shortcomings.[32] Even the FTAs mentioned above contain only a limited number of environmental provisions and lack robust mechanisms for climate governance.[33]

A commitment to uphold high levels of protection is currently applied to labour and environmental standards, requiring countries to provide and maintain stringent regulations. FTAs contain two types of clauses that enforce these commitments. The first is the non-derogation clause, which stipulates that contracting parties cannot derogate from their existing labour and environmental protection commitments.[34] The second is the non-regression clause, which requires countries to improve their domestic laws and regulations to ensure high levels of labour and environmental protection.[35]

The value of the commitment to high levels of protection lies in its ability to ensure the progressive improvement of countries' labour and environmental standards. Since existing labour and environmental standards are insufficient to meet sustainable development objectives, this commitment is crucial for mitigating the negative externalities of international trade.

---

[27] See United Nations (1992) Convention on Biological Diversity. 22 May 1992.

[28] See United Nations (1989) Basel Convention on the Control of Transboundary Movements of Hazardous Wastes and their Disposal. 22 March 1989.

[29] See United Nations (1981) Abidjan Convention. 23 March 1981; United Nations (1994) The Action Plan for the Protection and Development of the Marine Environment and Coastal Areas of the East Asian Seas Region. Adopted in April 1981; United Nations (1994) Northwest Pacific Action Plan. Adopted in 1994; United Nations (1976) The Convention for the Protection of the Mediterranean Sea Against Pollution (Barcelona Convention). 16 February 1976; United Nations (1975) The Mediterranean Action Plan. Adopted in 1975; United Nations (1983) The Convention for the Protection and Development of the Marine Environment of the Wider Caribbean Region (Cartagena Convention). 24 March 1983; United Nations (1981) The Caribbean Environment Programme. Adopted in 1981.

[30] See Convention on International Trade in Endangered Species of Wild Fauna and Flora, signed at Washington, D.C., on 3 March 1973, entered into force on 1 July 1975.

[31] See Articles 24.9–24.12 of the USMCA and Articles 20.5–20.6 of the CPTPP.

[32] See, for example, Harré (2021), pp. 15–27.

[33] See Laurens et al. (2019), pp. 672–675.

[34] See Article 23.4 of the USMCA and Article 19.4 of the CPTPP.

[35] See Article 23.2 of the CETA.

2.3 Taking Stock of Trade-Related Sustainable Development Commitments in WTO Law

2.3.1 Trade Liberalisation Commitments

2.3.1.1 General Agreement on Tariffs and Trade

Trade liberalisation is a dominant focus of the GATT. A typical example is the GATT 1994, which commits to liberalising trade in goods by significantly reducing tariffs among signatories and restricting the use of other trade barriers.[36]

Regarding tariffs, WTO members shall list the maximum tariff rates they can impose on other states for all products in their schedules of bound tariff commitments.[37] They shall also apply Most-Favoured-Nation (MFN) tariff rates that are not higher than these bound tariff rates.[38]

With respect to internal regulations, WTO members are prohibited from applying any internal regulations and taxes to favour domestic production over foreign products.[39] This rule is established by the National Treatment (NT) principle outlined in GATT Article III:1. Subparagraphs of GATT Article III clarify that internal restrictions include internal taxation, SPS measures, technical regulations, quotas, quantitative restrictions, transit restrictions, discrimination, and subsidies.[40] These provisions aim to eliminate non-tariff barriers and ensure a level playing field in domestic markets.[41]

The scope of application of the NT principle has been confirmed and clarified through various WTO decisions. For instance, in the *Italian Discrimination Against Imported Agricultural Machinery* case, the panel ruled that:

---

[36] Hudec (1993), p. 3.

[37] See GATT Article II; See WTO Panel Report, EC-Chicken Cuts, WT/DS269/R, adopted on 30 May 2005, para. 7.65; See also WTO Panel Reports, EC-IT Products, WT/DS375/R (WT/DS376R or WT/DS377/R), adopted 21 September 2010, para. 7.100.

[38] See GATT Article I; See also WTO Panel Reports, Brazil-Taxation, WT/DS472/R (WT/DS497/R), adopted on 11 January 2019, paras. 7.1041–7.1042.

[39] Cottier and Oesch (2005), p. 382.

[40] Hudec (1993), pp. 3–5.

[41] GATT Article III provides that importing goods and domestic goods shall be treated equally in terms of internal taxation. GATT Article IV sets out conditions for the use of screen quotas and thus limits the restriction on the importation of cinematograph films. GATT Article V ensures the freedom of transit, which is vital to the shipment of importing goods. GATT Article VIII prohibits the protection of domestic products in the form of fees and formalities. GATT Article XI explicitly requires that WTO members shall eliminate quantitative restrictions on importation. GATT Article XII prevents the abuse of safeguard measures for the balance of payments. GATT Article XVI requires that WTO members eliminate the use of subsidies. GATT Article XVII provides that state-owned companies shall not constitute discrimination to foreign competitors in their business.

> [. . .] the objective of this provision is to treat the imported products in the same way as the like domestic products once they had been cleared through customs.[42]

This decision implied that the NT principle should apply to all forms of internal restrictions. Later, in *Argentina-Hides and Leather*, the panel confirmed that the NT principle extends to tax measures, specifically a collection regime for income taxes related to import transactions.[43] The AB report in *EC-Bananas III* further affirmed that licensing systems affecting internal sales and purchases fall under the scope of GATT Article III:4 and are thus subject to the NT principle.[44] More recently, the AB report in *Brazil-Taxation* reiterated the broad scope of application of Article III:2, stating that any measures affecting competition conditions are also subject to the NT principle.[45]

Furthermore, any trade regulatory measure is subject to the Most-Favoured Nation (MFN) principle prescribed by GATT Article I.[46] In 1948, Mr. Wilgress, Chairman of the GATT contracting parties, affirmed this principle in his ruling on *Indian export rebates* during a meeting on August 24, 1948.[47] He ruled that:

> any advantage, favour, privilege or immunity granted with respect to internal taxes by any contracting party to any product destined for any other country shall be accorded immediately and unconditionally to the like product destined for the territories of all other contracting parties.[48]

Wilgress's interpretation not only extended the scope of application of the MFN principle to internal tax regulations but also implied that the principle should apply to all internal regulations. Subsequent WTO decisions followed this interpretation. For instance, in *EC-Bananas III*, the panel explained that:

> "advantages" in the sense of GATT Article I:1 are those that create 'more favourable import opportunities' or affect the commercial relationship between products of different origins.[49]

More directly, the AB clarified that the MFN principle applies to all categories of domestic regulatory measures. In *Canada-Autos*, the AB ruled that the term 'advantages' in Article I:1 encompasses any benefit for any products and for like products

[42] GATT Report of the Panel, Italian Discrimination Against Imported Agricultural Machinery, adopted on 23 October 1958, BISD 7S/60 (1961), para. 11.

[43] See WTO Panel Report, Argentina-Hides and Leather, WT/DS155/R, adopted on 16 February 2001, para. 11.145.

[44] See WTO Appellate Body Report, EC-Bananas III, WT/DS27/AB/R, adopted on 25 September 1997, para. 211.

[45] See WTO Appellate Body Reports, Brazil-Taxation, WT/DS472/AB/R (WT/DS497/AB/R), adopted on 11 January 2019, para. 5.15.

[46] Hudec (1993), p. 6; Mavroidis (2012), p. 129.

[47] Hudec (1990), p. 113.

[48] GATT (1948), p. 4.

[49] WTO Panel Report, EC-Bananas III, WT/DS27/R/ECU, adopted on 25 September 1997, para. 7.239.

originating in or destinated for all other members.[50] In *EC-Seal Products*, the AB reaffirmed that the objective of Article I:1 is to ensure equal competitive opportunities for like imported products from all members. The AB also ruled that a violation of Article I:1 does not depend on the actual trade effects of a measure.'[51] This ruling significantly broadened the scope of the MFN principle.

To date, the NT and MFN principles remain central to the liberalisation of trade in goods.[52] Scholars like Thomas Cottier even described them as the WTO's constitutional principles.[53]

2.3.1.2 Agreement on Technical Barriers to Trade

Similar to the three major WTO agreements,[54] Article 2.1 of the Agreement on Technical Barriers to Trade (TBT Agreement) upholds the MFN and NT principles by prohibiting discrimination between foreign like products.[55] Scholars, including Rüdiger Wolfrum, Peter-Tobias Stoll, and Anja Seibert-Fohr consider this provision to be a combined and abbreviated version of GATT Articles I and III.[56] Additionally, the TBT Agreement establishes detailed rules and procedures for assessing the conformity of technical regulations. These TBT rules promote free trade in goods and services in two ways.

First, the TBT Agreement provides a clear legal basis to prohibit any technical barrier to trade. Article 2.2 of the TBT Agreement provides that:

> Members' technical regulations are not prepared, adopted or applied with a view to or with effect to or with the effect of creating unnecessary obstacles to international trade.

In addition, this provision does not prohibit internal regulations designed to mitigate the externalities of international trade. It requires a risk assessment to determine whether a trade measure is necessary to achieve a legitimate objective, such as protecting the environment or public health.[57] In *EC-Sardines*, the panel interpreted this provision as aiming to balance the prevention of unnecessary trade barriers with

[50] WTO Appellate Body Reports, Canada-Autos, WT/DS139/AB/R (WT/DS142/AB/R), adopted on 19 June 2000, para. 79.

[51] See WTO Appellate Body Reports, EC-Seal Products, WT/DS400/AB/R (WT/DS401/AB/R), adopted on 18 June 2014, para. 5.87.

[52] Jackson (1997), p. 157.

[53] Cottier and Oesch (2005), pp. 346–428.

[54] Namely, the GATT, GATS, and the TRIPS Agreement.

[55] Kudryavtsev (2013), p. 20.

[56] Wolfrum et al. (2007), p. 214.

[57] See WTO Appellate Body Report, US-Tuna II (Mexico), WT/DS381/AB/R, adopted on 16 May 2012, para. 314; See also WTO Appellate Body Reports, US-COOL, WT/DS384/AB/R (WT/DS386/AB/R), adopted on 23 July 2012, paras. 371–372.

the need to allow policy space for domestic regulatory measures.[58] In *US-Tuna II*, the AB developed a necessity test under Article 2.2.[59] This approach was also adopted in subsequent leading cases, such as the *Korea-Beef* and *EC Asbestos* cases.

Moreover, Article 5.1.1 of the TBT Agreement prescribes that 'conformity assessment procedures are not prepared, adopted or applied with a view to or with the effect of creating unnecessary obstacles to international trade.' The first leading case regarding Article 5.1.1 is *Russia-Railway Equipment*. In this case, the panel explained the scope of conformity assessment[60] and identified three types of discrimination that members must avoid:

(1) The suppliers of another member who have been granted less favourable access are suppliers of products that are like products of domestic suppliers or suppliers from any other country who have been granted more favourable access.
(2) The importing Member (through the preparation, adoption or application of a covered conformity assessment procedure) grants access for suppliers of products from another member under conditions less favourable than those accorded to suppliers of domestic products or products from any other country.
(3) The importing member grants access under conditions less favourable for suppliers of like products in a comparable situation.[61]

Second, the TBT Agreement requires adherence to harmonised technical regulations. Members must strive to harmonise their technical regulations in all situations. According to Article 1.1 of the TBT Agreement, members should formulate their regulations and conduct conformity assessments in accordance with international standards.[62] If no international standard exists for a given technical regulation, Articles 2.9.1–2.9.3 require members to notify other members and provide their proposed technical regulations.[63] In such cases, Article 2.9.4 obliges members to negotiate with other members and consider their feedback on the proposed regulations. Article 2.6 requires members to actively participate in the work of

[58] See WTO Panel Report, EC-Sardines, WT/DS231/R, adopted on 23 October 2002, paras. 7.119–7.120.

[59] See WTO Appellate Body Report, US-Tuna II (Mexico), WT/DS381/AB/R, adopted on 16 May 2012, para. 322; and See also WTO Appellate Body Reports, US-COOL, WT/DS384/AB/R (WT/DS386/AB/R), adopted on 23 July 2012, paras. 374–378.

[60] In the light of Article 5.1.1, a conformity assessment must concern procedures for the assessment of conformity by central government bodies, and it must concern a situation where a positive assurance of conformity with technical regulations or standards is required (i.e., a mandatory conformity assessment procedure). See WTO Panel Report, Russia-Railway Equipment, WT/DS499/R, adopted on 5 March 2020, para. 7.249.

[61] See WTO Panel Report, Russia-Railway Equipment, WT/DS499/R, adopted on 5 March 2020, para. 7.251.

[62] See WTO Panel Reports, EC-Approval and Marketing of Biotech Products, WT/DS291/R (WT/DS292/R or WT/DS293/R), adopted on 29 September 2006, para. 7.300; See also WTO Panel Report, US-Tuna II (Mexico), WT/DS381/R, adopted on 13 June 2012, paras. 7.661–7.662.

[63] See WTO Panel Report, US-Clove Cigarettes, WT/DS406/R, adopted on 24 April 2012, para. 7.542.

international standardising bodies to promote the development of international standards. Additionally, Article 2.7 stipulates that members should consider accepting other members' technical regulations as equivalent, provided they adequately fulfil the objectives of their own regulations.[64]These provisions ensure the transparency of national technical regulations[65] and help prevent technical trade barriers.

2.3.1.3 General Agreement on Trade in Services

The GATS contributes to liberalising trade in services. While it upholds the core values of the GATT 1994, such as trade liberalisation, it provides members with greater policy space to restrict foreign service providers. During the negotiation of GATS in 1991, the US proposal for a general grant of MFN rights faced opposition from other negotiators. Finally, the US agreed to accept reservations to the MFN obligation in several areas to reach a consensus on liberalisation commitments.[66]

Therefore, the liberalisation of trade in services is limited. While GATS Article II provides for the MFN principle, it does not apply to all service sectors. Members can make reservations when acceding to the GATS, and the extension of these reservations depends on negotiations.[67] Some predicted that WTO members would maintain their reservations,[68] and so far, members have shown no willingness to increase their commitments to the MFN principle.

Likewise, the NT principle in the context of the GATS is slightly limited. GATS Article XVI 'Market Access' only requires members to apply national treatment to specific services and service suppliers listed in their schedule. This provision, in principle, narrows the scope of the NT principle. However, the panel in *China-Electronic Payment Services* clarified that the limitations of the market access obligation do not narrow the scope of the national treatment obligation. It stated that 'the scope of the national treatment obligation laid down in GATS Article XVII extends generally to all measures affecting the supply of services.'[69] This decision appears to modify members' obligations under Article XVI.

Despite the conflict between GATS Article XVI and XVII, other provisions clarify members' national treatment obligations. GATS Article VI (4) states that domestic regulations, such as qualification requirements and procedures, technical standards, and licensing requirements, shall not constitute unnecessary barriers to trade in services. GATS Article VII facilitates the recognition of professional

[64] TBT Agreement Article 2.7.

[65] Wolfrum et al. (2007), p. 230.

[66] See Hudec (1993), p. 191.

[67] See GATS Annex on Article II Exemptions.

[68] See Cottier and Oesch (2005), p. 355.

[69] See WTO Panel Report, China-Electronic Payment Services, WT/DS413/R, adopted on 31 August 2012, para. 7.652.

certifications between WTO members, allowing for the recognition of education, experience, requirements, and licenses or certifications granted in a particular country.[70] Harmonised professional qualification criteria help remove barriers to the circulation of labour between countries and liberalise services. GATS Article VIII explicitly limits service monopolies, requiring members to ensure that any monopoly supplier does not abuse its monopoly position.[71] GATS Article XII prohibits the misuse of safeguard measures to protect the balance of payments. GATS Article XV requires WTO members to avoid trade-distortive effects of subsidies.

Considering the limitations on trade in services, further negotiations on the liberalisation of services are imperative. Part IV of the GATS includes provisions for the progressive liberalisation of services. To achieve the GATS' objective of liberalising trade in services, members should extend the application of the MFN and NT principles to more service sectors. As noted by Cottier and Oesch, the NT principle is important in progressive and gradual liberalisation.[72] I argue that the MFN principle should also be crucial in the process.

2.3.1.4 Agreement on Trade-Related Aspects of Intellectual Property Rights

The Agreement on Trade-Related Aspects of Intellectual Property Rights (TRIPS Agreement) prevents trade barriers arising from regulations related to intellectual property rights (IPRs), thus contributing to free international trade.[73] To this end, the TRIPS Agreement establishes standards concerning the availability, scope, protection terms, and use of IPRs.[74] These standards supplement the traditional intellectual property regime by referencing pre-existing agreements, such as the World Intellectual Property Organisation (WIPO) convention.[75]

The TRIPS Agreement is not merely a duplication of the WIPO convention. It enhances the protection of IPRs in the context of international trade. First, the TRIPS Agreement extends the scope of national treatment. While national treatment has traditionally been a cornerstone of IPR protection,[76] it did not previously apply to all types of IPRs. The TRIPS Agreement notably extends national treatment to

---

[70] Article VII (1), GATS.

[71] See GATS Article VIII; See WTO Panel Report, Argentina-Textiles and Apparel, WT/DS56/R, adopted on 22 April 1998, paras. 6.74–6.75; See WTO Panel Report, US-Certain EC Products, WT/DS165/R, adopted on 10 January 2001, paras. 6.69–6.70; See also WTO Panel Reports, China-Raw Materials, WT/DS394/R (WT/DS395/R or WT/DS398/R), adopted on 22 February 2012, para. 7.839.

[72] Cottier and Oesch (2005), p. 385.

[73] TRIPs Agreement, Preamble.

[74] See Stoll et al. (2009), pp. 205–791.

[75] Stoll (2009), p. 2.

[76] See Cottier and Oesch (2005), p. 385.

copyright law, which was not covered under the Bern Convention.[77] Secondly, the TRIPS Agreement introduces the MFN principle into IPR law.[78] Article 4 mandates that any privileges granted to one member (including those exceeding TRIPS provisions) must be extended immediately and unconditionally to all other WTO members.[79] Although there are some exceptions,[80] this MFN obligation significantly enhances IPR protection. As Cottier and Oesch have noted, 'this improvement especially helps reduce inequalities between smaller countries and large trading powers on third markets.'[81] Moreover, the TRIPS Agreement incorporates IPR disputes into the WTO dispute settlement mechanism.[82] These legal arrangements explicitly define IPR rules and contribute to reducing distortions and impediments to international trade.

2.3.2 *Commitments to Reducing Economic Inequalities Between Developing and Developed Countries*

2.3.2.1 GATT Article XXXVI

The WTO Agreements include provisions aimed at reducing economic inequalities between developing and developed countries. The most explicit legal basis for this commitment is found in GATT Article XXXVI, which identifies the enhancement of living standards and the progressive development of members' economies as fundamental objectives of the GATT. This Article underscores the importance of fostering economic growth in less-developed countries, specifically requiring the WTO to assist in improving the export earnings of its less-developed country members.

---

77 Cottier (2005), pp. 1041–1120.

78 See Cottier and Oesch (2005), p. 357.

79 See WTO Panel Report, EC-Trademarks and Geographical Indications (US), WT/DS174/R, adopted on 20 April 2005, para. 7.704.

80 Article 4 of the TRIPS Agreement lists four categories of advantages, favours, privileges, and immunities which are exempted from the MFN obligation: (1) those deriving from international agreements on judicial assistance or law enforcement of a general nature and not particularly confined to the protection of intellectual property; (2) those granted in accordance with the provisions of the Bern Convention (1971) or the Rome Convention authorizing that the treatment accorded be a function not of national treatment but of the treatment accorded in another country; (3) those in respect of the rights of performers, producers of phonograms and broadcasting organizations or not provided under this Agreement; (4) those deriving from international agreements related to the protection of intellectual property which entered into force prior to the entry into force of the WTO Agreement, provided that such agreements are notified to the Council for TRIPs and do not constitute an arbitrary or unjustifiable discrimination against nationals of other Members.

81 Cottier and Oesch (2005), p. 358.

82 TRIPS Agreement, Article 64.

During the GATT era, panels twice upheld claims related to Article XXXVI, urging developed countries to help raise the export earnings of developing countries. In *European Communities-Refunds on Exports of Sugar*, Brazil argued that the European Communities (EC)' subsidies severely depressed world market prices of sugar, thereby displacing Brazilian exports, reducing sales opportunities, and diminishing Brazil's export earnings.'[83] The panel supported Brazil's claim and ruled that:

> [. . .] the European Communities had therefore not collaborated jointly with other contracting parties to further the principles and objectives set forth in Article XXXVI, in conformity with the guidelines given in Article XXXVIII.[84]

In *European Economic Community (EEC)-Restrictions on Imports of Dessert Apples*, Chile claimed that 'the EC made no conscious and purposeful efforts to ensure that Chile secure a share of growth in international trade in apples commensurate with the needs of its economic development.'[85] The panel found that 'the EEC's import measures on dessert apples did affect a product of particular export interest to less-developed contracting parties' and asked the EEC to remove protective measures on apples originating in Chile.[86]

There have been no cases regarding Article XXXVIII in the WTO era. However, during the WTO's 12th Ministerial Conference, WTO members reaffirmed their commitment to supporting small economies, particularly in trade facilitation.[87]

2.3.2.2 Special and Differential Treatment

The WTO Agreements include a range of special and differential treatment (SDT) provisions designed to support the economic development of developing and least-developed countries. In 2000, the WTO Committee on Trade and Development published an official document that comprehensively summarises these SDT provisions in WTO Agreements and decisions. These SDT provisions include:

(a) Provisions to increase the trade opportunities of developing country members.
(b) Provisions to safeguard the interests of developing country members.
(c) Provisions to offer flexibility in terms of commitments, action, and the use of policy instruments.
(d) Provisions to provide transitional time periods.

---

[83] See GATT Report of the Panel, European Communities-Refunds on Exports of Sugar, L/5011-27S/69, adopted on 10 November 1980, paras. 2.25 and 2.27–2.28.

[84] Ibid., para. (h).

[85] See GATT Report of the Panel, European Economic Community-Restrictions on Imports of Dessert Apples, L/6491-36S/93, adopted on 22 June 1989, paras. 8.4–8.5.

[86] Ibid., paras. 12.31–12.32.

[87] See WTO (2022d).

(e) Provisions to offer support and resources for technical assistance.
(f) Provisions to address the needs of least-developed country members.[88]

Although SDT provisions aim to mitigate economic inequalities between developing and developed countries, they face several challenges.

First, these provisions often lack details. For example, while Article 15 of the Anti-Dumping Agreement requires developed country members to consider the special circumstances of developing country members when imposing anti-dumping measures, it does not clearly define this obligation.[89] This lack of clarity was highlighted by the WTO panel in *US-Steel Plate*. The panel concluded that Article 15 did not impose any specific or general obligation on developed country members to undertake particular actions.[90] As a result, monitoring the implementation of these SDT provisions by developed countries can be challenging.

Second, the SDT is increasingly inadequate, particularly with the expiration of transition period clauses. WTO panels and the AB have identified two types of transition periods: those for general WTO commitments and those linked to obligations arising from international agreements or domestic laws of WTO members. While developing and least-developed countries have had some flexibility in their policy instruments,[91] they are no longer exempt from core WTO commitments. For example, the Agreement on Textiles and Clothing (ATC), which previously permitted quotas on textile and clothing products, ended on January 1, 2005. Similarly, the transition periods granted to developing countries under Articles 27.2 and 27.3 of the SCM Agreement have also been terminated.[92]

There is a diversity of opinions regarding the effectiveness of SDT provisions. Scholars from developing countries often argue that SDT provisions are crucial in mitigating inequalities between developing and developed countries.[93] Conversely, many scholars from developed countries contend that the current SDT provisions are insufficient for significantly boosting the economies of developing and least-developed countries.[94]

Third, the debate over the status of developing country within the WTO has intensified since 2019. The US proposed changing the current self-declaration approach, which allows WTO members to unilaterally declare their developing

---

[88] WTO (2000), p. 3.

[89] Anti-Dumping Agreement, Article 15.

[90] See WTO Panel Report, US-Steel Plate, WT/DS206/R, adopted on 29 July 2002, para. 7.110; See also WTO Panel Report, EC-Tube or Pipe Fittings, WT/DS219/R, adopted on 18 August 2003, para. 7.68.

[91] Developing and least developed countries have the right to enjoy a longer transition period than other countries when adapting to any domestic regulatory measures of Members. See, for example, WTO Appellate Body Report, US-Shrimp, WT/DS58/AB/R, adopted on 6 November 1998, paras. 174–175.

[92] See SCM Agreement, Article 27.

[93] See Mah (2012), pp. 2015–2017.

[94] See Conconi and Perroni (2015), p. 68.

country status.[95] Under the US proposal, the WTO would adopt a more objective and systematic approach to determining developing country status.[96] This approach involves a two-tier assessment process.

In the first step of this process, several categories of countries would be excluded from receiving developing country status. These categories include:

1. Members of the Organisation for Economic Cooperation and Development (OECD).
2. Countries that have begun the accession process to the OECD.
3. Members of the Group of 20 (G20).
4. Countries classified as 'high income' by the World Bank.
5. Countries that account for no less than 0.5 per cent of global merchandise trade (imports and exports).[97]

This initial assessment primarily focuses on national income levels, providing a broad measure of a country's overall economic development rather than targeting specific industries.

In the second step, the assessment shifts to a more detailed examination, focusing on specific industries within WTO members. The US proposal allows for the possibility that, in sector-specific negotiations, countries may also be deemed ineligible for SDT based on their performance in particular sectors.[98]

The US approach is based on a case-by-case analysis of the circumstances affecting different industries within WTO members. Under this proposal, a WTO member not listed among the exclusions based on its overall economic development may still be treated as a developed country in specific negotiations if its relevant industry is sufficiently advanced. However, the proposal does leave some ambiguity. It lacks precise standards for determining when a country should lose its developing country status in specific negotiations.

The US proposal has significantly tipped the balance between developing and developed countries within WTO negotiations. Major developing countries such as Korea, Brazil, and India have opted not to seek special treatment as developing countries in future WTO negotiations. While many developing countries that meet the US criteria still benefit from SDT, they now face an uncertain future. The withdrawal of large developing countries from the group of developing countries means that smaller developing and least-developed countries may be increasingly vulnerable in future WTO negotiations. This shifting dynamic makes anticipating the reintroduction of expanded SDT provisions into WTO Agreements more challenging. At the recent 12th WTO ministerial conference, while WTO members reaffirmed their commitment to existing SDT provisions for developing and

[95] WTO (2019).

[96] WTO General Council (2019).

[97] Ibid.

[98] Ibid.

least-developed countries,[99] they only provided some SDT provisions in the new Agreement on Fisheries Subsidies.[100] Apart from this, there was no additional clarification or expansion of SDT provisions in other WTO Agreements, such as those concerning green subsidies.[101]

2.3.3 Commitments to Industry Protection

2.3.3.1 Agreement on Implementation of Article VI of the General Agreement on Tariffs and Trade 1994 (Anti-Dumping Agreement)

Under Article 9 of the Anti-Dumping Agreement (AD Agreement), WTO members are permitted to impose anti-dumping duties on imported products that are sold at less than their normal value if this practice causes injury to domestic producers of similar products in the importing country.[102] This provision is designed to prevent trade distortions and safeguard domestic industries from unfair competition. Over the past decades, anti-dumping measures have become a common tool among WTO members, including some developing countries such as China and India.[103]

It is worth noting that the AD Agreement does not permit the use of anti-dumping duties for reasons other than addressing the specific issue of dumping. However, research has shown that many countries have initiated anti-dumping actions in response to broader macroeconomic factors.[104] For instance, some countries resort to anti-dumping measures when their exchange rates strengthen or their GDP growth rate declines.[105] Such economic conditions are not recognised as valid grounds for imposing anti-dumping duties under the AD Agreement. This misuse of anti-dumping measures highlights a gap in the existing legal framework, suggesting that countries may need additional or alternative legal tools to protect their industries effectively.

2.3.3.2 Agreement on Subsidies and Countervailing Measures

In addition to anti-dumping duties, the Agreement on Subsidies and Countervailing Measures (SCM Agreement) offers another mechanism for protecting domestic

[99] See WTO (2022c), para. 2.

[100] See WTO (2022a); Articles 6 and 7 of the Agreement on Fisheries Subsidies.

[101] See Ibid.

[102] Bentley and Silberston (2007), p. 10.

[103] Ibid., p. 119.

[104] Ibid., p. 120.

[105] Ibid., p. 120.

industries. It allows importing countries to seek remedies by either requesting that exporting countries raise the prices of their subsidised goods[106] or directly imposing countervailing duties on those imports.[107]

There is merely a slight difference between the applications of anti-dumping and countervailing measures. Many subsidised exports can also be classified as dumped products under Article 2 of the AD Agreement.[108] This overlap occurs because subsidies can prompt exporters to reduce their export prices. Consequently, the scope of countervailing and anti-dumping measures significantly overlaps. Importing countries may impose either anti-dumping duties or countervailing duties on subsidised products. However, it is important to note that a single product cannot be subjected to both types of duties for the same instance of dumping or export subsidisation.[109] Additionally, in countervailing investigations, representatives from importing countries must consult with their counterparts from exporting countries. This step is not required in anti-dumping investigations.[110] Therefore, diplomatic solutions play a crucial role in applying countervailing measures. If members want to address trade disputes through diplomacy, initiating countervailing duty investigations could be a favourable approach.

The most compelling issue regarding the SCM agreement is what subsidies WTO members can legally use. The SCM Agreement classifies subsidies into prohibited subsidies and actionable subsidies. Article 3 of the SCM Agreement explicitly prohibits WTO members from granting or maintaining export and import substitution subsidies. For actionable subsidies, WTO members must demonstrate their adverse trade effects to justify applying remedies.

Previously, WTO agreements included provisions for non-actionable subsidies, which were deemed acceptable under certain conditions. The SCM Agreement and the Agreement on Agriculture (AoA) allowed WTO members to provide subsidies for agriculture, research and development, regional assistance, and environmental adaptations, the latter often referred to as green subsidies. However, these exceptions have been significantly reduced over the past decade. Specifically, the provision for non-actionable subsidies related to research and development, regional assistance, and environmental adaptations expired on January 1, 2000. Additionally, under the 2015 Nairobi Ministerial Decision on Export Competition, WTO members agreed to phase out export subsidies for agricultural products gradually.[111]

Furthermore, at the WTO's 12th Ministerial Conference, members agreed to eliminate three categories of fisheries subsidies: (1) subsidies contributing to illegal,

---

[106] See SCM Agreement Article 18.

[107] See SCM Agreement Article 19; See also Bentley and Silberston (2007), p. 22.

[108] Ibid., p. 21.

[109] The GATT 1994, Article VI (5).

[110] Bentley and Silberston (2007), p. 32.

[111] WTO (2015).

unreported and unregulated fishing[112]; (2) subsidies regarding overfished stocks[113]; and (3) subsidies related to fishing outside the jurisdiction of a coastal member or non-member, and beyond the authority of relevant regional fisheries management organisations or arrangements.[114] It is worth noting that eliminating non-actionable subsidies, such as green subsidies, does not mean that WTO rules prohibit these subsidies outright. Instead, they become actionable, meaning that their legality can be challenged before WTO panels if other members request an examination.

The WTO's subsidy disciplines present both advantages and challenges for sustainable development. Restrictions on subsidies would promote free trade and drive economic growth. Specific subsidy restrictions are crucial for achieving certain SDGs. For instance, eliminating export subsidies for agricultural products would ensure members' food security, while removing fisheries subsidies would help conserve marine resources. Currently, overfishing by some countries threatens both fisheries resources and the food security of island nations. Addressing overfishing would help these fishery-dependent states manage their marine resources sustainably.

However, eliminating non-actionable subsidies for research and development, regional assistance, and environmental adaptations would diminish the WTO Agreements' support for sustainable development. The risk of litigation for subsidies aimed at poverty and eradication, economic growth, and green industries could impose significant pressure on members' sustainable development policies. Given this, WTO members may need to reconsider reintroducing non-actionable green subsidies and explore potential improvements to the original provisions. Additionally, as with anti-dumping measures, there is a concern about the potential abuse of countervailing measures.

2.3.3.3 Agreement on Safeguards

The Agreement on Safeguards (SG Agreement) allows members to implement safeguard measures to protect domestic industries from injury caused by increased imports.[115] These measures include *ad valorem* tariffs, tariff rate quotas, specific tariffs, and quantitative restrictions and quotas.[116]

Safeguard measures are more suitable for protecting domestic industries than anti-dumping and countervailing measures. First, the scope of safeguard measures is broader. Unlike anti-dumping and countervailing measures, which require evidence of unfair competition, such as dumping or subsidisation, safeguard measures can be

---

[112] See WTO (2022a), Article 3.

[113] Ibid., Article 4.

[114] Ibid., Article 5.

[115] The SG Agreement, Article 2.

[116] Mavroidis et al. (2010), p. 481.

implemented without such preconditions.[117] This allows importing countries to apply safeguard measures more widely to shield their domestic industries. Second, there is a lower risk of abuse of safeguard measures. These measures intend to be temporary and are designed to help domestic industries adjust to increased import competition.[118] For example, they can help preserve jobs and support unemployed workers in transitioning to new employment opportunities.[119]

However, safeguard measures have been implemented less frequently than anti-dumping and countervailing measures.[120] The unpopularity of safeguard measures is mainly due to two reasons.

First, safeguard measures are challenging to justify before WTO panels and the AB. According to GATT Article XIX (1) (a), safeguard measures must meet an 'unforeseen development' requirement,[121] meaning that the damage caused by increased imports must have been unexpected at the time of negotiations.[122] To date, only the US-Safeguard Measure on PV products has met this requirement.[123]

Second, many members are reluctant to use safeguard measures. GATT Article XIX mandates that safeguard measures must be non-discriminatory among WTO members.[124] This means that such measures must apply equally to all countries, including both trade partners and rivals, which can strain relationships with key trading partners and create undesirable diplomatic consequences.[125]

2.3.4 Commitments to Public Health, Environmental Protection, and Animal Protection

2.3.4.1 GATT Articles XX (b) & (g), Agreement on the Application of Sanitary and Phytosanitary Measures, and Agreement on Technical Barriers to Trade

Although WTO members do not explicitly commit to protecting public health, the environment, and animals, several WTO provisions support these objectives. GATT Article XX (b) allows members to enact and enforce measures necessary to protect

---

[117] Ibid., p. 468.

[118] Ibid., p. 468.

[119] Ibid., pp. 469–471.

[120] Ibid., p. 465; Bown (2002), pp. 47–62.

[121] WTO Appellate Body Reports, US-Lamb, WT/DS177/AB/R (WT/DS178/AB/R), adopted on 16 May 2001, paras. 72–3; See also WTO Panel Report, Ukraine-Passenger Cars, WT/DS468/R, adopted on 20 July 2015, paras. 7.83–7.84.

[122] Piérola (2014), p. 143.

[123] See Mavroidis et al. (2010), p. 508; Fang (2022), p. 242.

[124] See Mavroidis et al. (2010), p. 465.

[125] Ibid., p. 466.

human, animal, or plant life or health. GATT Article XX (g) permits the implementation of measures related to the conservation of exhaustible natural resources. Additionally, the SPS and TBT agreements ensure that members have the right to take SPS[126] or TBT measures to protect human, animal or plant life or health.[127] However, any domestic regulatory measures taken under these agreements must not constitute arbitrary or unjustifiable discrimination or a disguised restriction on international trade.[128]

These provisions form essential legal bases for fulfilling environmental and social rights within the framework of WTO Agreements. They provide WTO members with policy space to enact and enforce domestic laws and regulations for social and environmental purposes. Despite that, the implementation of internal regulations has often been controversial.

First, the content of existing commitments is not clear. GATT Article XX (b) does not explicitly define what is 'necessary to protect human, animal, or plant life or health'. WTO panels must determine the applicability of Article XX(b) on a case-by-case basis. Consequently, it is uncertain whether a domestic regulatory measure (e.g., a climate change policy) aligns with the legitimate objectives outlined in Article XX (b). WTO panels analyse the design and structure of internal regulations to ascertain the policy objective.[129] Therefore, members must explicitly state the policy objective in their laws or regulations to claim that their internal regulations aim to achieve such objectives.[130] In other words, members cannot rely on a general sustainable development commitment to justify a specific sustainable development-related policy. This approach narrows the scope of sustainable development commitments in WTO law and deters members from addressing sustainable development issues through the WTO dispute settlement system.

In addition, unclear commitments undermine the legitimacy of the WTO DSB. The AB's interpretation of GATT XX (g) is a good example. Paragraph (g) does not specify the scope of natural resources. In *US-Shrimp*, the AB interpreted the term 'natural resources' evolutionarily, considering contemporary global concerns about environmental protection and conservation.[131] This interpretation significantly broadened the meaning of natural resources. Following the AB Report in *US-Shrimp*, the panel in *US-Tuna II* further extended the scope of 'exhaustible

---

[126] SPS Agreement, Article 2 (1).

[127] TBT Agreement, Preamble.

[128] See GATT Article XX, the preamble to SPS Agreement, and the Preamble to TBT Agreement; See also WTO Appellate Body Report, US-Gasoline, WT/DS2/AB/R, adopted on 20 May 1996, pp. 28–29. '[...] The resulting discrimination must have been foreseen, and was not merely inadvertent or unavoidable.'

[129] See WTO Panel Report, EC-Tariff Preference, WT/DS246/R, adopted on 20 April 2004, para. 7.201.

[130] Ibid., paras. 7.201–7.202.

[131] See WTO Appellate Body Report, US-Shrimp, WT/DS58/AB/R, adopted on 6 November 1998, paras. 128–131.

natural resource' to include dolphins.[132] Although the evolutionary interpretation of Article XX (g) promotes environmental and animal protection, it reduces the predictability of sustainable development commitments in WTO law. This unpredictability, in turn, undermines the legitimacy of WTO panels and the AB. Many scholars, including Steve Charnovitz and Simon Lester, argue that members should avoid addressing sustainable development-related disputes within the WTO dispute settlement system.[133]

Second, unclear commitments adversely affect the review of members' domestic regulatory measures. Article 11 of the Understanding on Rules and Procedures Governing the Settlement of Disputes (DSU) mandates that WTO panels and the AB apply an objective assessment standard when evaluating the legality of members' domestic regulatory measures.[134] However, due to the lack of clear commitments, WTO panels and the AB often need to interpret WTO provisions during their assessments. This approach hinders the application of the objective assessment standard and unreasonably restricts the policy space for national regulators to enact and enforce regulatory measures. I will further discuss this issue in Chap. 3.

2.3.4.2 Environmental Subsidies

Article 8(2) of the Subsidies and Countervailing Measures (SCM) Agreement provided three categories of non-actionable subsidies, but this provision expired on January 1, 2000.[135] As stated above, these green subsidies are now actionable. There is a proposal to reintroduce the original 'green light' subsidies into the WTO Agreements.[136] If this were to happen, WTO members would not have to withstand the pressure of litigation risk when designing subsidy policies. They would gain more policy space to support domestic environmental industries like solar panel manufacturers. Studies have shown that subsidies are crucial for developing environmental industries.[137] Given this, it is important to consider whether these original provisions would need improvement if WTO members decided to reintroduce non-actionable green subsidies into the WTO Agreements.

Article 8(2)(a) of the SCM Agreement prescribes that WTO members can provide research subsidies to support development. Article 8(2)(b) stipulates that WTO members can provide subsidies to support the development of disadvantaged

---

[132] See WTO Panel Report, US-Tuna II (Mexico) (Article 21.5-Mexico), WT/DS381/RW, adopted on 3 December 2015, para. 7.521.

[133] See Charnovitz (2020); See also Lester (2021).

[134] See, for example WTO Appellate Body Report, Indonesia-Importation of Horticultural Products, Animals and Animal Products, WT/DS477/AB/R (WT/DS478/AB/R), adopted on 22 November 2017, para. 5.28 and WTO Appellate Body Report, Korea-Pneumatic Valves (Japan), WT/DS504/AB/R, adopted on 30 September 2019, paras. 5.156 and 5.158.

[135] SCM Agreement, Article 31; Wu (2015).

[136] Ibid., pp. 1–2.

[137] See Bougette and Charlier (2018), p. 182.

regions. Article 8(2)(c) provides a category of non-actionable subsidies that can be used to offset environmental adjustment costs. The application of these non-actionable subsidies is, of course, subject to specific conditions.

Concerning research subsidies, Article 8(2)(a) of the SCM Agreement sets out the maximum amount and scope of subsidisation. The subsidy shall cover at most 75 per cent of industrial research costs or 50 per cent of the costs of pre-competitive development activities.[138] The subsidies cover exclusively the following five items:

(1) Costs of personnel, including researchers, technicians, and other supporting staff employed exclusively in the research activity.
(2) Costs of instruments, equipment, land, and buildings used exclusively and permanently (except when disposed of on a commercial basis) for the research activity.
(3) Costs of consultancy and equivalent services used exclusively for the research activity, including bought-in research, technical knowledge, patents, etc.
(4) Additional overhead costs incurred directly as a result of the research activity.
(5) Other running costs (such as those of materials, supplies, and the like), incurred directly as a result of the research activity.[139]

Concerning subsidies to disadvantaged regions, Article 8(2)(b) of the SCM Agreement lays down conditions for eligibility. First, members can provide subsidies only to regions facing long-standing economic difficulties, as defined by criteria explicitly laid down in law, regulation, or other official documents.[140] Second, these subsidies must not target any specific enterprise, industry, or group of enterprises or industries.

The approach outlined in Article 8(2)(b) of the SCM Agreement is a double-edged sword. On the one hand, it prevents countries from misusing subsidies and upholds free trade commitments. On the other hand, it restricts the application of governmental financial support programmes, such as funds and tax incentives. Specifically, Article 8(2)(b) prohibits providing subsidies to specific industries or groups of industries, which can significantly limit the effectiveness of financial support programmes in addressing poverty. Environmental and geographical conditions often make particular industries, such as agriculture and mining, particularly suited to specific regions. Subsidising these industries can be a practical way to boost local economies. Prohibiting targeted subsidies under Article 8(2)(b) can prevent countries from adequately supporting industries crucial for the economic development of disadvantaged regions.

[138] SCM Agreement, Article 8 (2) (a).

[139] Ibid.

[140] See Article 8(2) (b) of the SCM Agreement. A region can be considered as disadvantaged if its income per capita or household income per capita, or GDP per capita, is not above 85 per cent of the average for the territory concerned. A region can also be considered as disadvantaged if its unemployment rate is at least 110 per cent of the average for the territory concerned. In addition, for a three-year assessment, countries may consider together income, unemployment rate, and other factors.

Furthermore, the use of research subsidies is limited quantitatively. The subsidy amount is determined as a percentage of industrial research costs or the costs of pre-competitive development activities. This calculation method can disadvantage developing and least-developed countries. While countries are capped at providing a subsidy of up to 75 per cent of industrial research costs, wealthier nations can invest more substantially in research programmes due to their greater economic resources. As a result, research institutions in developed countries can offer better working conditions, such as higher salaries and paid holidays, which can lead to a brain drain from developing and least-developed countries. This exacerbates the negative externalities of international trade.

Like other paragraphs of Article 8(2), paragraph (c) sets conditions for using environmental subsidies.[141] These conditions significantly limit the effectiveness of environmental subsidies in advancing green industrial policies, making it challenging for governments to address environmental adjustment costs efficiently. For instance, regarding renewable energy costs, the expenses associated with purchasing electricity, which are recurring and not directly related to facility adaptation, cannot be subsidised under these provisions.[142]

Furthermore, Article 8(2)(c) does not support the construction of facilities, which limits its applicability for promoting the development of various types of renewable energy,[143] such as biomass and wind power.[144] Developing renewable energy often relies on constructing new facilities rather than merely converting existing ones.[145]

Given these issues, WTO members should consider revising the subsidy rules if they decide to reintroduce non-actionable green subsidies into the WTO Agreements. As Mark Wu has noted:

> If the hope is to provide greater legal certainty to the enactment of certain subsidy policies to promote the development of renewable energy industries, the scope of the non-actionable category must be broadened beyond the original Article 8.[146]

[141] First, environmental subsidies are a one-time non-recurring measure. Second, the subsidy is limited to 20 per cent of the cost of adaptation. Third, the subsidy does not cover the cost of replacing and operating the assisted investment, which must be fully borne by firms. Fourth, the subsidy is directly linked to and proportionate to a firm's planned reduction of nuisances and pollution. Fifth, the subsidy does not cover any manufacturing cost savings which may be achieved. Sixth, the subsidy is available to all firms which can adopt the new equipment and/or production processes. See Article 8 (2) (c) of the SCM Agreement.

[142] Wu (2015), p. 8.

[143] Lee (2016), p. 626.

[144] Wu (2015), p. 9.

[145] Ibid., p. 9.

[146] Ibid., p. 9.

2.3.4.3 Public Health, Intellectual Property, and Medicinal Products

Intellectual property rules significantly impact countries' access to pharmaceutical products.[147] At the Doha Ministerial Conference in 2001, WTO members issued a joint ministerial statement on public health and IPRs for pharmaceutical products to ensure the availability of medicinal products.[148] Paragraph 5 of this statement allows for certain flexibilities in WTO members' commitments under the TRIPS Agreement. Specifically, paragraph 5(b) grants the right to issue compulsory licences and allows countries the freedom to determine the grounds for such licences.

Additionally, paragraph 6 acknowledges that WTO members with insufficient or no pharmaceutical manufacturing capacities may struggle to use compulsory licences under the TRIPS Agreement effectively. In response, the Council for TRIPS established detailed rules in 2003.[149] These rules aim to promote technology transfer and capacity building in the pharmaceutical sector while preventing the abuse of compulsory licenses.[150]

The importance of this commitment became particularly evident during the COVID-19 pandemic when developing countries urgently requested that developed countries waive patent rights for COVID-19 vaccines. At the WTO's 12th Ministerial Conference in 2022, WTO members agreed to such a waiver, a result that would have been inconceivable without the foundation laid by the Doha Declaration on the TRIPS Agreement and Public Health.[151]

2.4 Conclusion

This Chapter defines TSDCs and explores their implications for achieving the UN SDGs. TSDCs can be categorised into three groups: commitments to uphold inalienable, individual economic rights, commitments to fulfil social rights, and commitments to promote green economic activities. These commitments are crucial for achieving the UN SDGs. By adhering to these TSDCs, countries can, on the one hand, access the financial and material resources necessary for sustainable development and, on the other hand, prevent international trade from undermining environmental and social rights. Therefore, to achieve sustainable development, WTO members must recognise and ensure that their international trade policies align with TSDCs.

---

[147] See Cullet (2003), pp. 141–142.

[148] See WTO (2001).

[149] See WTO (2003), paras. 6–7.

[150] See Ibid., paras. 4–5.

[151] For 5 years, developing countries will be allowed to authorise the use of patented materials and ingredients to manufacture COVID-19 vaccines without the consent of the rights holder to produce vaccines for domestic and eligible markets. See WTO (2022b).

This Chapter also systematically outlines the TSDCs implicit in WTO rules, highlighting several areas lacking these commitments. First, WTO members' commitments to the freedom of trade in services are limited,[152] which diminishes the potential role of international trade in driving economic growth.

Second, the WTO agreements do not explicitly address reducing poverty and hunger through international trade. While GATT Article XXXVI indicates a commitment to reducing economic disparities between developed and developing countries, and some SDT provisions are provided to developing and least-developed countries, these measures fall short of meeting the SDGs related to poverty and hunger eradication.[153] Furthermore, objections from developed countries, such as the US, regarding the SDTs complicate the process of defining and negotiating the status of developing countries in future discussions.[154]

Third, there is a notable lack of protection for environmental industries within WTO rules. Although the WTO agreements permit the use of contingent protection measures, including anti-dumping, countervailing, and safeguard measures, to protect domestic environmental industries from dumped imports, the implementation of these measures is fraught with difficulties. Anti-dumping and countervailing measures are highly trade-restrictive and prone to abuse, while safeguards, although less restrictive,[155] present compliance challenges[156] and are less frequently used due to their non-discriminatory application.[157]

Moreover, WTO provisions related to public health, the environment, and animal protection are insufficient. Provisions in the GATT and SPS and TBT agreements do not explicitly address these issues, and WTO panels' interpretations limit their scope. This limitation adds to the uncertainty of rulings and restricts members' policy space for domestic regulation. Additionally, green subsidy provisions in the WTO Agreements have expired, leaving members without legal basis to support their environmental industries. The previous green subsidy provisions were also subject to significant limitations.

Finally, there are no explicit commitments to labour rights within the WTO Agreements. The only relevant provision is GATT Article XX(e), which allows WTO members to implement measures related to products of prison labour. There are no comprehensive labour standards or enforcement mechanisms within the WTO framework. At the Singapore Ministerial Conference in 1996, WTO members agreed that the ILO should be responsible for setting labour standards, thus leaving labour rights largely outside the scope of WTO regulations.[158]

[152] See Sect. 2.3.1.3.

[153] See Sect. 2.3.2.

[154] See Sect. 2.3.2.2.

[155] Mavroidis et al. (2010), pp. 468–469.

[156] Ibid., 508; Fang (2022), p. 242.

[157] Mavroidis et al. (2010), p. 465.

[158] See WTO (1996), para. 4.

The analysis presented in this Chapter reveals that the WTO Agreements have significantly contributed to advancing the UN SDGs. Since the establishment of the GATT, international trade rules have moved away from a laissez-faire liberalism that neglected social and environmental concerns. As noted by many scholars, the GATT recognises members' sovereign right to enact and enforce national laws and regulations. Theoretically, this framework allows national regulators to mitigate the negative externalities of international trade through well-designed trade policies.

Economic theory supports the idea that domestic interventions, when applied directly to the source of market distortions, often achieve better outcomes than import restrictions, which can reduce national gains from trade and introduce additional macroeconomic distortions.[159] For example, to address environmental issues arising from trade, implementing targeted environmental policies is generally more effective than trade restrictions. GATT Articles II and III, along with Article XX, provide WTO members with the means to optimise their trade policies in line with these objectives.

WTO panels and the AB have also worked to safeguard social and environmental interests when interpreting these provisions.[160] In some cases, such as the *EC-Asbestos* and *US-Clove Cigarettes*, the panels have implicitly referenced international standards related to environmental protection or public health to address trade disputes and achieve legitimate objectives beyond mere trade interests. This approach demonstrates the WTO's efforts to incorporate sustainability considerations within its framework. One could argue that, given the constraints of WTO rules, the DSB has maximised its role in advancing the SDGs. Despite the existing gaps and the need for more explicit and detailed TSDCs, the WTO has played a crucial role in promoting sustainable development within the boundaries of its current regulations.

However, it is important to acknowledge that the WTO Agreements have significant shortcomings in advancing sustainable development. To drive reform, this Chapter aims to spotlight these deficiencies. The absence of TSDCs within WTO rules obscures the sustainable development obligations that members are expected to uphold in international trade. This lack of clarity creates uncertainty and unpredictability regarding how WTO rules contribute to sustainable development. Additionally, the outcomes of WTO panel rulings often exhibit a high degree of contingency concerning their impact on sustainable development.

Moreover, this ambiguity undermines the legitimacy of WTO panel reports. For instance, the recognition of the sovereign right to regulate under GATT Articles II and III and the broad acknowledgement of sustainable development in the second recital of the GATT Preamble are notably vague. Although these provisions offer flexibility for WTO panels to support members' national sustainable development

[159] Petersmann (1991), p. 57.

[160] For example, see Philippe Sands (1999), para. 9. Philippe Sands stated that the decisions of the WTO AB as of 1999 showed that the compatibility of trade restrictions established in an MEA with WTO rules was less of a problem than previously thought.

policies through indirect references to international law or evolutionary interpretations, such interpretations may exceed the specific commitments made by WTO members upon joining the organisation. This approach to interpretation often leads to dissatisfaction and hampers the panels' ability to address negative trade externalities effectively. As a result, WTO panels struggle to enforce specific SDGs under the existing ambiguous rules.

Moreover, broad and hollow rules fail to provide WTO panels with the judicial powers to comprehensively balance equally important sustainability values while respecting WTO members' national regulatory autonomy. In a member-driven international organisation like the WTO, panels are constrained by members' commitments.

Achieving the UN SDGs requires a multilevel approach to governance that involves governments, regional organisations, and international institutions working together. These entities must establish and adhere to unified trade and sustainability criteria to promote sustainable development effectively. While the WTO Agreements offer a valuable framework for advancing the UN SDGs, their current implementation falls short of ensuring that international trade policies align with sustainable development requirements. In this regard, WTO law has significant room for improvement. There is a need for further progress in integrating sustainability into the WTO framework to fully realise the potential of international trade in supporting the SDGs.

Jurisprudence

GATT Report of the Panel, Italian Discrimination Against Imported Agricultural Machinery, adopted on 23 October 1958, BISD 7S/60 (1961)

GATT Report of the Panel, European Communities-Refunds on Exports of Sugar, L/5011-27S/69, adopted on 10 November 1980

GATT Report of the Panel, European Economic Community-Restrictions on Imports of Dessert Apples, L/6491-36S/93, adopted on 22 June 1989

WTO Appellate Body Report, US-Gasoline, WT/DS2/AB/R, adopted on 20 May 1996

WTO Panel Report, EC-Bananas III, WT/DS27/R/ECU, adopted on 25 September 1997

WTO Appellate Body Report, EC-Bananas III, WT/DS27/AB/R, adopted on 25 September 1997

WTO Panel Report, Argentina-Textiles and Apparel, WT/DS56/R, adopted on 22 April 1998

WTO Appellate Body Report, US-Shrimp, WT/DS58/AB/R, adopted on 6 November 1998

WTO Appellate Body Reports, Canada-Autos, WT/DS139/AB/R (WT/DS142/AB/R), adopted on 19 June 2000

WTO (2000) Implementation of Special and Differential Treatment Provisions in WTO Agreements and Decisions. WT/COMTD/W/7725. October 2000

WTO Panel Report, US-Certain EC Products, WT/DS165/R, adopted on 10 January 2001

WTO Panel Report, Argentina-Hides and Leather, WT/DS155/R, adopted on 16 February 2001

WTO Appellate Body Reports, US-Lamb, WT/DS177/AB/R (WT/DS178/AB/R), adopted on 16 May 2001

WTO Panel Report, US-Steel Plate, WT/DS206/R, adopted on 29 July 2002

WTO Panel Report, EC-Sardines, WT/DS231/R, adopted on 23 October 2002

WTO Panel Report, EC-Tube or Pipe Fittings, WT/DS219/R, adopted on 18 August 2003

WTO Panel Report, EC-Tariff Preference, WT/DS246/R, adopted on 20 April 2004

WTO Panel Report, EC-Trademarks and Geographical Indications (US), WT/DS174/R, adopted on 20 April 2005

WTO Panel Report, EC-Chicken Cuts, WT/DS269/R, adopted on 30 May 2005

WTO Panel Reports, EC-Approval and Marketing of Biotech Products, WT/DS291/R (WT/DS292/R or WT/DS293/R), adopted on 29 September 2006

WTO Panel Reports, EC-IT Products, WT/DS375/R (WT/DS376R or WT/DS377/R), adopted 21 September 2010

WTO Panel Reports, China-Raw Materials, WT/DS394/R (WT/DS395/R or WT/DS398/R), adopted on 22 February 2012

WTO Panel Report, US-Clove Cigarettes, WT/DS406/R, adopted on 24 April 2012

WTO Appellate Body Report, US-Tuna II (Mexico), WT/DS381/AB/R, adopted on 16 May 2012

WTO Panel Report, US-Tuna II (Mexico), WT/DS381/R, adopted on 13 June 2012

WTO Appellate Body Reports, US-COOL, WT/DS384/AB/R (WT/DS386/AB/R), adopted on 23 July 2012

WTO Panel Report, China-Electronic Payment Services, WT/DS413/R, adopted on 31 August 2012

WTO Appellate Body Reports, EC-Seal Products, WT/DS400/AB/R (WT/DS401/AB/R), adopted on 18 June 2014

WTO Panel Report, Ukraine-Passenger Cars, WT/DS468/R, adopted on 20 July 2015

WTO Panel Report, US-Tuna II (Mexico) (Article 21.5-Mexico), WT/DS381/RW, adopted on 3 December 2015

WTO Appellate Body Report, Indonesia-Importation of Horticultural Products, Animals and Animal Products, WT/DS477/AB/R (WT/DS478/AB/R), adopted on 22 November 2017

WTO Panel Reports, Brazil-Taxation, WT/DS472/R (WT/DS497/R), adopted on 11 January 2019

WTO Appellate Body Reports, Brazil-Taxation, WT/DS472/AB/R (WT/DS497/AB/R), adopted on 11 January 2019

WTO Appellate Body Report, Korea-Pneumatic Valves (Japan), WT/DS504/AB/R, adopted on 30 September 2019

WTO Panel Report, Russia-Railway Equipment, WT/DS499/R, adopted on 5 March 2020

References

Bibliography

Baumann-Pauly D, Posner M (2016) Making the business case for human rights: an assessment. In: Baumann-Pauly D, Nolan J (eds) Business and human rights from principles to practice. Routledge, London, pp 11–21

Bentley P, Silberston A (2007) Antidumping and countervailing action. Edward Elgar, Cheltenham

Bougette P, Charlier C (2018) Antidumping and feed-in tariffs as good buddies? Modeling the EU-China solar panel dispute. Energy J 39(6):171–188

Bown CP (2002) Why are safeguards under the WTO so unpopular? World Trade Rev 1(1):47–62

Breining-Kaufmann C (2005) The right to food and trade in agriculture. In: Cottier T et al (eds) Human rights and international trade. OUP, Oxford, pp 342–381

Bronckers M, Gruni G (2021) Retooling the sustainability standards in EU free trade agreements. J Int Econ Law 24(1):25–51

Bunn ID (2012) The right to development and international economic law. Hart, Oxford

Castellino J, Bradshaw S (2015) Sustainable development and social inclusion: why a changed approach is central to combating vulnerability. Washington Int Law J 24(3):459–494

Charnovitz S (2020) The Shadows of WTO Irrelevance. IELP blog. 14 December 2020. Available via https://ielp.worldtradelaw.net/2020/12/the-shadows-of-wto-irrelevance.html. Accessed 21 July 2024

Conconi P, Perroni C (2015) Special and differential treatment of developing countries in the WTO. World Trade Rev 14(1):67–86

Cottier T (2005) The agreement on trade-related aspects of intellectual property rights. In: Macrory PFJ et al (eds) The World Trade Organisation: legal, economic and political analysis. Springer, New York, pp 1041–1120

Cottier T, Oesch M (2005) International trade regulation. Cameron, London

Cullet P (2003) Patents and medicines: the relationship between TRIPS and the human right to health. Int Aff 79(1):139–160

Dimitropoulos G (2021) International commercial courts in the 'modern law of nature': adjudicatory unilateralism in special economic zones. J Int Econ Law 24(2):361–379

Fang MM (2022) Shedding any new light? The WTO's latest ruling in the US-China Solar Battle. Asian J WTO Int Health Law Policy 17(1):239–278

Harré L (2021) The risks and opportunities of free trade agreements for progressive labour law reform: the case of the TPP. Labour Ind 31(1):15–27

Harrison J et al (2019) Labour standards provisions in EU free trade agreements: reflections on the European Commission's reform agenda. World Trade Rev 18(4):635–657

Hudec RE (1990) The GATT legal system and world trade diplomacy. Butterworth Legal Publishers, Salem (New Hampshire)

Hudec RE (1993) Enforcing international trade law-the evolution of the modern GATT legal system. Butterworth Legal Publishers, Salem (New Hampshire)

Jackson JH (1997) The world trading system: law and policy of international economic relations. The MIT Press, Cambridge

Kudryavtsev A (2013) The TBT agreement in context. In: Epps T, Trebilcock MJ (eds) Research handbook on the WTO and technical barriers to trade. Edward Elgar, Cheltenham, pp 17–80

Laurens N et al (2019) NAFTA 2.0: the greenest trade agreement ever? World Trade Rev 18(4): 659–677
Lee J (2016) SCM agreement revisited: climate change renewable energy, and the SCM agreement. World Trade Rev 15(4):613–644
Lester S (2021) Soumaya Keynes' Great Special Report on World Trade (and Two Quibbles). IELP blog. 11 October 2021. Available via https://ielp.worldtradelaw.net/2021/10/soumaya-keynes-great-special-report-on-world-trade.html. Accessed 21 July 2024
Mah JS (2012) Special and differential treatment of developing countries and export promotion policies under the WTO. World Econ Wiley-Blackwell 34(12):1999–2018
Mavroidis PC (2012) Trade in goods. OUP, New York
Mavroidis PC et al (2010) The law and economics of contingent protection in the WTO. Edward Elgar, Cheltenham
Mbow C et al (2019) Food security. In: Shukla PR et al (eds) Climate change and land: an IPCC special report on climate change, desertification, land degradation, sustainable land management, food security, and greenhouse gas fluxes in terrestrial ecosystems. IPCC's special report. Available via https://www.ipcc.ch/srccl/chapter/chapter-5/. Accessed 21 July 2024
Petersmann E-U (1991) Constitutional functions and constitutional problems of international economic law. Routledge, New York
Petersmann E-U (2005) Human rights and international trade law: defining and connecting the two fields. In: Cottier T et al (eds) Human rights and international trade. OUP, Oxford, pp 30–94
Philippe Sands (1999) Foundation for International Environmental Law and Development (FIELD) (IEA 13). April 1999. Available via https://publications.parliament.uk/pa/cm199899/cmselect/cmenvtra/307r/307ap07.htm. Accessed 10 Nov 2022
Piérola F (2014) The challenge of safeguards in the WTO. CUP, Cambridge
Stoll, Peter-Tobias (2009) Introduction I. In: Stoll P-T et al (eds) WTO-trade-related aspects of intellectual property rights. Martinus Nijhoff Publishers 2009, Leiden, pp 1–11
Stoll P-T et al (eds) (2009) WTO-trade-related aspects of intellectual property rights. Martinus Nijhoff Publishers, Leiden
Temmerman F (2017) Trade in water under international law. Edward Elgar, Cheltenham
Wolfrum R et al (eds) (2007) WTO-technical barriers and SPS measures. Martinus Nijhoff Publishers, Leiden
Wu M (2015) Re-examining 'Green Light' Subsidies in the wake of new green industrial policies. E15 Expert Group on Reinvigorating Manufacturing New Industrial Policy and the Trade System. August 2015

Documents

GATT (1948) Summary Record of the Eleventh Meeting. GATT/CP.2/SR.11. 24 August 1948
The United Nations Environment Assembly (2018) Implementation Plan "Towards a Pollution-Free Planet". UNEP/EA.4/3. 21 December 2018. Available via https://undocs.org/UNEP/EA.4/3. Accessed 21 July 2024
UNGA (2015) Transforming Our World: the 2030 Agenda for Sustainable Development. UNGA Res/70/1. 21 October 2015
WTO (1996) Singapore Ministerial Declaration. WT/MIN (96)/DEC. 18 December 1996. Available via https://docs.wto.org/dol2fe/Pages/FE_Search/FE_S_S009-DP.aspx?language=E&CatalogueIdList=48267,32665&CurrentCatalogueIdIndex=0&FullTextHash=1&HasEnglishRecord=True&HasFrenchRecord=True&HasSpanishRecord=True. Accessed 21 July 2024

WTO (2001) Declaration on the TRIPS Agreement and Public Health. WT/MIN (01) DEC/2. 20 November 2001). Available via https://www.wto.org/english/thewto_e/minist_e/min01_e/mindecl_trips_e.htm. Accessed 21 July 2024

WTO (2003) Implementation of paragraph 6 of the Doha Declaration on the TRIPS Agreement and Public Health. Decision of the General Council of 30 August 2003. WT/L/540 and Corr.1. 1 September 2003), paras.6–7. Available via https://www.wto.org/english/tratop_e/trips_e/implem_para6_e.htm. Accessed 21 July 2024

WTO (2015) Export Competition. WT/MIN (15)/45, WT/L/980. Ministerial Decision of 19 December 2015. 21 December 2015

WTO (2019) An Undifferentiated WTO: Self-Declared Development Status Risks Institutional Irrelevance: Communication from the United States WT/GC/W/757/Rev. 1. 14 February 2019

WTO (2022a) Agreement on Fisheries Subsidies, Ministerial Decision of 17 June 2022. WT/ MIN (22) /33 WT/ L/ 1144. 22 June 2022

WTO (2022b) Draft Ministerial Decision on the TRIPS Agreement. WT/MIN (22) /W/15/Rev.2. 12–15 June 2022

WTO (2022c) MC 12 Outcome Document. WT/MIN (22)/24. 22 June 2022

WTO (2022d) Work Programme on Small Economies Ministerial Decision. WT/ MIN (22)/25 WT/L/1136. 17 June 2022

WTO General Council (2019) Draft General Council Decision: Procedures to Strengthen the Negotiating Function of the WTO. WT/GC/W/764. 15 February 2019

Chapter 3
WTO Law's Constraints on National Regulation

Abstract Under the current WTO framework, the sustainability of international trade hinges on the domestic regulation of WTO members. While WTO law acknowledges members' sovereign right to regulate, its effectiveness in supporting the implementation of these regulatory measures is debatable. Does the sovereign right imply that WTO members can effectively enact and enforce their national laws and regulations related to sustainable development? This Chapter will address this question. It begins by reviewing the evolution of regulatory autonomy within international trade. Then, it explains the concept of regulatory autonomy in WTO law and presents various scholarly perspectives on the topic. Notably, it focuses on the theory of embedded liberalism, which scholars have widely used in recent years to argue that WTO rules effectively guarantee the regulatory autonomy of members. Exploring this theory will inform my observations on the extent of WTO members' regulatory autonomy under current WTO rules. Finally, this Chapter systematically analyses the impact of the objective assessment standards of review and analytical frameworks under the WTO Agreements on the domestic regulatory autonomy of WTO members.

3.1 The Evolution of Regulatory Autonomy Within International Trade

Classical liberalism advocated for a *laissez-faire* approach, requiring governments to refrain from intervening in the economy and instead promote market deregulation and adherence to economic rationality.[1] A group of libertarian economists in the 1930s and 1960s significantly advanced this doctrine. Among them were scholars from the Mont Pelerin Society, the Freiburg Ordoliberalism school, and the Chicago

[1] Venugopal (2015), p. 172.

X. Zhao, *Integrating the UN SDGs into WTO Law*, EYIEL Monographs - Studies in European and International Economic Law 41,
https://doi.org/10.1007/978-3-031-73876-0_3

School. Due to their efforts, liberalism evolved from a radical concept into a sophisticated and inclusive ideology.[2]

Back in the late 1940s, some neoliberal economists began redefining the *laissez-faire* concept, deviating from the theories of orthodox liberalism. One of the pioneers in this movement was Henry Calvert Simons (1899–1946).[3] He argued that *laissez-faire* is not a 'do-nothing policy'. Instead, Simons claimed that *laissez-faire* necessitates a positive role for the state in maintaining competitive conditions, controlling currencies, protecting property rights, curbing monopoly power, and even maintaining social welfare.[4]

In addition, economists from the Freiburg School of Ordoliberalism and Friedrich Hayek (1899–1992)[5] contributed significantly to developing neoliberal theory. Walter Eucken (1891–1950),[6] one of the founders of the Freiburg School of Ordoliberalism, opined that:

> The state is to provide a constitutional framework, a liberal *ordo*, within which capitalist enterprises have the space to thrive in secured competition and that will prevent the "establishment of monopolistic power entities".[7]

Unlike his colleagues in the Mont Pelerin Society, Hayek valued the contribution of non-market spheres to a market expansion.[8] In the late years of his academic career, Hayek integrated more social values into his economic theory and contended that unlimited democracy would destroy a functioning market.[9]

In addition, as a non-neoliberalist, Karl Polanyi (1886–1964) widely discussed the relationship between markets and national regulation. He argued that markets must be embedded in society[10] because the emergence of a market economy can distort the relationship between capital and people.[11] Without rules, social processes and institutions will be shaped to meet market requirements rather than satisfy society's needs.[12] This market failure is detrimental to human welfare. To re-embed markets into society, national regulators must provide social regulations,

---

[2]Ibid., pp. 167–168.

[3]Henry Calvert Simons is one of essential scholars of the Chicago school of economics. See Stigler (1974), pp. 1–5.

[4]See Simons (1948); See also Peck (2008), p. 16.

[5]Scholars hold different views on whether Friedrich Hayek belongs to the Freiburg School of Ordoliberalism. See Kolev (2010); See also Rodrigues (2013), pp. 1001–1017.

[6]See Dathe (2009), pp. 53–100; See also Berghahn (2015), pp. 37–47.

[7]See Eucken (1952), p. 298; See also Berghahn (2015), p. 41.

[8]See Rodrigues (2013), p. 1015; See also Mirowski (2018), p. 898.

[9]Hayek (1979), p. 77; Mirowski (2018), p. 905.

[10]Kreide (2011), pp. 41–64.

[11]Mirowski (2018), p. 898.

[12]Kreide (2011), p. 41.

welfare services, and social protectionist measures.[13] Moreover, Polanyi opined that reforming the economic market should be subordinated to a democratic system.[14]

The development of neoliberalism has reshaped the post-WWII international trade order and promoted the reform of international trade rules. After WWII, the victorious nations established the Bretton Woods system in 1944 to prevent trade wars and preserve peace.[15] The system required the contracting parties of the Bretton Woods Agreement to peg their currencies to the US dollar to control inflation.[16] During this period, a balance between market liberalisation and government regulation characterised international trade rules. As stated in Chap. 2, the GATT 1947 recognised members' sovereign right to regulation. Members could enact and enforce national laws and regulations to prevent market failures. Hence, post-WWII international trade rules not only emphasised the importance of free trade but also aimed to prevent market failures from having spillover effects on world peace and development.

After the collapse of the Bretton Woods system in 1971,[17] the need for national regulatory autonomy became even stronger. The weakening of US economic leadership prompted a shift in US economic policies from support for free trade policies to trade protectionism.[18] The spillover effects of this policy change challenged the liberal international trading system, increasing government concerns about trade protectionism and the impact of international economic volatility on their domestic economies. Countries sought international trade rules that would provide more policy space for national regulation to address the new challenges of international trade during this period.[19]

Calls for increased national regulatory autonomy peaked during the negotiations to establish the WTO in Marrakech in 1994. Many states, especially developing countries, agreed to join the multilateral trading system only if the rules could ensure their national regulatory autonomy.[20] The WTO Agreements provide members with more explicit policy space than the GATT. Although WTO members did not incorporate new provisions into GATT to expand the scope of national regulatory

---

[13] Ibid., 41.

[14] Polanyi (1957), p. 234.

[15] See Jackson (1998), p. 15; See also Jackson (1997), p. 36.

[16] The parties set a specific fixed exchange rate for their currency with the US dollar and, as a result, cannot overly print their currency by delinking the value of their currency from gold standard. See Article IV of the Agreement of the International Monetary Fund.

[17] The US' national economic policy (i.e., the Great Society Programs) and military action (i.e., the Vietnam War) worsened the overvaluation of the US dollar by the early 1960s. On 15 August 1971, the incumbent US President Richard Nixon announced the temporary suspension of the dollar's convertibility into gold and thus ended the Bretton Woods system. See IMF, The End of the Bretton Woods System (1972–1981). Available via https://www.imf.org/external/about/histend.htm. Accessed 21 July 2024.

[18] See Hudec (1993), p. 21.

[19] Ibid., p. 23.

[20] See Sect. 2.3.1.2.

autonomy, they clarified the scope of their regulatory autonomy by concluding a series of new agreements, including the GATS, SPS Agreement, TBT Agreement, and TRIPS Agreement.[21]

3.2 National Regulation in WTO Law

3.2.1 Limited Regulatory Autonomy

While WTO law prohibits protectionist trade measures, it does not deprive members of the right to develop national disciplines.[22] WTO Agreements contain provisions that allocate regulatory jurisdiction among states, allowing members to regulate trade in goods and services.[23] For example, GATT Article XX provides policy space for WTO members to implement domestic regulatory measures on the condition that these measures do not constitute arbitrary discrimination or disguised trade barriers. These provisions grant members the right to enact and enforce domestic laws and regulations compatible with WTO law. Therefore, WTO law creates policy space for members to implement sustainable development-related measures.

The extent to which WTO members enjoy policy space remains uncertain due to the challenges in assessing the compatibility of sustainable development-related measures with WTO rules. The compatibility can vary based on the different interpretations of what constitutes WTO-consistent measures. Trachtman has observed that trade law traditionally imposed only a narrow set of limits on national regulatory autonomy, primarily to prevent states from applying discriminatory measures to imported goods.[24] If non-discrimination were the sole requirement for WTO compliance, members would indeed have considerable policy space to implement laws and regulations related to sustainable development because national measures can be non-discriminatory.

However, WTO law also stipulates that measures must not be more trade-restrictive than necessary to achieve their objectives.[25] The assessment of a measure's trade restrictiveness is done on a case-by-case basis, which complicates the determination of members' policy space.[26] Moreover, this test for trade restrictiveness raises the compliance threshold under WTO rules, further constraining members' policy space.[27]

---

[21] See Jackson (1998), p. 22.

[22] See Hudec (1993), p. 7; See also Du (2011), pp. 639–675; Reid (2010), pp. 877–901.

[23] See Trachtman (2007), pp. 633–643.

[24] Ibid., p. 632; See also Reid (2010), p. 889.

[25] I will discuss panels' analytical framework and panels' test of the trade-restrictiveness of a member's measure under different WTO provisions in Sect. 3.3.

[26] Reid (2010), pp. 894–895.

[27] Ibid., p. 895; See Wallach (2002), p. 823; See also Wagner (2000), p. 855.

3.2.2 *The Debate on Embedded Liberalism*

As noted above, the policy space provided by the WTO Agreements is limited. The question is whether WTO members' policy space for national regulation is sufficient to allow them to protect environmental and social interests from the negative externalities of international trade. For some scholars, the GATT regime had already provided an efficient way to ensure national regulatory autonomy back in 1947. They argue that the GATT regime was not characterised by *laissez-faire* liberalism but by embedded liberalism.[28] In their view, the GATT/WTO regime recognised from the very beginning the essential role of sufficient national policy space.[29] For instance, Du argues that the GATT regime secures a delicate political balance, ensuring sufficient policy space to achieve WTO members' regulatory purposes.[30] However, some scholars, such as Dunoff and Howse, contend that embedded liberalism does not accurately reflect the current nature and character of the WTO regime.[31]

I opine that the GATT regime was not entirely *laissez-faire* liberalism. Like other international economic institutions established in the post-war period, the GATT regime included interventionist elements that recognised the value of governmental intervention in economic activities. This was not the sole result of adopting a new economic theory like embedded liberalism. Indeed, Keynesian economic theory, with its interventionist ideas, played an essential role in reconstructing the post-war international economic regime. Keynes, as the delegate of the UK, significantly influenced the negotiations for establishing the Bretton Woods System, even though the US delegate, White, led the negotiations.[32]

Ruggie, the inventor of embedded liberalism,[33] also recognises the impact of Keynesian economic theory on reconstructing the post-war international economic regime. He wrote:

> True, the United States from the start of the negotiations was far less "Keynesian" in its positions than Great Britain. [...] But these differences among the industrialised countries concerned the forms and depth of state intervention to secure domestic stability, not the legitimacy of the objective.[34]

Although Polanyi's economic theory significantly influenced Ruggie's views and he, like Polanyi, addressed social purposes,[35] Ruggie did not interpret Keynes's

[28] For example, see Du (2011), p. 650.

[29] Ibid., p. 650.

[30] Ibid., p. 652.

[31] Ibid., p. 651; See Ehrlich (2010), p. 1013; See Dunoff (1999), p. 738; See also Howse (2002), p. 101.

[32] See Steil (2014).

[33] John G. Ruggie coined embedded liberalism in 1982. See Ruggie (1982), pp. 379–415.

[34] Ibid., p. 394.

[35] Ibid., pp. 385–388.

concept of 'intervention' in a new way. Regarding GATT provisions, he limited his analysis to issues related to balance-of-payments (GATT Article XII) and safeguard measures (GATT Article XIX).[36] He did not explore regulatory measures related to environmental protection, public health, or human rights. Thus, when Ruggie argued that states could use regulatory measures to ensure domestic stability,[37] he viewed national regulation as a tool to shield the domestic economy from external disruptions.[38]

Nevertheless, Ruggie's analysis of GATT Articles XII and XIX suggests that WTO members have similar regulatory autonomy within the scope of GATT Article XX. The absence of environmental and social issues in Ruggie's 1982 article is likely due to the article's focus on international transactions. Given its subject matter, which concentrated on trade and economic regulation, it is understandable that environmental and social concerns were not addressed in that context.[39]

Therefore, WTO rules theoretically allow WTO members to enforce national laws and regulations to protect environmental and social interests from the negative externalities of international trade. However, members must assert their regulatory autonomy through WTO dispute settlement proceedings. This means that WTO panels must assess the legality of members' domestic regulatory measures on a case-by-case basis. The outcome of these assessments determines the extent to which members can achieve their SDGs. Therefore, even though the GATT/WTO regime is characterised by embedded liberalism, we cannot assume that WTO members automatically have the autonomy to implement all types of sustainable development measures. Similarly, it cannot be taken for granted that the GATT/WTO regime provides sufficient policy space. How do WTO panels interpret regulatory autonomy under WTO rules? Do their panel reports offer members adequate policy space to restrict international trade for sustainable development? In Sects. 3.3 and 3.4, I will address these questions by analysing the standard of review and the analytical frameworks used to assess the legality of members' domestic measures.

3.3 Objective Assessment Standard of Review

3.3.1 Article 11 of the DSU

Article 11 of the DSU mandates that panels objectively assess the matters before them. In *Canada-Aircraft*, the AB clarified the interpretation of the word 'should' in Article 11 of the DSU, determining that an objective assessment is a duty incumbent

[36] Ibid., p. 411.
[37] Ibid., p. 405.
[38] Ibid., p. 405.
[39] Ibid., p. 383.

upon panels.[40] Subsequent panels and the AB have consistently upheld this interpretation.[41] Therefore, objective assessment is a fundamental rule that every WTO panel must adhere to when evaluating the legality of national regulatory measures.

This rule requires panels to assess both the facts of the case and the applicability of and conformity of the relevant covered agreements.[42] However, the DSU text does not provide detailed guidance on what constitutes an objective assessment. Instead, this concept has been developed through WTO panels and AB rulings in dispute settlement proceedings. According to established case law, the objective assessment standard of review requires a panel to independently evaluate the matter before it, acting as both a trier and weigher of facts.

3.3.2 *Independent Assessment*

Independent assessment requires that a panel must develop its own reasoning when evaluating the matter before it. This means a panel should not confine its analysis to the arguments or positions presented by the parties involved.[43] Cases such as *Argentina-Footwear (EC)* and *Australia-Automotive Leather II (Article 21.5-US)* illustrate the discretion panels have to develop their own reasoning. In *Argentina-Footwear (EC)*, the AB opined that a panel could not make an objective assessment if its reasoning was restricted to the specific provisions cited by the parties.[44] Similarly, in *Australia-Automotive Leather II (Article 21.5-US)*, the panel emphasised that its interpretation of a relevant WTO Agreement should not be limited by the particular arguments advanced by the parties.[45]

It is important to note that 'developing its own reasoning' is an obligation of WTO panels. Panels cannot fulfil this requirement by merely adopting the arguments presented by the parties. This principle was underscored in *US-Countervailing Measures (China)*, where the AB found that the panel had inappropriately relied solely on the language used by China's investigating authority when evaluating its

[40] See WTO Appellate Body Report, Canada-Aircraft, WT/DS70/AB/R, adopted on 20 August 1999, para. 187.

[41] See WTO Appellate Body Report, Mexico-Taxes on Soft Drinks, WT/DS308/AB/R, adopted on 24 March 2006, para. 51; See also WTO Panel Report, China-Publication and Audiovisual Products, WT/DS363/R, adopted on 19 January 2010, para. 6.46.

[42] See WTO Appellate Body Report, Mexico-Taxes on Soft Drinks, WT/DS308/AB/R, adopted on 24 March 2006, para. 51.

[43] See WTO Appellate Body Report, Brazil-Taxation, WT/DS472/AB/R, adopted on 11 January 2019, para. 5.171; WTO Appellate Body Report, Indonesia-Iron or Steel Products, WT/DS490/AB/R(WT/DS496/AB/R), adopted on 27 August 2018, paras. 5.32–5.33.

[44] See WTO Appellate Body Report, Argentina-Footwear (EC), WT/DS121/AB/R, adopted on 12 January 2000, para. 74.

[45] See WTO Panel Report, Australia-Automotive Leather II (Article 21.5-US), WT/DS126/RW, adopted on 11 February 2000, para. 6.19.

determinations in the countervailing duty proceedings. As a result, the panel failed to examine these determinations thoroughly. Consequently, the AB ruled that the panel had acted inconsistently with its obligations under Article 11 of the DSU in assessing China's claims under Article 12.7 of the SCM Agreement.[46] Additionally, this obligation requires panels to assess facts comprehensively, meaning that a panel must make efforts to obtain all evidence necessary for a complete evaluation of the matter before it.[47]

'Taking its own position when assessing the matter before it' is also a fundamental obligation for WTO panels. Jurisprudence has consistently demonstrated that an objective assessment requires panels to examine each legal requirement of WTO provisions based on the claims, arguments, and evidence presented by the parties, regardless of whether counter-arguments are made.[48] The panel report in *China-Publication and Audiovisual Products* encapsulates this obligation. The panel stated that:

> The function of panel is to make an 'objective assessment' of the matter before them. Consequently, the parties' common assessment in relation to a particular issue is, therefore, not in and of itself dispositive.[49]

3.3.3 Trier and Weigher of Facts

Panels have the right to examine all evidence essential to investigating facts. To date, panels have examined a wide range of evidence, including public statements by company executives and government officials,[50] findings from a member's

[46] See WTO Appellate Body Report, US-Countervailing Measures (China), WT/DS437/AB/R, adopted on 16 January 2015, paras. 4.188–4.190 and 4.196.

[47] See WTO Appellate Body Report, US-Large Civil Aircraft (2nd complaint), WT/DS353/AB/R, adopted on 23 March 2012, para. 1139; See also WTO Appellate Body Report, Russia-Commercial Vehicles, WT/DS479/AB/R, adopted on 9 April 2018, paras. 5.134 and 5.137.

[48] See WTO Panel Report, Mexico-Taxes on Soft Drinks, WT/DS308/R, adopted on 24 March 2006, para. 8.20; See also WTO Panel Report, US-Shrimp (Ecuador), WT/DS335/R, adopted on 20 February 2007, paras. 7.9 and 7.11; See also WTO Panel Report, US-Shrimp (Thailand), WT/DS343/R, adopted on 1 August 2008, para. 7.21; See also WTO Panel Report, United States-Certain Measures Affecting Imports of Poultry from China, WT/DS392/R, adopted on 29 September 2010, paras. 7.445–7.446; See also WTO Panel Report, China-TRQs, WT/DS517/R, adopted on 28 May 2019, para. 7.21.

[49] WTO Panel Report, China-Publication and Audiovisual Products, WT/DS363/R, adopted on 19 January 2010, para. 6.46.

[50] See WTO Panel Report, EC and Certain Member States-Large Civil Aircraft, WT/DS316/R, adopted on 1 June 2011, para. 7.1919; See also WTO Panel Reports, Argentina-Import Measures, WT/DS438/R (WT/DS444/R and WT/DS445/R), adopted on 26 January 2015, paras. 6.79–6.80.

domestic courts,[51] articles from newspapers and magazines,[52] scientific publications,[53] documents from market intelligence entities,[54] industry surveys,[55] and economic/statistical data.[56] Panels also must consider all the evidence presented before them. This obligation is well-established in WTO jurisprudence. In *EC-Hormones*, the AB explicitly stated that a panel has a duty to consider all evidence.[57] Consequently, deliberately ignoring or refusing to consider submitted evidence is inconsistent with a panel's obligation to assess facts objectively.[58] The AB also noted that wilfully distorting or misrepresenting evidence undermines the objectivity of the assessment.[59] This principle has been consistently upheld in subsequent cases.[60]

---

[51] See WTO Panel Report, US-Shrimp II (Viet Nam), WT/DS429/R, adopted on 22 April 2015, para. 7.308.

[52] See WTO Panel Reports, Argentina – Import Measures, WT /DS438/R (WT/DS444/R and WT/DS445/R), adopted on 26 January 2015, paras. 6.70–6.71.

[53] See WTO Panel Reports, Australia – Tobacco Plain Packaging (Honduras), WT/DS435/R, adopted on 29 June 2020, paras. 7.513–7.514, 7.516, 7.545, 7.549–7.550, 7.627 and 7.622.

[54] WTO Panel Reports, Argentina-Import Measures, WT/DS438/R (WT/DS444/R and WT/DS445/R), adopted on 26 January 2015, para. 6.106.

[55] Ibid., para. 6.103.

[56] WTO Appellate Body Reports, Australia – Tobacco Plain Packaging (Honduras), WT/DS435/AB/R, adopted on 29 June 2020, paras. 6.122–6.123.

[57] See WTO Appellate Body Report, EC – Hormones, WT/DS26/AB/R (WT/DS48/AB/R), adopted 13 February 1998, para. 133.

[58] Ibid.

[59] Ibid.

[60] See WTO Appellate Body Report, Japan-Apples, WT/DS245/AB/R, adopted 10 December 2003, para. 221; See also WTO Appellate Body Report, EC-Asbestos, WT/DS135/AB/R, adopted on 5 April 2001, para. 161; See also WTO Appellate Body Report, Australia-Salmon, WT/DS18/AB/R, adopted on 6 November 1998, para. 266; See also WTO Appellate Body Report, EC-Bed Linen (Article 21.5-India), WT/DS11/AB/R, adopted on 12 March 2001, paras. 170, 177, and 181; See also WTO Appellate Body Report, EC-Sardines, WT/DS231/AB/R, adopted on 23 October 2002, para. 299; See also WTO Appellate Body Report, EC-Tube or Pipe Fittings, WT/DS219/AB/R, adopted on 22 July 2003, para. 125; See also WTO Appellate Body Report, Japan-Measures Affecting Agricultural Products, WT/DS76/AB/R, adopted on 19 March 1999, paras. 141–142; See also WTO Appellate Body Report, Korea-Dairy, WT/DS98/AB/R, adopted on 12 January 2000, paras. 137 and 138; See also WTO Appellate Body Report, Korea – Alcoholic Beverages, WT/DS75/AB/R (WT/DS84/AB/R), adopted on 17 February 1999, paras. 161–162; See also WTO Appellate Body Report, US – Oil Country Tubular Goods Sunset Reviews, WT/DS268/AB/R, adopted on 17 December 2004, para. 313; See also WTO Appellate Body Report, US – Gambling, WT/DS285/AB/R, adopted on 20 April 2005, para. 363; See also WTO Appellate Body Report, EC – Selected Customs Matters, WT/DS315/AB/R, adopted on 11 December 2006, para. 258; See also WTO Appellate Body Report, US – Carbon Steel, WT/DS213/AB/R, adopted on 19 December 2002, para. 142; See also WTO Appellate Body Report, Brazil – Retreaded Tyres, WT/DS332/AB/R, adopted on 17 December 2007, para. 185.

However, the objective assessment standard of review does not oblige panels to examine every piece of evidence but rather grants them the discretion to do so.[61] In *EC-Hormones*, while emphasising the duty to consider presented evidence, the AB described panels as the trier and weigher of facts.[62] The AB acknowledged that panels have the discretion to determine which evidence to use in their findings,[63] recognising that it is not practical for panels to refer to every statement made by the expert advisors.[64] In *Australia-Salmon*, the AB further clarified that panels are not required to give the same weight to factual evidence as the parties.[65] In *EC-Bed Linen*, the AB distinguished between the admissibility of evidence and the weight given to it, helping to clarify the role of panels as trier and weigher of facts. According to the AB, panels may not consider evidence that is not relevant or necessary to its determinations because they are not probative to the issues at hand.[66]

Moreover, as triers and weighers of facts, panels have the discretion to make affirmative findings based on reasonable inferences drawn from circumstantial rather than direct evidence. In *Canada-Aircraft*, the AB affirmed this discretion, noting that drawing inferences is an inherent part of a panel's task of characterising the facts of a dispute.[67] In *US-Continued Zeroing*, the AB reiterated the necessity of drawing inferences, explaining that inferences are essential due to the varying design and operation of national regulatory systems.[68]

3.3.4 Violating the Obligation to Make an Objective Assessment

Although panels have the discretion to develop their own reasoning and serve as the trier and weigher of facts, this discretion is not unlimited. Panels must always adhere

[61] See WTO Appellate Body Reports, Australia – Tobacco Plain Packaging (Honduras), WT/DS435/AB/R, adopted on 29 June 2020, para. 6.210. The AB stated that '[A] panel could not be expected to reflect each and every aspect of every argument set forth in every exhibit submitted by every party in order to comply with the obligation to make an objective assessment of the matter before it.'

[62] See WTO Appellate Body Reports, EC – Hormones, WT/DS26/AB/R (WT/DS48/AB/R), adopted 13 February 1998, para. 132.

[63] Ibid., para. 135.

[64] Ibid., para. 138.

[65] See WTO Appellate Body Report, Australia-Salmon, WT/DS18/AB/R, adopted on 6 November 1998, para. 267.

[66] See WTO Panel Report, EC – Bed Linen, WT/DS141/R, adopted on 12 March 2001, para. 6.33.

[67] See WTO Appellate Body Report, Canada-Aircraft, WT/DS70/AB/R, adopted on 20 August 1999, para. 198.

[68] See WTO Appellate Body Report, US – Continued Zeroing, WT/DS350/AB/R, adopted on 19 February 2009, para. 357.

to their duty to render an objective assessment of the matters before them.[69] However, not every error of law or incorrect legal interpretation by a panel constitutes a violation of objective assessment.[70] As triers of fact, panels do not violate objective assessment by refusing to accord evidence the weight a party desires.[71] According to the AB, only egregious errors that undermine the objectivity of a panel's assessment constitute a violation.[72]

In *Australia-Tobacco Plain Packaging (Honduras)*, the AB identified four situations in which a panel fails to make an objective assessment:

1. The panel's reasoning was internally inconsistent or lacked coherence;
2. The panel disregarded, distorted, or misrepresented evidence;
3. The panel's findings lacked a sufficient evidentiary basis in the panel record; or
4. The panel was not even-handed in its treatment of the parties' evidence.[73]

[69] See WTO Appellate Body Reports, EC – Hormones, WT/DS26/AB/R (WT/DS48/AB/R), adopted 13 February 1998, para. 133; See also WTO Panel Report, Australia-Automotive Leather II (Article 21.5-US), WT/DS126/RW, adopted on 11 February 2000, para. 9.25; See also WTO Appellate Body Report, US – Carbon Steel, WT/DS213/AB/R, adopted on 19 December 2002, para. 153; See also WTO Appellate Body Report, EC – Bed Linen (Article 21.5 – India), WT/DS141/AB/RW, adopted on 24 April 2003, para. 181; See also WTO Appellate Body Report, Japan-Apples, WT/DS245/AB/R, adopted 10 December 2003, paras. 165–166; See also WTO Appellate Body Report, Canada – Wheat Exports and Grain Imports, WT/DS276/AB/R, adopted on 27 September 2004, para. 186; See also WTO Appellate Body Report, US – Zeroing (Japan), WT/DS322/AB/R, adopted on 23 January 2007, paras. 82 and 88; See also WTO Appellate Body Report, EC – Fasteners (China) (Article 21.5 – China), WT/DS397/AB/RW, adopted on 12 February 2016, para.5.61; See also WTO Appellate Body Reports, US – Tuna II (Mexico) (Article 21.5 – Mexico), WT/DS381/AB/RW/USA (WT/DS381/AB/RW2), adopted on 11 January 2019, paras. 7.219 and 7.223; See also WTO Appellate Body Report, US – Clove Cigarettes, WT/DS406/AB/R, adopted on 24 April 2012, para. 210; See also WTO Appellate Body Reports, US – Tuna II (Mexico) (Article 21.5 – Mexico), WT/DS381/AB/RW/USA (WT/DS381/AB/RW2), adopted on 11 January 2019, para. 254; See also WTO Appellate Body Report, India – Agricultural Products, WT/DS430/AB/R, adopted on 19 June 2015, para. 5.182; See also WTO Appellate Body Reports, US-COOL, WT/DS384/AB/R (WT/DS386/AB/R), adopted on 23 July 2012, paras. 299–300.

[70] See WTO Appellate Body Report, US – Steel Safeguards, WT/DS248/AB/R, adopted on 10 December 2003, para. 497; See also WTO Appellate Body Report, Japan – DRAMS (Korea), WT/DS336/AB/R, adopted on 17 December 2007, para. 184.

[71] WTO Appellate Body Report, EC – Bed Linen (Article 21.5 – India), WT/DS141/AB/RW, adopted on 24 April 2003, para. 177.

[72] See WTO Appellate Body Report, EC and certain member States – Large Civil Aircraft, WT/DS316/AB/R, adopted on 1 June 2011, para. 1318; See also WTO Appellate Body Report, Korea – Alcoholic Beverages, WT/DS75/AB/R (WT/DS84/AB/R), adopted on 17 February 1999, para. 164; See also WTO Appellate Body Report, Japan-Measures Affecting Agricultural Products, WT/DS76/AB/R, adopted on 19 March 1999, paras. 141–142.

[73] WTO Appellate Body Reports, Australia – Tobacco Plain Packaging (Honduras), WT/DS435/AB/R, adopted on 29 June 2020, para. 6.318.

These situations can be categorised into two groups of egregious errors: a failure to accord due process and a failure to provide a reasoned and adequate explanation or basic rationale.

The violation of due process has three forms. First, in *EC-Hormones*, the AB stated that a failure to accord due process constitutes a violation of objective assessment. Such a violation denies the fundamental fairness of parties submitting evidence. Therefore, if a panel disregards or distorts the evidence submitted, it violates its obligation of objective assessment.[74] Second, in *Chile-Price Band System*, the AB stated that a panel fails in its duty to respect due process if it makes a finding on a matter not before it,[75] as this denies a party a fair right of response.[76] Similarly, in *Thailand-Cigarettes (Philippines) (Article 21.5-Philippines II)*, the panel refused to rule on a complainant's claim,[77] highlighting that ensuring the respondent's right to comment is essential to respecting due process. Third, in *US/Canada-Continued Suspension*, the AB opined that appointing and consulting with scientific experts who were not independent or impartial violated due process.[78]

The second group of egregious errors involves a failure to provide a reasoned and adequate explanation or a basic rationale. While the DSU does not specify the elements for analysing such failures, existing case law provides guidance. A panel's reasoning and assessment of evidence must meet two requirements. First, it must be reasonable. In *Brazil-Taxation*, the panel recommended a 90-day period to remove prohibited subsidies under Article 4.7 of the SCM Agreement. The AB found that the panel had failed to provide a 'reasoned and adequate explanation' or 'basic rationale' for this recommendation.[79] The AB's analysis emphasised the reasonableness of the panel's assessment based on the specific circumstances of the case. Panels must employ a case-by-case analysis and consider all relevant factors to ensure the reasonableness of their assessments. Second, it must be consistent and coherent. The AB has examined the consistency and coherence of panels' reasoning in several cases. According to the AB, a panel violates its obligation to make an objective assessment if it treats differently pieces of evidence that share the same features.[80] Additionally, in *Thailand-Cigarettes (Philippines) (Article 21.5-*

---

[74]WTO Appellate Body Reports, EC – Hormones, WT/DS26/AB/R (WT/DS48/AB/R), adopted 13 February 1998, paras. 132–133, 135 and 138; See also WTO Appellate Body Report, Japan-Apples, WT/DS245/AB/R, adopted 10 December 2003, para. 222.

[75]WTO Appellate Body Report, Chile – Price Band System, WT/DS207/AB/R, adopted on 23 October 2002, para. 173.

[76]Ibid., para. 176.

[77]See WTO Panel Report, Thailand – Cigarettes (Philippines) (Article 21.5 – Philippines II), WT/DS371/RW, circulated on 12 November 2018, para. 7.61.

[78]See WTO Appellate Body Reports, US/Canada – Continued Suspension, WT/DS321/AB/R, adopted on 14 November 2008, para. 481.

[79]See WTO Appellate Body Report, Brazil-Taxation, WT/DS472/AB/R, adopted on 11 January 2019, para. 5.460.

[80]See WTO Appellate Body Report, US – Upland Cotton (Article 21.5 – Brazil), WT/DS267/AB/RW, adopted on 20 June 2008, paras. 294–295; See also WTO Appellate Body Report, EC and

Philippines II), the panel concluded that objectivity requires panels to refuse to re-examine arguments previously submitted by a party. Re-examining facts based on the same arguments undermines the consistency and coherence of a panel's reasoning. Therefore, under Article 11 of the DSU, a WTO Panel has a duty to re-examine earlier findings and reasonings only when new arguments are presented in a second recourse to Article 21.5.[81]

3.4 Analytical Frameworks for Assessing Members' Policy Space

3.4.1 GATT Article XX (a), (b), (d), (g), and (j)

In *US-Gasoline*, the AB introduced a two-tiered test under GATT Article XX for the first time.[82] This test requires a panel first to determine whether a measure falls under one of the exceptions listed in the various paragraphs of GATT Article XX. If it does, the panel must then assess whether the measure satisfies the requirements of the chapeau of GATT Article XX.[83]

To evaluate if a domestic regulatory measure meets the criteria of a paragraph under GATT Article XX, a panel must answer two questions: whether the measure is designed to protect a legitimate objective listed in a paragraph and whether the measure is necessary to protect that objective.[84] Existing case law indicates that this analytical framework applies to paragraphs (a), (b), (d), (g), and (j) of GATT Article XX.[85]

The second question involves a necessity test. In *Korea-Various Measures on Beef*, the AB defined 'necessity' and established the analytical framework for

certain member States – Large Civil Aircraft, WT/DS316/AB/R, adopted on 1 June 2011, para. 894; See also WTO Appellate Body Report, Russia-Commercial Vehicles, WT/DS479/AB/R, adopted on 9 April 2018, paras. 5.81–5.82.

[81] WTO Panel Report, Thailand – Cigarettes (Philippines) (Article 21.5 – Philippines II), WT/DS371/RW, circulated on 12 November 2018, para. 7.16.

[82] See WTO Appellate Body Report, US-Gasoline, WT/DS2/AB/R, adopted on 20 May 1996, p. 22.

[83] See Ibid., p. 22; See also WTO Appellate Body Report, US-Shrimp, WT/DS58/AB/R, adopted on 6 November 1998, paras. 119–120; See also WTO Appellate Body Report, Brazil – Retreaded Tyres, WT/DS332/AB/R, adopted on 17 December 2007, para. 139.

[84] See WTO Appellate Body Report, Colombia-Textiles, WT/DS461/AB/R, adopted on 22 June 2016, paras. 5.67–5.70; See also WTO Panel Report, US-Tariff Measures, WT/DS543/R, circulated on 15 September 2020, para. 7.125.

[85] See WTO Appellate Body Report, US-Gasoline, WT/DS2/AB/R, adopted on 20 May 1996, p. 17; See also WTO Appellate Body Report, China-Publications and Audiovisual Products, WT/DS363/AB/R, adopted on 19 January 2010, para. 310; See also WTO Appellate Body Report, US-Shrimp, WT/DS58/AB/R, adopted on 6 November 1998, para. 136; See also WTO Appellate Body Report, India-Solar Cells, WT/DS456/AB/R, adopted on 14 October 2016, paras. 5.58 and 5.60.

assessing the necessity of a member's regulatory measure.[86] The AB's definition of a necessity analysis involves weighing and balancing several factors, including the importance of the societal interest or value at stake, the contribution of the measure to its objective, and the trade restrictiveness of the measure. Typically, this analysis includes a comparison between the challenged measure and possible alternatives.[87]

After analysing a paragraph, a panel must examine the requirements in the chapeau of GATT Article XX. This involves determining whether a member's measure constitutes arbitrary or unjustifiable discrimination or a disguised restriction on international trade. The AB considers the chapeau test essential to prevent abuse of the exceptions under GATT Article XX.[88] In US-Shrimp, the AB emphasised the need to balance a member's right to invoke an exception with the duty to respect other members' treaty rights,[89] thereby minimising restrictions on international commerce similar to the trade restrictiveness test under the paragraphs of Article XX.

The AB's report in *US-Shrimp* outlines three elements for determining 'arbitrary or unjustifiable discrimination'[90]: the measure must result in discrimination, the discrimination must be arbitrary or unjustifiable, and the discrimination must occur between countries where the same conditions prevail.[91]

The core of determining 'arbitrary or unjustifiable discrimination' lies in demonstrating that the measure is arbitrary or unjustifiable. In *US-Tuna II (Mexico) (Article 21.5-Mexico)*, the AB stated that the analysis should focus on the cause of the discrimination or the rationale behind it.[92] Existing case law offers examples. In *US-Gasoline*, the AB found US measures to be unjustifiable discrimination and a disguised restriction on international trade because the resulting discrimination was foreseeable and not merely inadvertent.[93] In *US-Shrimp*, the AB ruled that the US measure was arbitrary discrimination as it failed to consider the regulatory appropriateness for conditions in exporting countries.[94] In *Brazil-Retreaded Tyres*, the AB concluded that Brazil's measure constituted arbitrary or unjustifiable discrimination

[86] See WTO Appellate Body Reports, Korea-Various Measures on Beef, WT/DS161/AB/R (WT/DS169/AB/R), adopted on 10 January 2001, paras. 161–164.

[87] See WTO Appellate Body Report, Colombia-Textiles, WT/DS461/AB/R, adopted on 22 June 2016, paras. 5.67–5.70; See also WTO Panel Report, US-Tariff Measures, WT/DS543/R, circulated on 15 September 2020, para. 7.125.

[88] See WTO Appellate Body Report, US-Gasoline, WT/DS2/AB/R, adopted on 20 May 1996, p. 22.

[89] See WTO Appellate Body Report, US-Shrimp, WT/DS58/AB/R, adopted on 6 November 1998, para. 157.

[90] Ibid., para. 150.

[91] See WTO Panel Report, EC-Tariff Preference, WT/DS246/R, adopted on 20 April 2004, paras. 7.228–7.229, 7.232 and 7.234.

[92] See WTO Appellate Body Reports, US – Tuna II (Mexico) (Article 21.5 – Mexico), WT/DS381/AB/RW/USA (WT/DS381/AB/RW2), adopted on 11 January 2019, para. 7.316.

[93] See WTO Appellate Body Report, US-Gasoline, WT/DS2/AB/R, adopted on 20 May 1996, pp. 28–29.

[94] See WTO Appellate Body Report, US-Shrimp, WT/DS58/AB/R, adopted on 6 November 1998, paras. 164–165.

since the rationale behind it was unrelated to the objective justified under Article XX.[95] These cases illustrate how national regulators can improve the design of laws and regulations to avoid arbitrary or unjustifiable discrimination.

Regarding disguised restrictions on international trade, the AB in *US-Gasoline* held that such a restriction could be identified using the same considerations for determining arbitrary or unjustifiable discrimination. The AB linked these concepts, ruling that a measure deemed unjustifiably discriminatory also constituted a disguised restriction on international trade.[96] Additionally, in *Brazil-Retreaded Tyres*, the AB noted that a panel should not base its determination on the impact of the measure on import volumes alone.[97] This implies that a disguised restriction must be unreasonable in nature, such as precluding competitive conditions in international trade.[98]

3.4.2 GATS Article XIV

The analytical framework for assessing GATT Article XX is also applicable to GATS Article XIV, as reflected in the AB report on *US-Gambling*. The AB found that previous decisions under GATT Article XX are relevant for the analysis under GATS Article XIV.[99] Accordingly, a panel must apply the same analytical framework used under GATT Article XX, which includes a two-tier analysis and the necessity test of a member's measure.[100]

This analytical framework requires a panel to determine whether the challenged measure falls within the scope of one of the paragraphs of GATS Article XIV and to consider whether that measure satisfies the requirements of the chapeau of GATS

[95] See WTO Appellate Body Report, Brazil – Retreaded Tyres, WT/DS332/AB/R, adopted on 17 December 2007, paras. 231–232.

[96] See WTO Appellate Body Report, US-Gasoline, WT/DS2/AB/R, adopted on 20 May 1996, p. 25.

[97] See WTO Appellate Body Report, Brazil – Retreaded Tyres, WT/DS332/AB/R, adopted on 17 December 2007, para. 239.

[98] Lo (2013) Journal of International Dispute Settlement 4(1): 111–137.

[99] See WTO Appellate Body Report, US – Gambling, WT/DS285/AB/R, adopted on 20 April 2005, para. 291. See also WTO Panel Report, Argentina-Financial Services, WT/DS453/R, adopted on 9 May 2016, para. 7.761. The panel made reference to the AB's statement in Brazil-Retreaded Tyres when examining the presence of arbitrary or unjustifiable discrimination; See also WTO Panel Report, EU-Energy Package, WT/DS476/R, circulated on 10 August 2018, para. 7.1244. The panel considered that the AB's reasoning about arbitrary or unjustifiable discrimination under Article XX are relevant for the assessment of arbitrary or unjustifiable discrimination under the chapeau of Article XIV of the GATS; See also WTO Appellate Body Report, US – Gambling, WT/DS285/AB/R, adopted on 20 April 2005, paras. 304–308. The AB followed its reasoning about GATT Article XX (d) in Korea-Various Measures on Beef to apply the necessity test under GATS Article XIV (a).

[100] See Appellate Body Report, US – Gambling, WT/DS285/AB/R, adopted on 20 April 2005, para. 292.

Article XIV. Specifically, the measure must not constitute arbitrary or unjustifiable discrimination or a disguised restriction on trade in services.[101]

Regarding the necessity test, a panel must first examine whether the measure is designed and necessary to achieve its regulatory objective.[102] The panel must then assess the trade restrictiveness of the measure and the availability of potential alternative measures. As with GATT Article XX, a potential alternative measure must achieve the same regulatory objective as the measure in dispute and be less trade-restrictive.[103]

3.4.3 Article 2.2 of the TBT Agreement

Article 2.2 of the TBT Agreement allows WTO members to implement technical regulations that may restrict international trade, provided these regulations serve legitimate purposes. These purposes include, but are not limited to, national security requirements, the prevention of deceptive practices, and the protection of human health or safety, animal or plant life or health, or the environment.[104] The crucial stipulation is that these measures should not be more trade-restrictive than necessary to achieve a legitimate objective. This requirement necessitates a panel to conduct a necessity test when evaluating the legality of a member's technical regulation.

In *US-Tuna II (Mexico)*, the AB summarised the analytical framework for determining whether a technical regulation is 'more trade-restrictive than necessary' under Article 2.2. According to the AB, a panel should begin by considering several factors: the degree of contribution the measure makes to the legitimate objective, the trade restrictiveness of the measure, and the nature and severity of the risks and consequences that would arise if the objective were not met.[105]

Subsequently, in most cases, a panel should compare the challenged measure with possible alternative measures.[106] This comparison involves assessing whether

---

[101] See Appellate Body Report, US – Gambling, WT/DS285/AB/R, adopted on 20 April 2005, para. 339; See also WTO Panel Report, Argentina-Financial Services, WT/DS453/R, adopted on 9 May 2016, paras. 7.745–7.746.

[102] See WTO Panel Report, US-Gambling, WT/DS285/R, adopted on 20 April 2005, para. 6.455; See also WTO Panel Report, EU-Energy Package, WT/DS476/R, circulated on 10 August 2018, paras. 7.229–7.231; See also WTO Panel Report, Argentina-Financial Services, WT/DS453/R, adopted on 9 May 2016, para. 7.661.

[103] See WTO Panel Report, US-Gambling, WT/DS285/R, adopted on 20 April 2005, paras. 304–308.

[104] See WTO Appellate Body Reports, US – Tuna II (Mexico) (Article 21.5 – Mexico), WT/DS381/AB/RW/USA (WT/DS381/AB/RW2), adopted on 11 January 2019, para. 313.

[105] See Ibid., para. 322; See also WTO Appellate Body Reports, US-COOL, WT/DS384/AB/R (WT/DS386/AB/R), adopted on 23 July 2012, paras. 374–378; and WTO Appellate Body Report, US-COOL (Article 21.5-Canada and Mexico), WT/DS384/AB/RW, adopted on 29 May 2015, para. 5.197.

[106] Ibid.

the proposed alternative is less trade-restrictive, whether it would make an equivalent contribution to the legitimate objective, considering the risks of non-fulfilment, and whether it is reasonably available.[107] In essence, the AB's analytical framework mirrors the necessity test under GATT Article XX.

3.4.4 Articles 5.5 and 5.6 of the SPS Agreement

Articles 5.5 and 5.6 of the SPS Agreement outline two conditions for implementing SPS measures that restrict international trade. Article 5.5 requires members to avoid arbitrary or unjustifiable distinctions in their levels of protection in different situations if such distinctions result in discrimination or a disguised restriction on international trade. Article 5.6 mandates that SPS measures should not be more trade-restrictive than necessary to achieve the appropriate level of sanitary or phytosanitary protection, considering technical and economic feasibility. These conditions require a panel to apply a test similar to that under GATT Article XX, although Articles 5.5 and 5.6 have different legal requirements.[108]

In *EC-Hormones*, the AB determined that a panel must demonstrate three elements to establish inconsistency with Article 5.5. First, the member imposing the measure must have adopted its own appropriate levels of sanitary protection against risks to human life or health in various situations. Second, these levels of protection must exhibit arbitrary or unjustifiable differences. Third, these differences must result in discrimination or a disguised restriction of international trade.[109] The analysis of the second and third elements is more complex than that of the first.

To analyse the second element, a panel must explain what constitutes an arbitrary or unjustifiable distinction under Article 5.5. In *US-Poultry (China)*, the panel referenced the AB's interpretation of 'arbitrary or unjustifiable' from GATT Article XX in *US-Gasoline*, *US-Shrimp*, and *US-Shrimp (Article 21.5-Malaysia)*.[110] The AB stated that the analysis should focus on the cause of the discrimination or the rationale behind it.[111] The panel in *US-Poultry (China)* further expanded on this by using dictionary definitions, interpreting 'arbitrary' as based on mere opinion or preference, capricious, unpredictable, inconsistent, unjustifiable, and indefensible.[112]

---

[107] Ibid.; See also WTO Panel Report, US-Clove Cigarettes, WT/DS406/R, adopted on 24 April 2012, para. 7.370.

[108] See WTO Appellate Body Reports, EC-Hormones, WT/DS26/AB/R (WT/DS48/AB/R), adopted 13 February 1998, paras. 238–239.

[109] See Ibid., para. 214.

[110] See WTO Panel Report, United States-Certain Measures Affecting Imports of Poultry from China, WT/DS392/R, adopted on 29 September 2010, paras. 7.260–7.261.

[111] Ibid.

[112] Ibid., para. 7.259.

The analysis of arbitrary or unjustifiable discrimination under Article 5.5 differs significantly from that under GATT Article XX because it relies on scientific evidence. In *US-Poultry (China)*, the panel required the respondent to demonstrate different levels of risk between comparable situations based on scientific evidence.[113]

The third element involves examining whether arbitrary or unjustifiable discrimination results in a disguised restriction on international trade. This assessment is unique to Article 5.5 and is not part of the analysis under the chapeau of GATT Article XX. The AB in *EC-Hormones* and *Australia-Salmon* provided guidance on this assessment, indicating that a panel should consider both the degree of difference and other factors, which vary case by case.[114] In *Australia-Salmon*, the panel identified three warning signals[115]: the arbitrary or unjustifiable nature of the differences in levels of protection,[116] the substantial difference in levels of protection between two different SPS measures,[117] and the inconsistency of the SPS measure with Articles 5.1 and 2.2 of the SPS Agreement. This last factor requires a member to base its SPS measure on a sufficient assessment of risks to human, animal, or plant life or health.[118]

Regarding Article 5.6, the AB stated that a panel must demonstrate three cumulative conditions to establish a violation[119]:

1. An SPS measure must be reasonably available, considering technical and economic feasibility.
2. The measure must achieve the member's appropriate level of sanitary or phytosanitary protection.
3. The measure must be significantly less restrictive to trade than the contested SPS measure.

In *India-Agricultural Products*, the AB emphasised that the analysis under Article 5.6 aims to determine whether a significantly less trade-restrictive alternative

[113] Ibid., para. 7.262.

[114] See WTO Appellate Body Reports, EC – Hormones, WT/DS26/AB/R (WT/DS48/AB/R), adopted 13 February 1998, para. 240.

[115] See WTO Appellate Body Report, Australia-Salmon, WT/DS18/AB/R, adopted on 6 November 1998, paras. 159, 161, 163 and 165.

[116] See WTO Panel Report, Australia-Salmon, WT/DS18/R, adopted on 6 November 1998, para. 8.150.

[117] Ibid.

[118] See WTO Appellate Body Report, Australia-Salmon, WT/DS18/AB/R, adopted on 6 November 1998, para. 165.

[119] See Ibid., para. 194; See also WTO Appellate Body Report, Japan-Measures Affecting Agricultural Products, WT/DS76/AB/R, adopted on 19 March 1999, para. 95; See also WTO Panel Report, United States-Certain Measures Affecting Imports of Poultry from China, WT/DS392/R, adopted on 29 September 2010, para. 7.331; See also WTO Panel Report, Australia-Apples, WT/DS367/R, adopted on 17 December 2010, para. 7.1098; See also WTO Panel Report, Japan-Apples (21.5), WT/DS245/RW, adopted on 20 July 2005, para. 8.162.

measure could meet the respondent's appropriate level of protection (ALOP).[120] This requires a necessity test similar to that under GATT Article XX.[121] A panel should first identify the member's ALOP, determine the level of protection achieved by the alternative measure(s), and compare it to the disputed measure's level of protection.[122] The panel must then decide if the alternative measure is significantly less trade-restrictive.[123] The main difference between the necessity test under Article 5.6 and GATT Article XX is the role of scientific evidence. The AB in *India-Agricultural Products* noted that the complainant must provide scientific evidence for all relevant elements under Article 5.6, including the respondent's ALOP and the protection level of the proposed alternative measure.[124]

3.5 Conclusion

The objective assessment standard of review and the analytical frameworks used by the WTO set a high threshold for compliance. First, although a member's measure may be non-discriminatory, it is often difficult to meet the requirement of less trade restrictiveness. As I have explained in Sect. 3.4, all existing analytical frameworks for assessing members' policy space for national regulation require a panel to apply a necessity test. Consequently, a member's measure can only be justified as necessary to achieve its regulatory objective if there is no less trade-restrictive alternative. The AB argues that this requirement does not constrain members' policy space because a panel must demonstrate that the alternative measure achieves the same regulatory objective. In theory, the necessity test balances a member's regulatory autonomy with free market values.[125] However, in practice, this balance is not always achieved.

Existing case law demonstrates that, despite the AB's instruction for panels to respect members' regulatory autonomy,[126] panels are not prohibited from examining the regulatory objectives of a member's measure. As the triers of fact, WTO panels must investigate the facts in light of the objective assessment standard of review and determine the actual regulatory objective of a member's measure. This duty is

---

[120] See WTO Appellate Body Report, India-Agricultural Products, WT/DS430/AB/R, adopted on 19 June 2015, para. 5.223.

[121] See Ibid., para. 7.611.

[122] See WTO Appellate Body Report, Australia-Apples, WT/DS367/AB/R, adopted on 17 December 2010, para. 368.

[123] See WTO Panel Report, Australia-Salmon, WT/DS18/R, adopted on 6 November 1998, para. 8.182.

[124] See WTO Appellate Body Report, India-Agricultural Products, WT/DS430/AB/R, adopted on 19 June 2015, para. 5.220.

[125] See WTO Appellate Body Report, China-Publications and Audiovisual Products, WT/DS363/AB/R, adopted on 19 January 2010, paras. 239 and 242–245.

[126] Neumann and Türk (2003), p. 209.

evident in the AB's findings on the analytical frameworks of Articles 5.5 and 5.6 of the SPS Agreement. The AB mandates that a panel determine the ALOP of a member's measure if the member fails to do so to prevent the member from evading its obligations under Articles 5.5 and 5.6.[127] This requirement compels a panel to scrutinise the actual ALOP of a member's SPS measure, regardless of whether the member has expressed that ALOP.[128] Similarly, under Article 2.2 of the TBT Agreement, the AB requires that a panel must not be bound by a member's characterisation of the objectives pursued through its measure when analysing the necessity of the member's technical regulation.[129]

Therefore, WTO panels are tasked with second-guessing the goals of a member's regulatory measure, which limits members' policy space for national regulation. When a panel finds the self-claimed regulatory aim incompatible with the actual regulatory goal, it must compromise or reject the member's regulatory objectives.[130] This analytical framework prioritises trade liberalisation over sustainable development. Otherwise, a panel should assess the necessity of a member's measure based on its original purpose rather than the so-called actual purpose, finding an alternative regulation that achieves the member's original goal. Thus, panels' analysis of the trade-restrictiveness of members' measures skews the balance between trade liberalisation and regulatory autonomy.[131]

Another problem with second-guessing a member's regulatory objective is identifying viable alternative measures. Although feasible alternative measures are necessary for members to achieve their regulatory objectives, the DSU does not mandate this obligation. According to Article 3(4) and Article 19 of the DSU, a panel's responsibility for achieving a satisfactory settlement of a dispute is vague, as it need not provide any feasible alternative measure to achieve the objective originally claimed by members in their submissions. The only caveat is that a panel's recommendations and rulings must not add to or diminish the rights and obligations provided in the covered agreements.

Second, panels' fact-finding for SPS measures heavily relies on scientific evidence, significantly increasing the burden of proof for WTO members. As I have explained in Sect. 3.4.4, a member must provide a sufficient risk assessment to meet the legal requirements of Articles 5.5 and 5.6, demonstrating that its different levels of sanitary protection do not constitute arbitrary or unjustifiable differences that result in discrimination or a disguised restriction on international trade. Additionally,

[127] See WTO Appellate Body Report, Australia-Salmon, WT/DS18/AB/R, adopted on 6 November 1998, para. 207.

[128] See WTO Panel Report, United States-Certain Measures Affecting Imports of Poultry from China, WT/DS392/R, adopted on 29 September 2010, para. 7.244.

[129] See WTO Appellate Body Report, US-Tuna II (Mexico), WT/DS381/AB/R, adopted on 16 May 2012, para. 314; See also WTO Appellate Body Reports, US-COOL, WT/DS384/AB/R (WT/DS386/AB/R), adopted on 23 July 2012, paras. 371–372.

[130] Kapterian (2010), pp. 125–126.

[131] Ibid., 90.

the member must provide evidence of the respondent's ALOP and the proposed alternative measure's level of protection.

However, it is challenging for members to complete a sufficient risk assessment for all risks with the current level of science. In cases where scientific evidence is inadequate for a comprehensive risk assessment, members must rely on Article 5.7 of the SPS Agreement to justify their SPS measures.[132] This Article outlines four requirements for adopting and maintaining a provisional SPS measure:

1. An SPS measure is imposed in respect of a situation where relevant scientific information is insufficient.
2. An SPS measure is adopted on the basis of available pertinent information.
3. The member who adopted the SPS measure seeks to obtain the additional information necessary for a more objective risk assessment.
4. The member who adopted the SPS measure reviews the measure accordingly within a reasonable period of time.[133]

In practice, it is difficult for members to demonstrate that relevant scientific evidence is insufficient. In *US/Canada-Continued Suspension*, the AB stated that the possibility of further research or analysis alone does not render the relevant scientific evidence insufficient.[134] The AB's reasoning is controversial as it underestimates the impact of scientific uncertainty on achieving the regulatory objective of a member's SPS measure. According to the AB, scientific uncertainty does not impede the completion of a sufficient risk assessment.[135] However, this interpretation is not conducive to protecting human, animal, or plant life or health, as it prohibits members from regulating potential risks. The danger of this interpretation was illustrated in *Korea-Radionuclides*, where the panel declined to consider the insufficiency of scientific information regarding potential future radioactive contamination, posing a threat to the health of Korean consumers.[136]

If a member fails to demonstrate that relevant scientific evidence is insufficient, it cannot invoke Article 5.7 to derogate from its obligation to complete a sufficient risk assessment. This makes it difficult for a member to meet the legal requirements of Articles 5.5 and 5.6, as the analysis of these requirements relies on the member's risk assessment.

---

[132] See WTO Appellate Body Reports, US/Canada-Continued Suspension, WT/DS321/AB/R, adopted on 14 November 2008, para. 701.

[133] See WTO Panel Report, Japan-Agricultural Products II, WT/DS76/R, adopted on 19 March 1999, paras. 8.56–8.57 and 8.60; See also WTO Appellate Body Report, Japan-Measures Affecting Agricultural Products, WT/DS76/AB/R, adopted on 19 March 1999, para. 89.

[134] See WTO Appellate Body Reports, US/Canada-Continued Suspension, WT/DS321/AB/R, adopted on 14 November 2008, para. 702.

[135] See WTO Appellate Body Report, Japan-Apples, WT/DS245/AB/R, adopted 10 December 2003, paras. 183–184. The AB stated that the application of Article 5.7 is triggered not by the existence of scientific uncertainty, but rather by the insufficiency of scientific evidence.

[136] See WTO Panel Report, Korea-Radionuclides, WT/DS495/R, adopted on 26 April 2019, para. 7.95.

Finally, in some cases, members cannot ask a panel to consider their sustainable development purposes because WTO law lacks sufficient TSDCs.[137] As a result, a panel does not need to apply a necessity test to balance trade liberalisation and regulatory autonomy in these cases. Consequently, members face greater challenges in defending the legality of their regulatory measures, thereby having less policy space for national regulation.

Jurisprudence

WTO Appellate Body Report, US-Gasoline, WT/DS2/AB/R, adopted on 20 May 1996

WTO Appellate Body Reports, EC-Hormones, WT/DS26/AB/R (WT/DS48/AB/R), adopted 13 February 1998

WTO Panel Report, Australia-Salmon, WT/DS18/R, adopted on 6 November 1998

WTO Appellate Body Report, Australia-Salmon, WT/DS18/AB/R, adopted on 6 November 1998

WTO Appellate Body Report, US-Shrimp, WT/DS58/AB/R, adopted on 6 November 1998

WTO Appellate Body Report, Korea-Alcoholic Beverages, WT/DS75/AB/R (WT/DS84/AB/R), adopted on 17 February 1999

WTO Panel Report, Japan-Agricultural Products II, WT/DS76/R, adopted on 19 March 1999

WTO Appellate Body Report, Japan-Measures Affecting Agricultural Products, WT/DS76/AB/R, adopted on 19 March 1999

WTO Appellate Body Report, Canada-Aircraft, WT/DS70/AB/R, adopted on 20 August 1999

WTO Appellate Body Report, Korea-Dairy, WT/DS98/AB/R, adopted on 12 January 2000

WTO Appellate Body Report, Argentina-Footwear (EC), WT/DS121/AB/R, adopted on 12 January 2000

WTO Panel Report, Australia-Automotive Leather II (Article 21.5-US), WT/DS126/RW, adopted on 11 February 2000

WTO Appellate Body Reports, Korea-Various Measures on Beef, WT/DS161/AB/R (WT/DS169/AB/R), adopted on 10 January 2001

WTO Appellate Body Report, EC-Bed Linen (Article 21.5-India), WT/DS11/AB/R, adopted on 12 March 2001

WTO Panel Report, EC-Bed Linen, WT/DS141/R, adopted on 12 March 2001

WTO Appellate Body Report, EC-Asbestos, WT/DS135/AB/R, adopted on 5 April 2001

[137] See Sect. 2.3.

WTO Appellate Body Report, Chile-Price Band System, WT/DS207/AB/R, adopted on 23 October 2002
WTO Appellate Body Report, EC-Sardines, WT/DS231/AB/R, adopted on 23 October 2002
WTO Appellate Body Report, US-Carbon Steel, WT/DS213/AB/R, adopted on 19 December 2002
WTO Appellate Body Report, EC-Bed Linen (Article 21.5 – India), WT/DS141/AB/RW, adopted on 24 April 2003
WTO Appellate Body Report, EC-Tube or Pipe Fittings, WT/DS219/AB/R, adopted on 22 July 2003
WTO Appellate Body Report, Japan-Apples, WT/DS245/AB/R, adopted 10 December 2003
WTO Appellate Body Report, US-Steel Safeguards, WT/DS248/AB/R, adopted on 10 December 2003
WTO Panel Report, EC-Tariff Preference, WT/DS246/R, adopted on 20 April 2004
WTO Appellate Body Report, Canada-Wheat Exports and Grain Imports, WT/DS276/AB/R, adopted on 27 September 2004
WTO Appellate Body Report, US-Oil Country Tubular Goods Sunset Reviews, WT/DS268/AB/R, adopted on 17 December 2004
WTO Panel Report, US-Gambling, WT/DS285/R, adopted on 20 April 2005
WTO Appellate Body Report, US-Gambling, WT/DS285/AB/R, adopted on 20 April 2005
WTO Panel Report, Japan-Apples (21.5), WT/DS245/RW, adopted on 20 July 2005
WTO Panel Report, Mexico-Taxes on Soft Drinks, WT/DS308/R, adopted on 24 March 2006
WTO Appellate Body Report, Mexico-Taxes on Soft Drinks, WT/DS308/AB/R, adopted on 24 March 2006
WTO Appellate Body Report, EC-Selected Customs Matters, WT/DS315/AB/R, adopted on 11 December 2006
WTO Appellate Body Report, US-Zeroing (Japan), WT/DS322/AB/R, adopted on 23 January 2007
WTO Panel Report, US-Shrimp (Ecuador), WT/DS335/R, adopted on 20 February 2007
WTO Appellate Body Report, Brazil-Retreaded Tyres, WT/DS332/AB/R, adopted on 17 December 2007
WTO Appellate Body Report, Japan-DRAMS (Korea), WT/DS336/AB/R, adopted on 17 December 2007
WTO Appellate Body Report, US-Upland Cotton (Article 21.5 – Brazil), WT/DS267/AB/RW, adopted on 20 June 2008
WTO Panel Report, US-Shrimp (Thailand), WT/DS343/R, adopted on 1 August 2008
WTO Appellate Body Reports, US/Canada-Continued Suspension, WT/DS321/AB/R, adopted on 14 November 2008
WTO Appellate Body Report, US-Continued Zeroing, WT/DS350/AB/R, adopted on 19 February 2009

WTO Panel Report, China-Publication and Audiovisual Products, WT/DS363/R, adopted on 19 January 2010
WTO Appellate Body Report, China-Publications and Audiovisual Products, WT/DS363/AB/R, adopted on 19 January 2010
WTO Panel Report, United States-Certain Measures Affecting Imports of Poultry from China, WT/DS392/R, adopted on 29 September 2010
WTO Panel Report, Australia-Apples, WT/DS367/R, adopted on 17 December 2010
WTO Appellate Body Report, Australia-Apples, WT/DS367/AB/R, adopted on 17 December 2010
WTO Panel Report, EC and Certain Member States-Large Civil Aircraft, WT/DS316/R, adopted on 1 June 2011
WTO Appellate Body Report, EC and certain member States-Large Civil Aircraft, WT/DS316/AB/R, adopted on 1 June 2011
WTO Appellate Body Report, US-Large Civil Aircraft (2nd complaint), WT/DS353/AB/R, adopted on 23 March 2012
WTO Panel Report, US-Clove Cigarettes, WT/DS406/R, adopted on 24 April 2012
WTO Appellate Body Report, US-Clove Cigarettes, WT/DS406/AB/R, adopted on 24 April 2012
WTO Appellate Body Report, US-Tuna II (Mexico), WT/DS381/AB/R, adopted on 16 May 2012
WTO Appellate Body Reports, US-COOL, WT/DS384/AB/R (WT/DS386/AB/R), adopted on 23 July 2012
WTO Appellate Body Report, US-Countervailing Measures (China), WT/DS437/AB/R, adopted on 16 January 2015
WTO Panel Reports, Argentina-Import Measures, WT/DS438/R (WT/DS444/R and WT/DS445/R), adopted on 26 January 2015
WTO Panel Report, US-Shrimp II (Viet Nam), WT/DS429/R, adopted on 22 April 2015
WTO Appellate Body Report, US-COOL (Article 21.5-Canada and Mexico), WT/DS384/AB/RW, adopted on 29 May 2015
WTO Appellate Body Report, India-Agricultural Products, WT/DS430/AB/R, adopted on 19 June 2015
WTO Appellate Body Report, EC-Fasteners (China) (Article 21.5 – China), WT/DS397/AB/RW, adopted on 12 February 2016
WTO Panel Report, Argentina-Financial Services, WT/DS453/R, adopted on 9 May 2016
WTO Appellate Body Report, Colombia-Textiles, WT/DS461/AB/R, adopted on 22 June 2016
WTO Appellate Body Report, India-Solar Cells, WT/DS456/AB/R, adopted on 14 October 2016
WTO Appellate Body Report, Russia-Commercial Vehicles, WT/DS479/AB/R, adopted on 9 April 2018
WTO Panel Report, EU-Energy Package, WT/DS476/R, circulated on 10 August 2018

WTO Appellate Body Report, Indonesia-Iron or Steel Products, WT/DS490/AB/R (WT/DS496/AB/R), adopted on 27 August 2018
WTO Panel Report, Thailand-Cigarettes (Philippines) (Article 21.5 – Philippines II), WT/DS371/RW, circulated on 12 November 2018
WTO Appellate Body Reports, US-Tuna II (Mexico) (Article 21.5 – Mexico), WT/DS381/AB/RW/USA (WT/DS381/AB/RW2), adopted on 11 January 2019
WTO Appellate Body Report, Brazil-Taxation, WT/DS472/AB/R, adopted on 11 January 2019
WTO Panel Report, Korea-Radionuclides, WT/DS495/R, adopted on 26 April 2019
WTO Panel Report, China-TRQs, WT/DS517/R, adopted on 28 May 2019
WTO Panel Reports, Australia-Tobacco Plain Packaging (Honduras), WT/DS435/R, adopted on 29 June 2020
WTO Appellate Body Reports, Australia-Tobacco Plain Packaging (Honduras), WT/DS435/AB/R, adopted on 29 June 2020
WTO Panel Report, US-Tariff Measures, WT/DS543/R, circulated on 15 September 2020

References

Berghahn VR (2015) Ordoliberalism, Ludwig Erhard, and West Germany's "Economic Basic Law". Eur Rev Int Stud 2(3):37–47
Dathe U (2009) Walter Euckens Weg zum Liberalismus (1918–1934). Ordo 60:53–100
Du MM (2011) The rise of national regulatory autonomy in the GATT/WTO regime. J Int Econ Law 14(3):639–675
Dunoff J (1999) The death of the trade regime. Eur J Int Law 10(4):733–762
Ehrlich D (2010) The fair trade challenge to embedded liberalism. Int Stud Q 54:1013–1033
Eucken W (1952) Grundsätze der Wirtschaftspolitik. A. Franke, Sankt Augustin and Berlin
Hayek F (1979) Law, legislation and liberty. University of Chicago Press, Chicago
Howse R (2002) From politics to technocracy-and Back again: the fate of the multilateral trading regime. Am J Int Law 96:94–117
Hudec RE (1993) Enforcing international trade law-the evolution of the modern GATT legal system. Butterworth Legal Publishers, Salem (New Hampshire)
Jackson JH (1997) The world trading system: law and policy of international economic relations. The MIT Press, Cambridge
Jackson JH (1998) The World Trade Organization: constitution and jurisprudence. Cassell Imprint, London
Kapterian G (2010) A critique of the WTO jurisprudence on necessity. Int Comp Law Q 59(1): 89–127
Kolev S (2010) F.A. Hayek as an Ordo-liberal. HWWI Research Paper, No.5-11. Available via https://www.econstor.eu/bitstream/10419/48214/1/664099874.pdf. Accessed 15 Aug 2022
Kreide R (2011) Re-embedding the market through law? The ambivalence of Juridification in the international context. In: Joerges C, Falke J (eds) Karl Polanyi globalisation and the potential of law in transnational markets. Hart, Oxford, pp 41–64
Lo C-F (2013) The proper interpretation of 'disguised restriction on international trade' under the WTO: the need to look at the protective effect. J Int Disp Settlement 4(1):111–137
Mirowski P (2018) Polanyi vs Hayek? Globalizations 15(7):894–910

Neumann J, Türk E (2003) Necessity revisited: proportionality in World Trade Organization law after Korea-Beef, EC-Asbestos and EC-sardines. J World Trade 37(1):199–233
Peck J (2008) Remaking Laissez-Faire. Prog Hum Geogr 32(1):3–43
Polanyi K (1957) The great transformation. Beacon Press, Boston
Reid E (2010) Regulatory autonomy in the EU and WTO: defining and defending its limits. J World Trade 44(4):877–901
Rodrigues J (2013) The political and moral economies of neoliberalism: Mises and Hayek. Camb J Econ 37:1001–1017
Ruggie JG (1982) International regimes, transactions, and change: embedded liberalism in the postwar economic order. Int Organ 36(2):379–415
Simons HC (1948) Economic policy for a free society. University of Chicago Press, Chicago
Steil B (2014) The Battle of Bretton woods: John Maynard Keynes, Harry Dexter White, and the making of a New World order. Princeton University Press, Princeton
Stigler GJ (1974) Henry Calvert Simons. J Law Econ 17(1):1–5
Trachtman JP (2007) Regulatory jurisdiction and the WTO. J Int Econ Law 10(3):631–651
Venugopal R (2015) Neoliberalism as concept. Econ Soc 44(2):165–187
Wagner JM (2000) The WTO's interpretation of SPS agreement has undermined right of governments to establish appropriate level of protection against risk. Law Policy Int Bus 31:855–859
Wallach LM (2002) Accountable governance in the era of globalization: the WTO, NAFTA and international harmonization of standards. Univ Kansas Law Rev 50(4):823–866

Part II
Reconciling WTO Law with the United Nations Sustainable Development Goals

Chapter 4
The Sustainability Test

Abstract Existing WTO rules overly protect economic interests, often at the expense of sustainable development. To address this issue, I recommend that WTO panels adopt the sustainability test. This test, which I have developed as an academic concept, aims to reconcile WTO law with the UN SDGs. The core of the sustainability test is to balance all elements of sustainable development associated with domestic measures, ensuring that WTO panels' rulings do not undermine these goals. Balancing different elements associated with legal issues is a common challenge. Judges have developed methodologies for balancing conflicting legal values, known as the balancing approach, especially in the context of constitutional and human rights law. This approach varies across national legal systems and is known as the proportionality test in Germany. Although the balancing approach originated in national law, it is now widely used in international dispute settlements, including those of the WTO. What balancing approach do WTO panels currently take? Can this approach effectively balance different sustainable development interests? If not, what approach should WTO panels adopt when applying the sustainability test? This chapter addresses these critical questions.

4.1 Definition of the Sustainability Test

The sustainability test is an academic concept that I propose to reconcile WTO law with the UN SDGs.[1] This test should replace the necessity test used by WTO panels. It has two essential missions: first, to examine whether a member's domestic regulatory measure achieves the UN SDGs; second, to balance various sustainable development elements within a member's domestic regulatory measure if that measure aims to promote sustainable development.

The first mission of the sustainability test involves identifying the regulatory objective of a member's measure. This step should focus solely on assessing the

[1] The UN SDGs include 17 goals and 169 targets, and here I refer to all these goals and targets. See UNGA (2015).

X. Zhao, *Integrating the UN SDGs into WTO Law*, EYIEL Monographs - Studies in European and International Economic Law 41,
https://doi.org/10.1007/978-3-031-73876-0_4

measure's objective to ensure it aligns with the UN SDGs. In WTO dispute settlement proceedings, panels and the AB have developed an approach for determining the purpose of a member's domestic regulatory measure under GATT Article XX (a), (b), (d), and (j). This approach hinges on the design of the measure. For instance, in *Colombia-Textiles*, the AB determined that a measure aimed to protect public morals if it was designed to do so.[2] Similarly, the *EC-Tariff Preferences* panel analysed the design and structure of the EC's regulation to identify its objective under GATT Article XX(b).[3] When addressing GATT Article XX(d), the AB emphasised the rational relationship between the design and structure of laws or regulations and their regulatory objectives.[4] For GATT Article XX(j), the objective of measures also depended on the design of the member's domestic regulatory measure.[5] As detailed in Sect. 3.4.2, GATT Article XX, particularly paragraphs (a), (b), and (d), significantly influences the analytical framework under GATS Article XIV.[6] The justification of a member's regulatory measure under GATS Article XIV relies on demonstrating compliance with the objectives listed in its paragraphs.

WTO jurisprudence concerning these provisions has contributed to developing a balancing approach within WTO law, also known as the WTO proportionality test. WTO members have accepted this approach and have not challenged its evolutionary analysis before the AB, indicating that it is a sound and useful legal tool for identifying a member's regulatory measure. Therefore, the sustainability test should follow this approach in identifying the objective of a member's measure. More precisely, reviewing statutes, legislative history, and other evidence related to the structure of a member's domestic regulations or laws helps determine whether a measure aims to achieve the UN SDGs.

To avoid misunderstanding, I must clarify that the test methodology discussed here solely refers to a measure's 'design test' as applied under GATT Article XX (a), (b), (d), (j), and relevant GATS provisions. In my concept of the sustainability test, the 'objective test' does not include any 'necessity test' regarding the achievement of a measure's objective.[7] It also differs from the 'objective test' under Article 5.6 of the SPS Agreement and Article 2.2 of the TBT Agreement.[8] As discussed in Chap. 3,

---

[2]See WTO Appellate Body Report, Colombia-Textiles, WT/DS461/AB/R, adopted on 22 June 2016, paras. 5.67–5.70.

[3]See WTO Panel Report, EC-Tariff Preference, WT/DS246/R, adopted on 20 April 2004, paras. 7.201–202.

[4]See WTO Appellate Body Report, India-Solar Cells, WT/DS456/AB/R, adopted on 14 October 2016, para. 5.58.

[5]See Ibid., paras. 5.58 and 5.60.

[6]See WTO Panel Report, US-Gambling, WT/DS285/R, adopted on 20 April 2005, para. 6.455; See also WTO Panel Report, EU-Energy Package, WT/DS476/R, circulated on 10 August 2018, paras. 7.229–7.231.

[7]When applying the WTO necessity test, a panel also needs to examine whether a member's measure is necessary to achieve the objective of that measure. Please see Sect. 3.3.1.

[8]Please see Sects. 3.3.3 and 3.3.4, and the conclusion of Chap. 3.

the 'objective tests' under these articles differ from those under GATT Article XX and GATS Article XIV.[9] Under Article 5.6 or Article 2.2, a panel must analyse the measure's application in addition to its design.[10] This requirement is not conducive to protecting sustainable development interests because it allows for second-guessing the objective of a member's measure, thereby limiting members' policy space for national regulation.[11] Given this, the sustainability test should not include this 'necessity test' in identifying the objective of members' measures.

Upon completing the first mission, a panel should balance various sustainable development elements within a measure if it finds that the measure is designed to achieve the UN SDGs. Otherwise, it must rule that a member should comply with its trade obligations under WTO Agreements. Typically, this could involve balancing interests related to different SDGs. For instance, conflicts might arise between the eradication of poverty (an economic interest) and environmental protection (an environmental interest), or between economic growth (an economic interest) and the protection of labour rights (a social interest).[12] Other conflicts could include balancing environmental protection with traditional cultural practices or economic activities like hunting and fishing. Additionally, panels may need to resolve conflicts within a single dimension of sustainable development.[13] In all cases, the objective of this second assessment is to protect sustainable development interests as comprehensively as possible. Currently, there is no detailed framework for balancing various sustainable development interests. In the following sections, I will conceptualise the balancing approach of the sustainability test by referencing the development of balancing approaches in national law, international law in general, and WTO jurisprudence.

4.2 Development History of Balancing Approach

4.2.1 Balancing Approach in National Law

4.2.1.1 Balancing Approach as a Legal Methodology

A balancing approach is a traditional legal methodology in national law, applied differently across various states. This variation manifests in two main aspects. First, balancing approaches are termed differently in national legal systems. For instance, the approach in European countries is commonly referred to as the proportionality

[9] See the conclusion of Chap. 3.

[10] See WTO Appellate Body Report, Australia-Salmon, WT/DS18/AB/R, adopted on 6 November 1998, para. 207; See also WTO Appellate Body Reports, US-COOL, WT/DS384/AB/R (WT/DS386/AB/R), adopted on 23 July 2012, paras. 391, 396, and 424.

[11] See the conclusion of Chap. 3.

[12] Giovanella (2017), p. 8.

[13] Ibid, p. 8.

test,[14] which originated in German law,[15] while in many other countries, such as the US, it is known as the balancing test.[16] In jurisdictions employing the proportionality test, the balancing test constitutes the final step.[17] In other cases, the balancing test stands as an independent method.[18]

Secondly, the analytical framework for the balancing approach varies between legal systems. For example, the German Constitutional Court employs a proportionality test defined by Alexy to resolve conflicts between constitutional rights.[19] This German concept, deeply rooted in European culture and influenced by ancient Greek philosophers such as Aristotle, Cicero, Augustine, and Aquinas, serves as a reference for many national courts analysing the proportionality of restrictive measures.[20] Although these courts adopt a framework similar to Alexy's proportionality analysis, their approaches often differ to some extent from the original German model.[21] This variation, influenced by different cultural backgrounds, can reshape the European idea and offer other countries a perspective aligned with the EU's. The difference is especially notable in countries with diverse judicial cultures, such as common law countries, which continue to apply the balancing test distinct from the civil law systems of European continental countries.

Despite the diverse definitions, a balancing approach must ensure the effective functioning of national legal systems by establishing an axiological hierarchy among conflicting norms. Legislators play a crucial role in creating this hierarchy, determining the priority of constitutional rights through legislation, a process known as constitutional balancing.[22] However, legislators' role is limited by judicial uncertainties, as some issues prompting legal disputes may be unforeseeable.[23] While legislators can predict that certain absolute rights must take precedence, they cannot pre-determine the hierarchy among most constitutional rights due to their general

---

[14] See Çalı (2018), p. 2; See also Judgment by the European Court of Human Rights (Fifth Section), case of Renaud v France No. 13290/07 of 25 February 2010, para 38; See also Judgment by the European Court of Human Rights (First Section), case of Constantinescu v Romania No. 28871/95 of 27 June 2000.

[15] See De Sena and Acconciamessa (2021), p. 3.

[16] Ibid., p. 3.

[17] See Rachovitsa (2020), p. 2; See also Christoffersen (2009), p. 194; See also Layrysen (2018), p. 316.

[18] See De Sena and Acconciamessa (2021), p. 2.

[19] See Rachovitsa (2020), p. 2; See also Kleinlein (2011), p. 1149.

[20] Emily Crawford considers that the origin of proportionality can be traced back to the thoughts of ancient Greek philosophers, including Aristote's distributive justice and the opinion of just war of Cicero, Augustine, and Aquinas. See Crawford (2020), p. 2; See also De Sena and Acconciamessa (2021), p. 3.

[21] Concerning Robert Alexy's proportionality analysis, see Möller (2007), pp. 453–467. See also further discussion below.

[22] See Giovanella (2017), p. 14; See also Barak (2012b), pp. 345–346.

[23] Giovanella (2017), pp. 6–7.

equality and harmony, only occasionally conflicting in specific circumstances.[24] Thus, determining the priority of rights requires a case-by-case analysis.[25]

In this context, national courts bear the responsibility of balancing contradictory constitutional rights by interpreting laws and regulations, a process called interpretative balancing.[26] The value of this approach lies in its ability to help judges perform interpretative balancing, thereby protecting conflicting values to the greatest extent possible. This method aims to protect a constitutional right without necessarily invalidating its opposite,[27] allowing for the preservation of one right with minimal sacrifice to the other.[28]

4.2.1.2 Various Balancing Approaches

As previously discussed, there are various definitions of the balancing approach. In this section, I will explore some of the most representative methods. The first is Alexy's proportionality test, which he proposed for balancing constitutional rights in cases where conflicting constitutional principles are at stake.[29] Alexy's test divides the principle of proportionality into three subprinciples:[30] suitability, necessity, and proportionality in the narrow sense.[31] National courts must evaluate these three elements when applying the proportionality test.[32] Among these, the proportionality in the narrow sense involves the actual balancing activity,[33] where judges compare the severity of interference with one constitutional right (i.e., the intensity of interference) against the importance of upholding another constitutional right.[34] A measure is considered proportionate if it imposes the least restriction on a constitutionally protected right while effectively protecting another constitutional right.[35]

[24] See Zucca (2018), pp. 26–27. Zucca introduces the concept of a partial conflict. In a partial conflict, neither of the two conflicting rights must necessarily be totally disregarded; See also Giovanella (2017), p. 8.

[25] See Giovanella (2017), p. 13.

[26] See Barak (2012b), pp. 345–346.

[27] See Giovanella (2017), pp. 10–11.

[28] See Pino (2010), p. 183; See also Alexy (2002), p. 45ff.

[29] See Alexy (2003a), p. 135ff. Concerning the meaning of conflicting constitutional principles, please see Dworkin (1977).

[30] See Alexy (2005), p. 572; See also Alexy (2002), p. 66ff.

[31] See Giovanella (2017), p. 17. Giovanella mentions that Stone Sweet and Mathews named the proportionality in the narrow sense as balancing in the strict sense. See Sweet and Mathews (2008), p. 75.

[32] To know the details of Alexy's proportionality test, please see Möller (2007), pp. 453–467; See also Giovanella's summary in Giovanella (2017), pp. 17–18.

[33] Pino (2010), p. 207; See also Barak (2010), p. 7.

[34] See Alexy (2003a), p. 136; See also Alexy (2003b), pp. 440–441.

[35] Ibid., 440–445.

Alexy's proportionality test is widely regarded as one of the most influential balancing approaches globally, having impacted judicial reasoning in both civil law and common law countries.[36] Interestingly, rather than harmonising balancing approaches across jurisdictions, Alexy's test has led to the development of various adaptations.[37] For example, Barak has proposed an enhancement to Alexy's framework by incorporating social importance alongside the intensity of interference. According to Barak, judges should assess the social importance of the benefits gained from enforcing a particular law against the social significance of limiting the constitutional right in question.[38]

Another prominent balancing approach is the US balancing test. The US Supreme Court applies this test to resolve conflicts between constitutional rights,[39] such as those under the Eighth Amendment (protection against excessive bail, fines, and cruel and unusual punishment), the Fourth Amendment (protection against unreasonable searches and seizures), and the First Amendment (freedom of speech).[40] In practice, the Supreme Court employs varying tests, standards, and rules depending on the rights involved.[41] Vincent Liuzzi identified three scenarios in which the Court applies the balancing test[42]: (1) when judges determine that a predetermined 'compelling' or 'substantial' interest outweighs other interests; (2) when judges use an explicit rule, test, or standard to balance conflicting constitutional rights; and (3) when judges do not adhere to any specific rule or guide.[43] Despite these variations, all approaches fundamentally assess the proportionality of a measure's impact on fundamental rights,[44] mirroring the core elements of Alexy's proportionality test: the need for a sufficiently important government purpose, a rational connection between means and ends, and the use of less restrictive means to achieve the same goal.[45]

Canada's Oakes test is another notable balancing approach.[46] The Canadian Supreme Court uses this test to address conflicts between constitutional rights protected by Section 1 of the Canadian Charter of Rights and Freedoms.[47] Chief Justice Dickson initially formulated the Oakes test in *R. v. Oakes* and refined it in

---

[36] See Barak (2012b), pp. 198–199; See also Sweet and Mathews (2008), pp. 148–152; See also De Sena and Acconciamessa (2021), p. 3.

[37] See Giovanella (2017), p. 20.

[38] See Barak (2012a), pp. 745–746.

[39] See Aleinikoff (1987), pp. 943–1005.

[40] See Giovanella (2017), p. 24.

[41] Luizzi (1980), pp. 377–386.

[42] Ibid., 377–386.

[43] Vincent Luizzi considers that this sort of balancing approach is genuine balancing because courts first search for the prevailing interest. See Luizzi (1980), pp. 388–389.

[44] See Bin (1992), pp. 125–126.

[45] See Jackson (2015), p. 3099; See also Giovanella (2017), p. 24.

[46] See Choudhry (2006), p. 502.

[47] Ibid., p. 501; See also Giovanella (2017), p. 23.

subsequent cases.[48] The Oakes test requires the party seeking to justify a limitation on a Charter right or freedom[49] to demonstrate that the limitation is reasonable and demonstrably justified in a free and democratic society.[50] To justify such a limitation, the party must establish two criteria: first, the objective of the measure must be of sufficient importance to warrant overriding a constitutionally protected right or freedom,[51] addressing pressing and substantial concerns in a democratic society[52]; second, the judge must apply a proportionality test to ensure that the means chosen are reasonable and justified.[53] This test involves three steps: examining whether the measure is carefully designed to achieve its objective (i.e., rationally connected and not arbitrary),[54] whether it impairs the right or freedom as little as possible,[55] and whether the measure's adverse effects are proportionate to the importance of the objective.[56] The objective of this test is to balance the interests of society with those of individuals and groups.[57]

The three aforementioned approaches have significantly influenced balancing methods in other jurisdictions. For example, the Italian Constitutional Court employs a reasonableness criterion to balance conflicting interests,[58] using concepts such as *contemperamento* (tempering opposite interests), *minimomezzo* (least

---

[48] See Supreme Court of Canada, R. v. Oakes [1986] S.C.J. No.7, [1986] 1 S.C.R. 103; See also Supreme Court of Canada, R. v. Edwards Books & Art Ltd., [1986] S.C.J. No. 70, [1986] 2 S.C.R. 713; See also Supreme Court of Canada, Irwin Toy Ltd. v. Quebec (Attorney General) [1989] S.C.J. No. 36, [1989] 1 S.C.R. 927.

[49] Brian Dickson C.J. provided a non-exclusive list of underlying values and principles of a free and democratic society that are the genesis of the rights and freedoms guaranteed by the Canadian Charter of Human Rights and Freedoms. These values and principles include but no limited to respect for the inherent dignity of the human person, commitment to social justice and equity, accommodation of a wide variety of beliefs, respect for cultural and group identity, and faith in social and political institutions which enhance the participation of individuals and groups in society. See Supreme Court of Canada, R. v. Oakes [1986] S.C.J. No.7, [1986] 1 S.C.R. 103, p. 136.

[50] Ibid., pp. 136–137.

[51] Ibid., p. 138.

[52] Ibid., pp. 138–139; See also Giovanella (2017), p. 23.

[53] Ibid., p. 139; See also Supreme Court of Canada, R. v. Big M Drug Mart Ltd. [1985] S.C.J. No. 17, [1985] 1 S.C.R. 295, p. 352.

[54] Supreme Court of Canada, R. v. Oakes [1986] S.C.J. No.7, [1986] 1 S.C.R. 103, p. 139.

[55] Ibid., p. 139; See also Supreme Court of Canada, R. v. Big M Drug Mart Ltd. [1985] S.C.J. No. 17, [1985] 1 S.C.R. 295, p. 352.

[56] Supreme Court of Canada, R. v. Oakes [1986] S.C.J. No.7, [1986] 1 S.C.R. 103, p. 139; See also Giovanella (2017), p. 23; See also Choudhry (2006), p. 505.

[57] The Oakes test resolves conflicts between collective rights and individual rights. Although these two rights are contradictory in a given case, they both serve to protect the underlying values and principles of a free and democratic society. In this sense, the rights that the Oakes test balances do not constitute a total conflict. Rather, they constitute a partial clash. See Supreme Court of Canada, R. v. Oakes [1986] S.C.J. No.7, [1986] 1 S.C.R. 103, p. 139; See also Giovanella (2017), p. 8.

[58] Scaccia (1998), p. 3975.

invasive means), and *coessenzialità del limite* (coessential limits).[59] Although the Italian Constitutional Court does not explicitly recognise the principle of proportionality,[60] scholars like Scaccia argue that the Italian approach is similar to Alexy's test. Indeed, the restriction on interference with constitutional rights is central to any balancing approach. The Italian Constitutional Court's analysis aligns with Alexy's classical proportionality test by ensuring that neither conflicting constitutional right is wholly sacrificed,[61] excessive or unnecessary means are avoided,[62] and a constitutional right is minimally infringed upon to realise other constitutional values or interests.[63]

Lastly, the English necessity test, a more traditional balancing approach used by British courts, emphasises the reasonableness and fairness of decisions rather than directly balancing conflicting interests.[64] This test assesses whether there is a viable alternative to the regulatory measure in dispute. For instance, in *Regina v. Secretary of State for the Home Department*, Lord Bingham of Cornhill prioritised inmates' fundamental rights by asserting that an alternative measure could have replaced the prison's search policy without comparing the conflicting interests directly.[65]

4.2.2 Balancing Approach in International Law in General

The balancing approach originated within national legal systems and was crucial in resolving conflicts between various values, principles, legal interests, or policies.[66] Today, this approach is widely used in international dispute settlement. The development of the balancing approach in international law can be viewed from two perspectives: the transplant of law from national legal systems to international law and the evolution of the balancing approach from a nascent, immature methodology to a well-recognised and sound legal framework accepted by the international community.[67] The balancing approach is no longer confined to the municipal law

---

[59] Ibid., pp. 3973–3985. Scaccia's terminologies are Italian in origin. English translation of these terms is done by Giovanella. See Giovanella (2017), p. 22.

[60] See Ibid., p. 22.

[61] Ibid., p. 22; See also Bin (1992), p. 81ff.

[62] See Giovanella (2017), p. 22.

[63] See Scaccia (1998), p. 3392.

[64] The EU's proportionality test has replaced the English necessity test due to the integration of EU law in UK law. See Crawford (2020), p. 5.

[65] Namely, the security of the prison and the fundamental rights of inmates. See House of Lords, Judgments-Regina v. Secretary of State for the Home Department, ex parte Daly, on 23 May 2001, [2001] UKHL 26, paras. 19 and 20(2).

[66] See De Sena and Acconciamessa (2021), p. 2; See also Pino (2014), p. 541.

[67] We can understand the development of balancing approach in international law by reference to a procedure of the revolutionary creation of norms of international law. Please see Kunz (1947), p. 121.

of specific geographical regions but is applied by national courts in diverse legal systems.[68]

The spread of the balancing approach among states is partly attributed to its development in international law, reflecting its transition from a national or regional methodology to an international one. The emergence of international institutions, with broad acceptance at the national level, has been essential in institutionalising international law and providing a forum where this legal methodology could be applied globally. International institutions facilitate the spread of international law, especially since judges in civil law countries typically lack the discretion to reference foreign legal methodologies embodied in foreign judgments and decisions.[69] Incorporating the balancing approach into municipal law often results from accepting superior international norms, such as those in judgments or decisions of international courts or tribunals.[70] Key institutions promoting the spread of the balancing approach include the International Court of Justice (ICJ),[71] United Nations Human Rights Bodies,[72] the European Court of Justice (CJEU), the European Court of Human Rights (ECtHR),[73] the Inter-American Court of Human Rights (IACtHR),[74] the African Court on Human and Peoples' Rights (ACtHPR),[75] the WTO Dispute Settlement Body,[76] and various investment tribunals.[77]

Additionally, international law played a crucial role in developing the balancing approach as a methodology. This approach was first used in international humanitarian law (the law of war)[78] and subsequently applied in other fields of international law.[79] Grotius was the first to introduce the ancient Greek philosophers' thoughts on proportionality regarding 'just war' into international law, although his theories on the initiation of war and activities during the war do not explicitly mention 'proportionality' or 'balancing'.[80] In landmark cases concerning the use of force, such as the

---

[68] See Sect. 4.2.1.

[69] See Hodge (2019).

[70] See Kunz (1947), p. 121; Peters (2016), pp. 1–24.

[71] Concerning the use of balancing test in the ICJ, see Cannizzaro (2019).

[72] Concerning the use of balancing test in United Nations Human Rights Bodies, see Keller and Walther (2018).

[73] Concerning the use of balancing test in the European Court of Human Rights, see Çalı (2018).

[74] Concerning the use of balancing test in Inter-American Court of Human Rights, see Lixinski (2019).

[75] Concerning the use of balancing test in the African Court on Human and Peoples' Rights, see Rachovitsa (2020).

[76] I will discuss the application of balancing approach in the WTO dispute settlement proceedings in Sect. 4.2.3.

[77] Concerning the use of balancing test in investment arbitration, see Henckels (2018).

[78] Many scholars use 'the law of wars' when discussing international humanitarian law. See, for example, Kalshoven (2007).

[79] See Crawford (2020), pp. 2–3.

[80] Grotius (2001); See also Crawford (2020), p. 2.

Caroline Incident[81] and the *Naulilaa Arbitration (Portugal v Germany)*,[82] judges referred to Grotius' legal philosophy to measure the proportionality of states' military actions and judge their legitimacy.[83]

After WWII, international judicial institutions further developed the balancing approach. Milestones in this development include the establishment of the UN, the use of proportionality analysis in ICJ judgments, and the rise of UN human rights bodies and various regional human rights courts. The UN's establishment significantly contributed to integrating the balancing approach in international dispute settlement, with the ICJ frequently using the concept of proportionality to resolve conflicts between sovereign states.[84] For instance, in *Nicaragua v United States of America*, the court affirmed proportionality as a fundamental rule of customary international law,[85] stating that the use of force in self-defence must be proportional to the armed attack and necessary to respond to it.[86] This analysis was followed in subsequent cases, including the *Oil Platforms case (Iran v. US)*[87] and *Armed Activities on the Territory of the Congo (DRC v Uganda)*.[88] In these cases, the ICJ assessed the proportionality of military actions to determine their legitimacy.[89]

Maritime delimitation is another area where the ICJ employs proportionality analysis in its judgments.[90] In this context, the proportionality analysis serves to ensure a just and equitable division of disputed maritime areas.[91] The court aims to

---

[81] To know more details of this case, please see Wood (2018).

[82] To know more details of this case, please see Pfeil (2007).

[83] See Crawford (2020), p. 3.

[84] Ibid., p. 4.

[85] See Advisory Opinion by the ICJ, Legality of the Threat or Use of Nuclear Weapons, 8 July 1996, para. 41.

[86] See Judgment by the ICJ, Case Concerning Military and Paramilitary Activities in and against Nicaragua (Nicaragua v United States of America), 27 June 1986, paras. 176 and 194; See also Wilmshurst (2005), p. 10.

[87] See Judgment by the ICJ, Case Concerning Oil Platforms (Islamic Republic of Iran v. United States of America), 6 November 2003, para. 74.

[88] See Judgment by the ICJ, Case Concerning Armed Activities on the Territory of the Congo (Democratic Republic of the Congo v Uganda), 19 December 2005.

[89] See Judgment by the ICJ, Case Concerning Oil Platforms (Islamic Republic of Iran v. United States of America), 6 November 2003, paras. 75–78; See Judgment by the ICJ, Case Concerning Armed Activities on the Territory of the Congo (Democratic Republic of the Congo v Uganda), 19 December 2005, para. 147.

[90] See Crawford (2020), p. 4.

[91] See Judgement by the ICJ, North Sea Continental Shelf Cases (Federal Republic of Germany/ Denmark; Federal Republic of Germany/Netherlands), 20 February 1969; To know more details about this case, please see Elferink (2013); See also Judgement by the ICJ, Case Concerning the Continental Shelf (Tunisia/Libyan Arab Jamahiriya), 24 February 1982; To know more details about this case, please see Stoll (2008); See also Judgement by the ICJ, Case Concerning Delimitation of the Maritime Boundary in the Gulf of Maine Area (Canada/United States of America), 12 October 1984; To know more details about this case, please see Nelson (2007); See Judgement by the ICJ, Case Concerning the Continental Shelf (Libyan Arab Jamahiriya/Malta), 3 June 1985; To know more details about this case, please see also Elferink (2006).

balance the interests of states with straight coastlines against those with markedly concave or convex coasts and to reduce highly irregular coastlines to their true proportions.[92] To achieve this, the court must eliminate disproportionately distorting effects on otherwise acceptable boundary lines caused by the presence of islets and protrusions,[93] ensuring these factors do not unduly impact the equitable division.[94]

In *Gabcíkovo-Nagymaros*, the ICJ analysed the proportionality of peaceful countermeasures. Here, Czechoslovakia unilaterally diverted water from the Danube, prompting Hungary to accuse Czechoslovakia of depriving it of its right to an equitable and reasonable share of the river's natural resources. The court held that the effects of a countermeasure must be commensurate with the injury suffered, taking account of the rights in question.[95] Consequently, it found that Czechoslovakia's countermeasure disproportionately affected the ecology of the riparian area of Szigetköz.[96]

The rise of human rights courts further propelled the development of the balancing approach. International human rights law protects both absolute and qualified human rights.[97] For the latter, there is a need to balance individual rights against public interests and conflicting individual rights.[98] In European human rights law, these conflicts are categorised as vertical and horizontal conflicts.[99] The ECtHR applies a three-pronged proportionality test, as defined by Alexy, to address these conflicts,[100] though the court does not always apply the test consistently.[101] Sometimes, the court employs a balancing test to ensure domestic authorities strike a fair balance between competing interests transparently and structurally.[102]

---

[92] See Judgement by the ICJ, North Sea Continental Shelf Cases (Federal Republic of Germany/Denmark; Federal Republic of Germany/Netherlands), 20 February 1969, para. 98; See also Judgement by the ICJ, Case Concerning Delimitation of the Maritime Boundary in the Gulf of Maine Area (Canada/United States of America), 12 October 1984, para. 185; See also Judgement by the ICJ, Case Concerning the Continental Shelf (Libyan Arab Jamahiriya/Malta), 3 June 1985, para. 56.

[93] See Judgement by the ICJ, North Sea Continental Shelf Cases (Federal Republic of Germany/Denmark; Federal Republic of Germany/Netherlands), 20 February 1969, para. 13; See Judgement by the ICJ, Case Concerning the Continental Shelf (Libyan Arab Jamahiriya/Malta), 3 June 1985, para. 64.

[94] See Judgement by the ICJ, North Sea Continental Shelf Cases (Federal Republic of Germany/Denmark; Federal Republic of Germany/Netherlands), 20 February 1969, para. 57.

[95] See Judgment by the ICJ, Case Concerning the Gabcíkovo-Nagymaros Project (Hungary/Slovakia), 25 September 1997, para. 85.

[96] Ibid., para. 85.

[97] Qualified rights are rights that are subject to limitations and thus must be assessed against the proportionality requirements. Instead, absolute human rights are not subject to the principle of proportionality or balancing test. See Rachovitsa (2020), p. 2.

[98] To know more details, please see Keller and Walther (2018), pp. 2–17.

[99] See Çalı (2018), p. 2.

[100] See Ibid., p. 2.

[101] See Ibid., p. 2.

[102] See Mowbray (2010), pp. 289–317.

Other regional human rights courts, such as the African Court on Human and Peoples' Rights (ACtHPR) and the Inter-American Court of Human Rights (IACtHR), have adopted the EU balancing approach mutatis mutandis. Although the ACtHPR does not provide a legal basis for restricting human rights, the ACtHPR's judges refer to the ECtHR's approach and creatively apply the proportionality test to delineate human rights.[103] For example, in *Tanganyika Law Society*, the ACtHPR established a three-part test under the ACHPR,[104] ruling that any human rights restriction must be prescribed by law, serve a legitimate aim, and be proportionate to the aim pursued.[105]

Similarly, the IACtHR closely follows the ECtHR's proportionality analysis.[106] However, the IACtHR's approach is distinguished by Article 30 of the ACHPR, which explicitly provides the legal basis for a three-pronged test. This article sets three cumulative conditions for limiting human rights: the restriction must be required by law, pursue a general interest, and be tailored to achieve that interest.[107]

4.2.3 Balancing Approach in WTO Dispute Settlement

In WTO law, proportionality or balancing is an academic concept rather than an explicitly written principle.[108] Although scholars like Hilf believed that proportionality was a fundamental principle underlying the multilateral trading system,[109] the GATT/WTO agreements have never explicitly recognised it.[110] Moreover, WTO panels and the AB have not officially referred to any balancing approach,[111] such as the proportionality test or a balancing test, to resolve conflicts between competing

---

[103] See Rachovitsa (2020), p. 3.

[104] See Judgment by the African Court on Human and Peoples' Rights, Tanganyika Law Society and Legal and Human Rights Centre and Reverend Christopher R Mtikila v United Republic of Tanzania (Tanganyika Law Society), Application No. 011 of 2011, para. 106.

[105] See Ibid., para. 106; See also Judgment by the African Court on Human and Peoples' Rights, Lohé Issa Konaté v Burkina Faso (Lohé Issa Konaté), Application No. 004/2013, para. 125.

[106] See Lixinski (2019), p. 2.

[107] See Judgment by the Inter-American Court of Human Rights, Usón Ramírez v. Venezuela, 20 November 2009, Serie C No.207, para. 79.

[108] See Hilf (2001), pp. 111–130.

[109] Ibid., p. 130. Meinhard Hilf defined that the principle of proportionality underlying the multilateral trading system shall rule any process of interpretation and application of WTO law with a view to obtaining a due relation between the different interests at stake.

[110] See Marceau and Trachtman (2002), p. 851; See also Kennett et al. (2003), p. 1.

[111] See Desmedt (2001), p. 480.

rights or interests.[112] As a result, some scholars have denied the existence of the principle of proportionality in WTO law.[113]

Nevertheless, it is undisputed that the principle of proportionality exists in some sectors of the WTO Agreements. Additionally, WTO panels and the AB have, in certain cases, analysed whether members' measures were proportionate to their regulatory objectives through the 'least trade-restrictive test'.[114] These cases involved WTO countermeasure rules, the chapeau of GATT Article XX, paragraphs (a), (b), and (d) of GATT Articles XX, GATS Article XIV, Article 2.2 of the TBT Agreement, and Article 5.6 of the SPS Agreement.[115] Scholars have also identified the use of the principle of proportionality based on the language in relevant provisions.[116] For example, Mitchell argues that the terms 'equivalent', 'temporary', and 'to the extent necessary' demonstrate the qualitative and quantitative requirements of the principle of proportionality as defined by international law.[117] Similarly, Cottier and others contend that terms like 'unjustifiable', 'arbitrary', 'necessary', 'rationally', 'shall not be more trade-restrictive than necessary', 'reasonably', 'appropriate level', and 'less restrictive to trade' also indicate the principle of proportionality.[118]

Furthermore, case law supports the existence of the principle of proportionality in WTO law, with three milestones marking the development of the analytical framework for the WTO proportionality test.[119] First, the panel and AB in *US-Gasoline* modified the WTO's old-fashioned necessity test into a two-prong test,[120] which can be seen as the prototype of the WTO proportionality test as conceived by scholars. This analytical framework requires panels to examine whether a measure is applied

[112] In the context of WTO law, competing interests may include trade liberalization, non-discrimination, sovereignty, sustainable development, and the principles of public international law. Please see Hilf (2001), p. 111.

[113] See Van den Bossche (2008), p. 283; See also Desmedt (2001), pp. 443 and 477–478. Desmedt opposes the existence of an overarching proportionality principle in WTO law not because the principle itself does not appear in WTO rules but because the WTO has not prepared sufficiently to use this principle; See also Neumann and Türk (2003), p. 231. Neumann and Türk consider that neither do the current rules on "necessity" incorporate an explicit reference to proportionality, nor have WTO tribunals adopted a true fully-fledged proportionality test in their case law.

[114] See Cottier et al. (2012), p. 22.

[115] See Desmedt (2001), p. 477. Exceptional clauses include those under the GATT, GATS, the TBT Agreement, and the SPS Agreement; See also Marceau and Trachtman (2014), pp. 368–382; See also Mitchell (2006), pp. 993–1008; See also Mitchell (2008), pp. 201–202; See also Cottier et al. (2012), pp. 13–21.

[116] See Mitchell (2008), p. 191.

[117] See Mitchell (2006), pp. 994–998.

[118] See Cottier et al. (2012), pp. 16–21.

[119] See Marceau and Trachtman (2014), pp. 368–371.

[120] See WTO Panel Report, United States-Standards for Reformulated and Conventional Gasoline, WT/DS2/R, adopted 29 January 1996, paragraph 6.20; See also WTO Appellate Body Report, US-Gasoline, WT/DS2/AB/R, adopted on 20 May 1996, p. 22.

in a manner that constitutes arbitrary or unjustifiable discrimination or a disguised restriction on international trade.

Second, in Korea-Various Measures on Beef, the panel and AB explicitly began considering a set of variables when assessing the necessity of members' measures.[121] The AB asserted that WTO panels should consider the contribution of members' measures to achieving the pursued objective, the importance of the common interests or values pursued, and the measures' restrictive effects on international trade.[122] According to existing WTO jurisprudence, it is easier for a panel to consider a measure with a relatively slight impact on imported products as necessary than one with intense or broader restrictive effects.[123]

Third, the panel and AB in *EC-Asbestos* further established a two-tiered analytical framework for the proportionality test, examining members' measures under the paragraphs of GATT Article XX and the chapeau of that article separately.[124] WTO panels must first determine whether a measure is necessary to achieve the legitimate objectives listed in the paragraphs of GATT Article XX. Then, panels must examine whether the measure meets the requirements of the chapeau of GATT Article XX, namely, not constituting discrimination or disguised regulatory trade barriers.[125] Due to the impact of GATT Article XX on panels' analysis of other provisions, this analytical framework, referred to as the proportionality test, also applies mutatis mutandis to other disputes regarding the balancing of trade and non-trade interests.[126]

4.3 Balancing Test in the Sustainability Test

In applying a sustainability test, a WTO panel should adopt a balancing approach that adequately addresses the complexities of sustainable development. Many scholars advocate for the stricto sensu proportionality test, the final step of the WTO proportionality test.[127] This test is a well-established method for resolving conflicts between competing rights or interests and is likely to meet the requirements

[121] WTO Panel Report, Korea-Measures Affecting Imports of Fresh Chilled and Frozen Beef, WT/DS161/R (WT/DS169/R), 31 July 2000, para. 654; WTO Appellate Body Reports, Korea-Various Measures on Beef, WT/DS161/AB/R (WT/DS169/AB/R), adopted on 10 January 2001, para. 156; See also Kapterian (2010), p. 107.

[122] WTO Appellate Body Reports, Korea-Various Measures on Beef, WT/DS161/AB/R (WT/DS169/AB/R), adopted on 10 January 2001, para. 161–164.

[123] See Ibid., paras. 161–164.

[124] See Sect. 3.3.1.

[125] There are many articles discussing the use of disguised regulatory trade barriers. See, for example, Kogan (2004); See also Kim (2012), pp. 426–475; See also Kim (2016), pp. 283–310; See also Lo (2013), pp. 111–137.

[126] To know more details, please see Sect. 3.3.

[127] Please see citations used in Sect. 4.2.3.

for balancing the diverse and sometimes conflicting interests related to sustainable development.[128] For example, Reid has argued that the proportionality test could serve as a sustainable development-based approach to WTO dispute settlement.[129] She believes that, as demonstrated in the EU legal order,[130] this three-pronged proportionality test can help panels balance interests across the three dimensions of sustainable development.[131]

However, the stricto sensu proportionality test, often used as a 'least trade restrictive test,[132]' has notable drawbacks.[133] First, it downplays the importance of sustainable development by not prioritising it over other interests.[134] Second, WTO panels apply this test only in limited cases. As discussed in Part I, WTO provisions do not provide sufficient legal bases for TSDCs nor ensure members' policy space for domestic regulatory measures. Consequently, the proportionality test's contribution to achieving the UN SDGs is limited due to its restricted reach and analytical framework. Therefore, the balancing approach of the sustainability test should be reframed to better protect sustainable development interests.

---

[128] Please see Sect. 4.2.

[129] See Reid (2015), p. 306.

[130] See ECJ's Judgment, United Kingdom v Commission, Case C-180/96, 5 May 1998, para. 96. The ECJ defined that the proportionality test is a three-pronged assessment: (1) the measure has to be in a causal relationship to the aim pursued, (2) it must be the least onerous measure available, and (3) it should not cause disadvantages that are disproportionate to the aims pursued. Other relevant case law include ECJ's Judgment, Commission v Malta, Case C-76/08, 10 September 2009; ECJ's Judgment, Azienda Agro-Zootecnica Franchin Sarl Eolica di Altamura Srl v Regione Puglia Case, C-2/10, 21 July 2011; ECJ's Judgment, The Queen v. Secretary of State for the Environment, Minister of Agriculture, Fisheries and Food, Ex parte: H.A. Standley and others, D.G.D. Meston and Others, Intervener: National Farmers' Union, Case C-293/97, 29 April 1999; ECJ's Judgment, Stichting Natuur en Milieu and Others v College van Gedeputeerde Staten van Groningen and College van Gedeputeerde Staten van Zuid-Holland, Case C-165/09 to C-167/09, 26 May 2011.

[131] See Reid (2015), pp. 303–311.

[132] See Marceau and Trachtman (2014), p. 368.

[133] See, for example, Ellis (2008), p. 653. Ellis criticises that those current understandings of what is reasonable and equitable do not take sufficient account of long-term impacts on society and environment, and concerns about environmental impact are given too little weight when set against promises of economic growth. As a result, sustainable development and related concepts serve to disrupt the current consensus about the appropriate balance to strike among environment, society and development. Another example is that the panel in *China-Rare Earths* considered that the principles of international law had no direct effect on WTO adjudication. See WTO Panel Report, China-Rare Earths, WT/DS431/R, adopted on 29 August 2014, para. 7.261.

[134] Only equally important interests are subject to the principle of proportionality or a balancing test. Because WTO panels use the proportionality test conceptualised by scholars to balance trade interests not related to sustainable development and interests relating to sustainable development (e.g., environmental interests, public health, and animal health), the panels must consider that these two groups of interests are of equal importance. In this way, panels deny the priority of interests relating to sustainable development. Regarding the use of the principle of proportionality or a balancing test, please see Rachovitsa (2020), p. 2; See also Keller and Walther (2018), pp. 2–17.

The balancing approach must meet two conditions: it should cover any trade-related UN SDGs, sufficiently protecting the relevant interests,[135] and it should efficiently balance various sustainable development interests.[136] Ideally, this approach should safeguard as many sustainable development interests as possible. Unfortunately, its application scope is limited due to the lack of TSDCs in WTO law.[137] It is hoped that WTO members will expand the scope of this approach by integrating more TSDCs into WTO law through negotiations, covering free trade obligations, commitments to reducing economic inequalities, industry protection, public health, environmental protection, animal protection, and labour right commitments.[138] With these commitments in place, WTO Agreements would encompass a broader range of competing or conflicting interests.[139] WTO members could then use the sustainability test to balance sustainable development interests rather than merely trade and non-trade interests.[140]

When weighing and balancing conflicting interests, a panel should prioritise sustainable development interests and find a balance among them.[141] This means the balancing approach should focus on the interests within the three dimensions of sustainable development. This argument is supported by existing jurisprudence, which requires courts and tribunals, whether national or international, to balance rights or interests of equal importance.[142] In the context of WTO law, if all TSDCs are incorporated into WTO Agreements, there would be two groups of interests: sustainable development interests[143] and those irrelevant to sustainable development. Within each group, interests are of equal importance horizontally,[144] but there

135 I defined trade-related sustainable development commitments in Sect. 2.1 of this book. I further took stock of trade-related sustainable development commitments in WTO law. To cover any trade-related UN SDG, the balancing approach I discuss here must be applicable to all the WTO's trade-related sustainable development commitments.

136 These interests are associated with trade-related sustainable development commitments discussed in Sect. 2.3.

137 I discussed the shortfall in trade-related sustainable development commitments in Sect. 2.3.

138 See Sect. 2.3.

139 There would be different groups of competing interests: (1) there would be conflict between two interests not related to sustainable development; (2) between interests relating to sustainable development; and (3) between interest not related to sustainable development and interest relating to sustainable development.

140 It is noteworthy that economic interests could also be an interest relating to sustainable development, for example economic growth may help reduce poverty in the very circumstances.

141 Balancing the interests relating to sustainable development means that panel should protect one of interests relating to sustainable development without sacrificing the conflicting interest relating to sustainable development. Besides, the damage to that conflict interest must be as little as possible.

142 See Rachovitsa (2020), p. 2; See also Çalı (2018), pp. 3–7.

143 Here, I refer to interests relating to sustainable development, whether included in WTO Agreements or not. Please see Sect. 2.3.

144 The equality between the interests relating to sustainable development is clear-cut. See UNGA (2012), para. 75; See also Purvis and Mao (2019), p. 685.

is a hierarchy between the two groups. Sustainable development interests are higher ranked due to members' international obligations on sustainable development.[145] Therefore, sustainable development interests should be prioritised over other interests.

Given the above considerations, the balancing test in the sustainability test should consist of two steps. First, a panel should identify which group the conflicting interests belong to.[146] If the conflicting interests include one sustainable development interest and several unrelated interests, the panel should end the test and protect the sustainable development interest. If there are multiple conflicting sustainable development interests, the panel should continue the balancing test to weigh these interests, considering all relevant factors. With the enrichment of sustainable development interests, the factors a panel should consider may vary from case to case.

4.4 Conclusion

To protect sustainable development interests, I suggest that WTO panels replace the necessity test with a sustainability test.[147] This sustainability test should complete two essential missions. First, it should determine whether a member's domestic regulatory measure aligns with the UN SDGs.[148] Second, it should balance various sustainable development elements within a member's domestic regulatory measure, provided that the measure aims to promote sustainable development.[149]

Observing the balancing approach in national laws reveals many different methods for reconciling conflicting legal values.[150] Despite these differences, the balancing approach must require that judges consider all relevant constitutional values without prioritising one at the expense of others.[151]

International judicial institutions have extensively used the balancing approach to analyse issues in international humanitarian law and international human rights law,[152] substantially following the practices of national judges.[153] However, WTO panels slightly modify this approach when balancing the values associated with WTO members' domestic measures.[154]

---

[145] See, for example, UNGA (2015), Preamble.

[146] Member States should prove to which group the interest in question belongs. In the procedure, a panel may refer to the assessment of the objective of members' measure. Please see Sect. 3.3.

[147] See Sect. 4.1.

[148] Ibid.

[149] Ibid.

[150] See Sect. 4.2.1.

[151] See Sect. 4.2.1.2.

[152] See Sect. 4.2.2.

[153] Ibid.

[154] See Sect. 4.2.3.

The balancing approach adopted by WTO panels, known as the stricto sensu proportionality test, fails to effectively balance the elements of sustainable development, including the economy, environment, and social rights. The stricto sensu proportionality test, often used as the 'least trade restrictive test,' has significant drawbacks.[155] First, it downplays the importance of sustainable development by denying its priority over other interests. Second, WTO panels apply this test only in limited cases.

The WTO panels' balancing approach should meet two conditions: (1) it should encompass any trade-related UN SDG, thereby sufficiently protecting relevant interests, and (2) it should efficiently balance various sustainable development interests.[156] A panel must first identify the groups to which the conflicting interests belong. If the conflicting interests include two groups, and only one belongs to the sustainable development group, the panel must protect the sustainable development interest. If more than one conflicting interest belongs to the sustainable development group, the panel must balance all sustainable development elements to protect these interests. Given the complexity of sustainable development interests, the factors a panel must consider will vary from case to case. In the next chapter, I will discuss these factors in terms of each TSDC.

Jurisprudence

WTO Panel Report, United States-Standards for Reformulated and Conventional Gasoline, WT/DS2/R, adopted 29 January 1996

WTO Appellate Body Report, US-Gasoline, WT/DS2/AB/R, adopted on 20 May 1996

WTO Appellate Body Report, Australia-Salmon, WT/DS18/AB/R, adopted on 6 November 1998

WTO Panel Report, Korea-Measures Affecting Imports of Fresh Chilled and Frozen Beef, WT/DS161/R (WT/DS169/R), 31 July 2000

WTO Appellate Body Reports, Korea-Various Measures on Beef, WT/DS161/AB/R (WT/DS169/AB/R), adopted on 10 January 2001

WTO Panel Report, EC-Tariff Preference, WT/DS246/R, adopted on 20 April 2004

WTO Appellate Body Reports, US-COOL, WT/DS384/AB/R (WT/DS386/AB/R), adopted on 23 July 2012

WTO Panel Report, China-Rare Earths, WT/DS431/R, adopted on 29 August 2014

WTO Appellate Body Report, Colombia-Textiles, WT/DS461/AB/R, adopted on 22 June 2016

WTO Appellate Body Report, India-Solar Cells, WT/DS456/AB/R, adopted on 14 October 2016

[155] See Sect. 4.3.

[156] Ibid.

Supreme Court of Canada, R. v. Big M Drug Mart Ltd. [1985] S.C.J. No. 17, [1985] 1 S.C.R. 295
Supreme Court of Canada, R. v. Oakes [1986] S.C.J. No. 7, [1986] 1 S.C.R. 103
Supreme Court of Canada, R. v. Edwards Books & Art Ltd., [1986] S.C.J. No. 70, [1986] 2 S.C.R. 713
Supreme Court of Canada, Irwin Toy Ltd. v. Quebec (Attorney General) [1989] S.C. J. No. 36, [1989] 1 S.C.R. 927
ECJ's Judgment, United Kingdom v Commission, Case C-180/96, 5 May 1998
ECJ's Judgment, The Queen v. Secretary of State for the Environment, Minister of Agriculture, Fisheries and Food, Ex parte: H.A. Standley and others, D.G.D. Meston and Others, Intervener: National Farmers' Union, Case C-293/97, 29 April 1999
ECJ's Judgment, Commission v Malta, Case C-76/08, 10 September 2009
ECJ's Judgment, Stichting Natuur en Milieu and Others v College van Gedeputeerde Staten van Groningen and College van Gedeputeerde Staten van Zuid-Holland, Case C-165/09 to C-167/09, 26 May 2011
ECJ's Judgment, Azienda Agro-Zootecnica Franchin Sarl Eolica di Altamura Srl v Regione Puglia Case, C-2/10, 21 July 2011
Judgment by the European Court of Human Rights (First Section), case of Constantinescu v Romania No. 28871/95 of 27 June 2000.
House of Lords, Judgments-Regina v. Secretary of State for the Home Department, ex parte Daly, on 23 May 2001, [2001] UKHL 26
Judgment by the European Court of Human Rights (Fifth Section), case of Renaud v France No. 13290/07 of 25 February 2010
Judgment by the Inter-American Court of Human Rights, Usón Ramírez v. Venezuela, 20 November 2009, Serie C No. 207
Judgment by the African Court on Human and Peoples' Rights, Tanganyika Law Society and Legal and Human Rights Centre and Reverend Christopher R Mtikila v United Republic of Tanzania (Tanganyika Law Society), Application No. 011 of 2011
Judgment by the African Court on Human and Peoples' Rights, Lohé Issa Konaté v Burkina Faso (Lohé Issa Konaté), Application No. 004/2013
Judgement by the ICJ, North Sea Continental Shelf Cases (Federal Republic of Germany/Denmark; Federal Republic of Germany/Netherlands), 20 February 1969
Judgement by the ICJ, Case Concerning the Continental Shelf (Tunisia/Libyan Arab Jamahiriya), 24 February 1982
Judgement by the ICJ, Case Concerning Delimitation of the Maritime Boundary in the Gulf of Maine Area (Canada/United States of America), 12 October 1984
Judgement by the ICJ, Case Concerning the Continental Shelf (Libyan Arab Jamahiriya/Malta), 3 June 1985
Judgment by the ICJ, Case Concerning Military and Paramilitary Activities in and against Nicaragua (Nicaragua v United States of America), 27 June 1986
Advisory Opinion by the ICJ, Legality of the Threat or Use of Nuclear Weapons, 8 July 1996

Judgment by the ICJ, Case Concerning the Gabcíkovo-Nagymaros Project (Hungary/Slovakia), 25 September 1997
Judgment by the ICJ, Case Concerning Oil Platforms (Islamic Republic of Iran v. United States of America), 6 November 2003
Judgment by the ICJ, Case Concerning Armed Activities on the Territory of the Congo (Democratic Republic of the Congo v Uganda), 19 December 2005

References

Bibliography

Aleinikoff TA (1987) Constitutional law in the age of balancing. Yale Law J 96(5):943–1005
Alexy R (2002) A theory of constitutional rights. OUP, New York
Alexy R (2003a) Constitutional rights, balancing, and rationality. Ratio Juris 16(2):131–140
Alexy R (2003b) On balancing and subsumption. A structural comparison. Ratio Juris 16(4): 433–449
Alexy R (2005) Balancing, constitutional review, and representation. Int J Const Law 3(4):572–581
Barak A (2010) Proportionality and principled balancing. Law Ethics Human Rights 4(1):1–16
Barak A (2012a) Proportionality (2). In: Rosenfeld M, Sajó A (eds) The Oxford handbook of comparative constitutional law. OUP, New York, pp 738–755
Barak A (2012b) Proportionality: constitutional rights and their limitations. CUP, New York
Bin R (1992) Diritti e argomenti. Il bilanciamento degli interessi nella girurisprudenza costituzionali. Giuffrè, Milano
Çalı B (2018) Balancing test: European Court of Human Rights (ECtHR). Max Planck Encyclopedia of International Law. Available via https://opil.ouplaw.com/view/10.1093/law-mpeipro/e3426.013.3426/law-mpeipro-e3426?prd=OPIL. Accessed 21 July 2024
Cannizzaro E (2019) Balancing test: International Court of Justice (ICJ). Max Planck Encyclopedias of International Law. Available via https://opil.ouplaw.com/view/10.1093/law-mpeipro/e3424.013.3424/law-mpeipro-e3424. Accessed 21 July 2024
Choudhry S (2006) So what is the real legacy of oakes? Two decades of proportionality analysis under the Canadian Charter's section 1. Supreme Court Law Rev 34:501–535
Christoffersen J (2009) Fair balance: proportionality, subsidiarity and primarity in the European convention on human rights. Martinus Nijhoff Publishers, Leiden
Cottier T et al (2012) The Principle of Proportionality in International Law. NCCR Trade Working Paper. 2012/38. Available via https://www.wti.org/media/filer_public/9f/1b/9f1bd3cf-dafd-4e14-b07d-8934a0c66b8f/proportionality_final_29102012_with_nccr_coversheet.pdf. Accessed 21 July 2024
Crawford E (2020) Proportionality. Max Planck Encyclopedia of Public International Law. Available via https://opil.ouplaw.com/display/10.1093/law:epil/9780199231690/law-9780199231690-e1459. Accessed 21 July 2024
De Sena P, Acconciamessa L (2021) Balancing test. Max Planck Encyclopedia of International Law. Available via https://opil.ouplaw.com/display/10.1093/law-mpeipro/e1257.013.1257/law-mpeipro-e1257. Accessed 21 July 2024
Desmedt A (2001) Proportionality in WTO law. J Int Econ Law 4(3):441–480
Dworkin R (1977) Taking rights seriously. Duckworth, London
Elferink AGO (2006) Continental Shelf Case (Libyan Arab Jamahiriya/Malta). Max Planck Encyclopedias of International law. Available via https://opil.ouplaw.com/view/10.1093/law:epil/9780199231690/law-9780199231690-e116?prd=OPIL. Accessed 21 July 2024

Elferink AGO (2013) North Sea Continental Shelf Cases. Max Planck Encyclopedias of International Law. Available via https://opil.ouplaw.com/view/10.1093/law:epil/9780199231690/law-9780199231690-e180. Accessed 21 July 2024
Ellis J (2008) Sustainable development as a legal principle: a rhetorical analysis. In: Fabri HR et al (eds) Select proceedings of the European society of international law, Vol. 2. Bloomsbury, London, pp 641–660
Giovanella F (2017) Copyright and information privacy: conflicting rights in balance. Edward Elgar, Cheltenham
Grotius H (2001) On the law of war and peace, translated from the original Latin De Jure Belli ac Pacis and slightly abridged by A.C. Campbell, A. M. Batoche Books, Kitchener
Henckels C (2018) Balancing test: investment arbitration. Max Planck Encyclopedias of International Law. Available via https://opil.ouplaw.com/view/10.1093/law-mpeipro/e3420.013.3420/law-mpeipro-e3420. Accessed 21 July 2024
Hilf M (2001) Power, rules and principles-which orientation for WTO/GATT law? J Int Econ Law 4(1):111–130
Hodge P (2019) Justice of the Supreme Court of the United Kingdom, the scope of judicial law-making in the common law tradition. Max Planck Institute of Comparative and International Private Law. 28 October 2019. Available via https://www.supremecourt.uk/docs/speech-191028.pdf. Accessed 21 July 2024
Jackson VC (2015) Constitutional law in an age of proportionality. Yale Law J 124(8):3094–3196
Kalshoven F (2007) Reflections on the law of war. Martinus Nijhoff Publishers, Leiden
Kapterian G (2010) A critique of the WTO jurisprudence on necessity. Int Comp Law Q 59(1): 89–127
Keller H, Walther R (2018) Balancing test: United Nations Human Rights Bodies. Max Planck Encyclopedias of International Law. Available via https://opil.ouplaw.com/view/10.1093/law-mpeipro/e3422.013.3422/law-mpeipro-e3422. Accessed 21 July 2024
Kennett M et al (2003) Second guessing national level policy choices: necessity, proportionality and balance in the WTO services negotiations. Centre for International Environmental Law Working Paper. Available via https://www.ciel.org/reports/second-guessing-national-policy-choices-necessity-proportionality-and-balance-in-the-wto-services-negotiations-august-2003-kennett-neumann-tuerk-presented-at-the-world-trade-organizations-5-2/. Accessed 21 July 2024
Kim M (2012) Disguised protectionism and linkages to the GATT/WTO. World Polit 64(3): 426–475
Kim M (2016) Enduring trade disputes: disguised protectionism and duration and recurrence of international trade disputes. Rev Int Org 11:283–310
Kleinlein T (2011) Judicial law making by judicial restraint? The potential of balancing in international economic law. German Law J 12:1141–1174
Kogan LA (2004) Enlightened environmentalism or disguised protectionism? Assessing the Impact of EU Precaution-Based Standards on Developing Countries. US National Foreign Trade Council Working paper. Available via https://www.wto.org/english/forums_e/ngo_e/posp47_nftc_enlightened_e.pdf. Accessed 21 July 2024
Kunz JL (1947) Revolutionary creation of norms of international law. Am J Int Law 41(1):119–126
Layrysen L (2018) System of restrictions. In: Van Dijk P et al (eds) Theory and practice of the European convention on human rights. Intersentia, Cambridge, pp 307–330
Lixinski L (2019) Balancing test: Inter-American Court of Human Rights (IACtHR). Max Planck Encyclopedias of International Law. Available via https://opil.ouplaw.com/view/10.1093/law-mpeipro/e3425.013.3425/law-mpeipro-e3425. Accessed 21 July 2024
Lo C-F (2013) The proper interpretation of "disguised restriction on international trade" under the WTO: the need to look at the protective effect. J Int Disp Settlement 4(1):111–137
Luizzi V (1980) Balancing of interests in courts. Jurimetrics 20(4):373–404
Marceau G, Trachtman JP (2002) The technical barriers to trade agreement, the sanitary and phytosanitary measures agreement, and the general agreement on tariffs and trade-a map of

the World Trade Organization law of domestic regulation of goods. J World Trade 36(5): 811–881
Marceau G, Trachtman JP (2014) A map of the World Trade Organization law of domestic regulation of goods: the technical barriers to trade agreement, the sanitary and phytosanitary measures agreement, and the general agreement on tariffs and trade. J World Trade 48(2): 351–432
Mitchell AD (2006) Proportionality and remedies in WTO disputes. Eur J Int Law 17(5):985–1008
Mitchell AD (2008) Legal principles in WTO disputes. CUP, New York
Möller K (2007) Balancing and the structure of constitutional rights. Int J Const Law 5(3):453–468
Mowbray A (2010) A study of the principle of fair balance in the jurisprudence of the European court of human rights. Hum Rights Law Rev 10(2):289–317
Nelson D (2007) Gulf of Maine case. Max Planck Encyclopedias of International Law. Available via https://opil.ouplaw.com/view/10.1093/law:epil/9780199231690/law-9780199231690-e144?rskey=Cbix21&result=327&prd=OPIL. Accessed 21 July 2024
Neumann J, Türk E (2003) Necessity revisited: proportionality in World Trade Organization law after Korea-beef, EC-Asbestos and EC-sardines. J World Trade 37(1):199–233
Peters A (2016) Proportionality as a global constitutional principle. MPIL Research Paper Series. 22 April 2016. Available via https://papers.ssrn.com/sol3/papers.cfm?abstract_id=2773733#. Accessed 21 July 2024
Pfeil J (2007) Naulilaa Arbitration (Portugal v Germany). Max Planck Encyclopedias of International Law. Available via https://opil.ouplaw.com/view/10.1093/law:epil/9780199231690/law-9780199231690-e178. Accessed 21 July 2024
Pino G (2010) Diritti e Interpretazione. Il Ragionamento Giuridico Nello Stato Costituzionale. Il Mulino, Bologna
Pino G (2014) Diritti Fondamentali e Principio di Proporzionalità. Ragion Pratica 34:541–556
Purvis B, Mao Y (2019) Three pillars of sustainability: in search of conceptual origins. Sustain Sci 14:681–695
Rachovitsa A (2020) Balancing test: African Court on Human and Peoples' Rights (ACtHPR). Max Planck Encyclopedia of International Procedural Law. Available via https://opil.ouplaw.com/view/10.1093/law-mpeipro/e3636.013.3636/law-mpeipro-e3636. Accessed 21 July 2024
Reid E (2015) Balancing human rights, environmental protection and international trade: lessons from the EU experience. Hart, Oxford
Scaccia G (1998) Il bilanciamento degli interessi come tecnica di controllo Costituzionale. Giurisprudenza Costituzionale 3953–4000
Stoll P-T (2008) Continental shelf. Max Planck Encyclopedias of International Law. Available via https://opil.ouplaw.com/view/10.1093/law:epil/9780199231690/law-9780199231690-e1152?prd=OPIL. Accessed 21 July 2024
Sweet AS, Mathews J (2008) Proportionality balancing and global constitutionalism. Columbia J Transnatl Law 47:68–149
Van den Bossche P (2008) Looking for proportionality in WTO law. Legal Iss Econ Integr 35(3): 283–294
Wilmshurst E (2005) Principles of international law on the use of force by states in self-defence. Chatham House. October 2005. Available via https://www.chathamhouse.org/sites/default/files/public/Research/International%20Law/ilpforce.doc#:~:text=The%20ICJ%20has%20confirmed%20that,necessary%20to%20respond%20to%20it.%E2%80%9D&text=This%20requires%20that%20the%20level,attack%20or%20remove%20the%20threat. Accessed 21 July 2024
Wood M (2018) The Caroline Incident-1837. In: Ruys T et al (eds) The use of force in international law: a case-based approach. OUP, New York, pp 5–16
Zucca L (2018) Conflicts of fundamental rights as constitutional dilemmas. In: Brems E (ed) Conflicts between fundamental rights. Intersentia, Antwerp, pp 19–38

Documents

UNGA (2012) The Future We Want. Resolution adopted by the General Assembly. A/RES/66/288. 27 July 2012

UNGA (2015) Transforming Our World: the 2030 Agenda for Sustainable Development. UNGA Res/70/1. 21 October 2015

Chapter 5
Balancing Sustainability Elements

Abstract To protect sustainable development interests, WTO panels must consider all relevant elements. These elements vary according to different sustainable development interests. This chapter outlines the factors that WTO panels must consider when balancing these diverse interests. To avoid redundancy, I categorise these factors into four groups based on the TSDCs. They involve (1) poverty eradication and inequality reduction, (2) food security and sustainable agriculture, (3) human health and life, and (4) climate change, the ocean, and clean energy. This chapter explains how WTO panels avoid undermining any SDGs when reviewing the legality of domestic measures related to various SDGs. In other words, it addresses how to ensure that WTO members pursue one TSDC without violating others using the sustainability test. The findings in this chapter are based on a critical analysis of WTO jurisprudence. Therefore, each section will first explore the SDGs related to a TSDC, their relevance to WTO dispute settlement, and the issues in existing cases before indicating how WTO panels should balance these factors when implementing the sustainability test.

5.1 Poverty Eradication and Inequality Reduction

5.1.1 SDG 1, SDG 9, and SDG 10

SDG 1 is dedicated to eradicating poverty,[1] aiming to increase daily available wealth to above $1.25 per person.[2] This goal calls on states to maximise individuals' living standards through specific targets: reducing by at least half the proportion of men, women, and children of all ages living in poverty in all its dimensions according to national definitions[3] and implementing nationally appropriate social protection

[1] See UNGA (2015), targets 1.1–1. b.

[2] Ibid., target 1.1.

[3] Ibid., target 1.2.

X. Zhao, *Integrating the UN SDGs into WTO Law*, EYIEL Monographs - Studies in European and International Economic Law 41,
https://doi.org/10.1007/978-3-031-73876-0_5

systems to achieve substantial coverage of the poor and the vulnerable.[4] Moreover, SDG 1 emphasises the importance of gender equality in ending poverty. It mandates that countries ensure all men and women, particularly the poor and vulnerable, have equal rights to economic resources and basic services. These include ownership and control over land and other forms of property, inheritance, natural resources, appropriate new technology, and financial services, including microfinance.[5] Another focus of SDG 1 is to mitigate the impact of climate change on individual well-being and build the resilience of the poor and those in vulnerable situations. The goal aims to reduce their exposure and vulnerability to climate-related extreme events and other economic, social, and environmental shocks and disasters.[6] SDG 1 also calls for global cooperation in mobilising financial resources to provide adequate and predictable means for developing countries, particularly least-developed countries.[7] This will support the implementation of programmes and policies to end poverty in all its dimensions. Finally, creating a sound policy framework is a crucial focus of SDG 1. The goal, recognising the vital role of policy in achieving these targets, requires the inclusion of pro-poor and gender-sensitive development strategies at various levels of government policies, with a focus on promoting investment in poverty eradication actions.[8]

SDG 9 is dedicated to building resilient infrastructure, promoting inclusive and sustainable industrialisation, and fostering innovation. This goal encompasses several critical targets. The first priority is to develop high-quality, reliable, sustainable, and resilient infrastructure.[9] This target aims to upgrade infrastructure and retrofit industries to ensure their sustainability.[10] Additionally, enhancing the technological capabilities of industrial sectors is a crucial target.[11] The goal, recognising the lack of technological capacities in developing countries, emphasises technology development in these regions and mandates that developed countries assist developing nations in advancing research and innovation.[12] A specific focus of this target is to ensure universal and affordable internet access in the least-developed countries, enabling their populations to access information and communication technology.[13] Finally, SDG 9 calls for enhancing small-scale industrial enterprises' access to financial services, empowering them to achieve the targets outlined in SDG 9.[14]

[4] Ibid., target 1.3.
[5] Ibid., target 1.4.
[6] Ibid., target 1.5.
[7] Ibid., target 1.a.
[8] Ibid., target 1.b.
[9] Ibid., target 9.1.
[10] Ibid., target 9.4.
[11] Ibid., target 9.5.
[12] Ibid., targets 9.a-b.
[13] Ibid., target 9.c.
[14] Ibid., target 9.3.

SDG 10 aims to achieve and sustain income growth for the bottom 40 per cent of the population at a rate exceeding the national average,[15] ensuring that this growth is free from discrimination based on age, sex, disability, race, ethnicity, origin, religion, or any other status.[16] The goal also addresses economic disparities between developing and developed countries by enhancing the representation and influence of developing countries in global economic and financial institutions.[17] Countries should facilitate orderly, safe, regular, and responsible migration and mobility of people.[18] They are also required to implement the SDT principle for developing countries under WTO Agreements,[19] which involves promoting official development assistance and financial flows, such as direct investments, to countries with the greatest need.[20] Another essential criterion for achieving this target is to reduce transaction costs for migrant remittances to less than 3 per cent.[21]

5.1.2 *Relevance of the WTO Dispute Settlement*

SDG 1, SDG 9, and SDG 10 are crucial for economic growth, and the WTO plays a pivotal role in their achievement. To address these goals, WTO members must commit to reducing the economic disparity between developed and developing countries and incorporate rules within WTO dispute settlement processes that safeguard the interests related to these SDGs. Essentially, WTO law should ensure a fair and equitable international trading system, enabling smaller economies to compete on equal footing with larger ones and improving their poverty status.

WTO regulations must support domestic efforts to eradicate poverty and reduce inequality while fostering the development of emerging economic sectors, particularly e-commerce. To protect these economic interests, WTO members can implement domestic regulatory measures. They can use tariffs, contingent trade protective measures, and administrative procedures to adjust the costs associated with cross-border commercial activities and market access.

If domestic measures by a WTO member unreasonably diminish the revenue of exporters and service providers, such as e-commerce traders, and consequently impact the foreign trade income of developing countries,[22] exporting countries may call for WTO panels to review and potentially rule against such trade-protective

[15] Ibid., target 10.1.
[16] Ibid., target 10.2.
[17] Ibid., target 10.6.
[18] Ibid., target 10.7.
[19] Ibid., target 10.a.
[20] Ibid., target 10.b.
[21] Ibid., target 10.c.
[22] Hudec (1993), pp. 3–4.

measures. This process ensures fair and equitable participation in global trade and supports the achievement of SDG 1, SDG 9, and SDG 10.

However, protectionism is not inherently negative when viewed through the lens of poverty eradication and inequality reduction. The complexities of international trade mean that its effects on sustainable development can vary based on specific circumstances. At times, countries may need to use contingent trade protective measures to achieve these SDGs. For instance, protecting pillar and labour-intensive industries from unfair competition is crucial for providing employment opportunities and ensuring economic stability in developing countries. Industries like cash crop production are vital for economic growth and food security in agricultural developing countries. Unlike developed countries with more diversified and resilient economies, these countries often depend heavily on a few key industries for economic development. The collapse of key industries can lead to severe economic repercussions and humanitarian crises. Therefore, developing and least-developed countries need to use contingent trade protective measures to safeguard their economic interests related to SDG 1, SDG 9, and SDG 10. WTO panels should be empowered to balance the values of free trade with the needs for poverty alleviation and inequality reduction while ensuring that protectionist measures do not become excessive.

5.1.3 Existing WTO Rules and Jurisprudence

5.1.3.1 No Exception Clause

The inclusion of exception clauses might be a critical step toward balancing economic interests with goals related to poverty eradication and reducing inequality. As discussed in Chap. 3, exception clauses provide members with the policy space needed for domestic regulatory measures. These provisions can help reconcile various interests within the framework of WTO law. In principle, WTO agreements could incorporate exception clauses concerning SDG 1 (No Poverty), SDG 9 (Industry, Innovation, and Infrastructure), and SDG 10 (Reduced Inequality), allowing members to deviate from their trade obligations under WTO rules. The existing general exception clause, GATT Article XX, serves as a reference for designing the language and structure of these potential clauses.

Currently, WTO agreements do not have specific exception clauses for SDG 1, SDG 9, and SDG 10. Existing clauses do not provide the necessary policy space for domestic regulatory measures aimed at eradicating poverty. This issue has two dimensions. First, there is no link between existing exception clauses and the protection of key or labour-intensive industries. General exception clauses do not clearly indicate that members can deviate from their trade obligations to eradicate poverty or reduce economic inequalities. Additionally, security exception provisions, including GATT Article XXI and GATS Article XIV bis, explicitly exclude economic security from their scope, defining 'security' solely as the absence of threats to peace.

Second, WTO rules establish a dual system: one for justifying contingent trade protective measures and another for justifying domestic regulatory measures based on exception clauses. Since WTO members agreed to use contingent trade protective measures to protect industries and included these provisions in specific agreements, they did not incorporate the protection of key or labour-intensive industries into GATT Article XX. As a result, general and security exception clauses have no reason to address poverty eradication and inequality reduction—two critical interests for achieving SDG 1, SDG 9, and SDG 10.

5.1.3.2 Public Moral Provisions

Although no exception clauses explicitly address poverty eradication and inequality reduction, panels can sometimes use certain exception clauses to uphold members' regulatory measures in these areas. One notable type of provision that panels may use is public moral provisions. However, WTO Agreements only include such clauses in GATT Article XX (a) and GATS Article XIV (a), thereby limiting their scope of application in WTO law.

The case of *US-Gambling* illustrates how a panel could use public moral provisions to support poverty eradication or inequality reduction. This leading case involved the prohibition of cross-border supply of gambling and betting services, where US regulations impacted Antigua and Barbuda's online gambling industry.[23] The conflict arose between Antigua and Barbuda's economic interests and the US's social interests. The US argued, and the panel and AB acknowledged, that the measures aimed to protect society from threats such as money laundering, organised crime, fraud, underage gambling, and pathological gambling.[24] From a sustainable development perspective, these measures safeguarded crucial societal interests, including public morals, public order, and public health. The panel characterised these societal interests as 'vital and important in the highest degree,' akin to protecting human life and health against life-threatening risks.[25]

However, the panel did not consider the impact of US measures on Antigua and Barbuda's online gambling industry from a sustainable development perspective. This industry was closely tied to poverty eradication and inequality reduction, both key aspects of economic sustainability. As the second-largest employer on the island, the cross-border gambling industry was crucial for economic development and social progress.[26] The industry's income could alleviate poverty and reduce

---

[23] See WTO Panel Report, US-Gambling, WT/DS285/R, adopted on 20 April 2005, pp. 169–195.

[24] See Ibid., para. 6.489.

[25] See Ibid., para. 6.492.

[26] To know the importance of the cross-border gambling industry to Antigua and Barbuda, one can see Kanter and Rivlin (2007); See also the statement of Antigua before the WTO dispute settlement

social crime in Antigua and Barbuda.[27] Thus, the US prohibition on cross-border gambling services undermined the economic interests related to sustainable development, an aspect the panel should have considered when reviewing the US claim.

The panel ruled that the US failed to prove the necessity of its measure to protect its social interests from Antigua and Barbuda's online gambling industry, suggesting the dispute could be resolved through international consultations outside the WTO dispute settlement process.[28] The AB upheld this finding,[29] noting that US domestic legislative provisions discriminated against Antigua and Barbuda's cross-border gambling suppliers by allowing US domestic suppliers to offer remote gambling services for interstate horse racing.[30] This decision highlighted that the panel did not review the US claim from a sustainable development perspective, as it failed to balance the various related interests.

The core issue in *US-Gambling* was identifying the interests related to Antigua and Barbuda's online gambling industry. If the panel had determined that the industry was not significant for poverty eradication or inequality reduction, the conflict would have been between the US's social interests and Antigua and Barbuda's economic interests. In this scenario, from a sustainable development perspective, the panel should have prioritised sustainable development interests and upheld the US prohibition. However, if the online gambling industry was found significant for poverty eradication or reducing inequality in Antigua and Barbuda, the conflict would have been between the economic and social dimensions of sustainable development. In this case, the panel needed to balance these interests, ensuring that the US measures protected its social interests without entirely sacrificing the economic interests of Antigua and Barbuda's cross-border gambling suppliers.

5.1.3.3 Provisions Preventing Deceptive Practices

Another kind of exception clause that a panel may use to uphold members' domestic regulatory measures on poverty eradication or inequality reduction is provisions preventing deceptive practices, such as GATT Article XX (d). The leading case concerning this exception is *Korea-Various Measures on Beef*, which suggests that a panel may use this clause to promote the achievement of SDG 1 and SDG 10.

---

panel at WTO Panel Report, US-Gambling, WT/DS285/R, adopted on 20 April 2005, para. 3.2. ('The Antiguan government has taken steps since the mid-1990s to build up a primarily internet-based, "remote-access" gaming industry as part of its economic development strategy'); See also Wohl (2009).

[27] See Grant (2015).

[28] See WTO Panel Report, US-Gambling, WT/DS285/R, adopted on 20 April 2005, paras. 6.529–6.531.

[29] See WTO Appellate Body Report, US-Gambling, WT/DS285/AB/R, adopted on 20 April 2005, paras. 317–318 and 326–327.

[30] See Ibid., para. 369.

Before delving deeper, let us quickly review the panel's report. The panel's reasoning centres on GATT Article XX (d), which allows members to regulate trade activities to prevent deceptive practices. In this case, the Korean government established a dual retail system under its Unfair Competition Act,[31] requiring domestic and imported beef to be sold in separate stores to prevent fraudulent practices by butcher shops.[32] However, the complainants, the US and Australia, argued that this measure constituted less favourable treatment compared to that accorded to similar domestic products, thus violating GATT Article III:4.[33] The panel ruled that Korea failed to justify its deviation from GATT Article III:4 because the measure was not necessary to avoid fraudulent practices.[34] The panel noted that alternative measures could achieve the same regulatory objective. Supporting the panel's ruling were facts indicating that (1) the Korean measure did not prevent fraud between domestic beef products,[35] and (2) less trade-restrictive alternatives, such as labelling and record-keeping, could achieve the same goal.[36] The AB upheld the panel's ruling, albeit with some nuances.[37]

Korea's claim that its measure regulated deceptive practices without focusing on other social concerns like public morals reduced the dispute to a pure trade conflict between WTO members. The importance of interests protected by a regulatory measure determines the extent to which the measure can restrict the freedom of trade. The more critical the social interest, the more tolerable the measure's impact on trade will be. Therefore, Korea might have justified its measure if it protected sustainable development interests. Indeed, regulations on deceptive practices may protect more than just economic interests; they can also safeguard consumers from fake products and foster a culture of fraud prevention. This, in turn, contributes to consumer health and public morals.

Moreover, a regulation on deceptive practices may help eradicate poverty or reduce economic inequality by protecting a developing country's pillar industries or labour-intensive industries from the threat of fake imported goods. From a sustainable development perspective, the panel should have considered these elements, potentially using GATT Article XX (d) to contribute to achieving SDG 1 and SDG 10. However, the scope of this exception clause is limited, as it specifically applies to the prevention of fraud.

Another relevant exception clause is Article 2.2 of the TBT Agreement, which allows members to use labelling schemes to declare the origin of products or prevent

---

[31] See WTO Panel Report, Korea-Measures Affecting Imports of Fresh Chilled and Frozen Beef, WT/DS161/R (WT/DS169/R), 31 July 2000, pp. 4–7.

[32] See Ibid., para. 237.

[33] See Ibid., pp. 12–13.

[34] See Ibid., paras. 675–676.

[35] See Ibid., paras. 661–662.

[36] See Ibid., paras. 666–674.

[37] See WTO Appellate Body Reports, Korea-Various Measures on Beef, WT/DS161/AB/R (WT/DS169/AB/R), adopted on 10 January 2001, para. 186.

deceptive practices. According to the findings in *Korea-Various Measures on Beef*, a labelling scheme is likely easier to satisfy panels' necessity test because it is less trade-restrictive than a dual retail scheme like the Korean regulation. Thus, members could potentially use this exception clause to justify regulatory measures aimed at poverty eradication or inequality reduction.

However, the *US-COOL* case demonstrates that a panel might not interpret poverty eradication or inequality reduction as legitimate objectives under Article 2.2 of the TBT Agreement. Article 2.2 provides a non-exhaustive, open-ended list of legitimate objectives, suggesting it could protect more purposes than other provisions. In *US-COOL*, the US invoked this provision to justify its mandatory labelling scheme for informing consumers about the origin of meat products.[38] While the US claimed that the objective of its labelling scheme was preventing deceptive practices,[39] the complainant, Canada, accused the US of illegally protecting its meat products.[40] The panel interpreted Article 2.2's language to imply that any measure aiming to eradicate poverty or reduce economic inequality could be considered legitimate if based on social norms.[41] However, the AB held that the panel's statements about social norms were ultimately inconsequential for its conclusion on legitimacy, indicating that the AB did not prioritise social norms over economic interests unrelated to sustainable development.[42]

The cases of *Korea-Various Measures on Beef* and *US-COOL* suggest that members cannot use provisions preventing deceptive practices to justify regulatory measures regarding poverty eradication or inequality reduction. The problem is that even the least trade-restrictive measures, like the US labelling scheme in *US-COOL*, are seen as imposing too many restrictions on trade freedom according to the panel's standard. The panel in *US-COOL* stated that a labelling scheme is considerably trade-restrictive if it limits the competitive opportunities for imported products compared to the pre-measure situation.[43] This stringent standard makes it difficult for members to justify their regulatory measures on deceptive practices.

This standard is overly stringent because it requires a panel to dismiss any regulatory measure that restricts international trade, making it difficult for members to justify measures against deceptive practices. For these provisions to help achieve SDG 1 and SDG 10, WTO panels must lower this criterion. Any regulatory measure contributing to the UN SDGs is valuable, and the legitimacy of such a measure should not depend on its ability to fully achieve a specific SDG. It is nearly

[38] See WTO Appellate Body Reports, US-COOL, WT/DS384/AB/R (WT/DS386/AB/R), adopted on 23 July 2012, para. 442 and footnote 895.

[39] See WTO Panel Report, United States-Certain Country of Origin Labelling (COOL) Requirements, WT/DS384/R (WT/DS386/R), adopted on 23 July 2012, para. 7.626.

[40] See Howse and Levy (2013), p. 334.

[41] See WTO Panel Report, United States-Certain Country of Origin Labelling (COOL) Requirements, WT/DS384/R (WT/DS386/R), adopted on 23 July 2012, para. 7.632.

[42] See WTO Appellate Body Reports, US-COOL, WT/DS384/AB/R (WT/DS386/AB/R), adopted on 23 July 2012, paras. 448.

[43] See Ibid., para. 477.

impossible for a single measure to accomplish one SDG alone. The *US-COOL* AB supported the idea that a panel should lower its standard of trade restrictiveness.[44] The judge argued that the fulfilment of a labelling measure's objective should be based on the degree of contribution it makes rather than whether it can entirely achieve the regulatory objective.[45] Unfortunately, the AB did not complete the legal analysis under Article 2.2 of the TBT Agreement due to a lack of relevant factual findings.[46] Consequently, the US labelling scheme remained inconsistent with WTO rules.

5.1.4 Key Considerations for Balancing Sustainability Elements

To effectively achieve SDG 1, SDG 9, and SDG 10, WTO panels must first thoroughly consider all relevant sustainability elements. They should then balance these potentially conflicting elements according to the principle of proportionality. The comprehensiveness of these elements determines the extent to which the panel can protect sustainable development interests in a given situation. Therefore, the primary question for the panel to address is identifying these relevant sustainability elements.

There are two groups of sustainability elements related to SDG 1, SDG 9, and SDG 10: elements concerning poverty eradication or inequality reduction and elements related to conflicting SDGs. Elements linked to poverty eradication or inequality reduction include economic growth, employment, salaries, and foreign trade income. Improvements in these areas reflect a society progressing from poverty to prosperity. In practice, pillar industries and labour-intensive industries often determine a country's performance in these elements. Disputes regarding poverty eradication or inequality reduction are inevitably tied to the protection of these industries. Therefore, a panel must assess the impact of members' domestic industries on their economic prosperity. The more significant the industry's impact on the economy, the more relevant it is to eradicate poverty or reduce economic inequality.

WTO jurisprudence offers valuable examples for assessing this impact. In *US-Wool Shirts and Blouses*, the panel examined the importance of an industry to a member's economy under the Agreement on Textiles and Clothing (terminated on January 1, 2005).[47] The US government imposed a temporary safeguard measure to limit textile imports from India to protect its domestic manufacturers of woven wool

[44] See Ibid., para. 461.

[45] See Ibid., para. 468.

[46] See Ibid., para. 491.

[47] See WTO Panel Report, United States-Measures Affecting Imports of Woven Wool Shirts and Blouses from India, WT/DS33/R, adopted on 23 May 1997, p. 69.

shirts and blouses.[48] Despite side effects on trade, the measure was deemed legitimate because Article 6 of the Agreement allowed for safeguard action if increased imports caused severe damage or an actual threat. The panel considered economic variables such as output, productivity, capacity utilisation, inventories, market share, exports, wages, employment, domestic prices, profits, and investment to determine the relevance of a member's industry to poverty eradication or inequality reduction.

Regarding sustainability elements related to other SDGs, WTO jurisprudence suggests that panels consider social and environmental factors. Social concerns often arise from economic activities, and members have cited numerous social elements associated with regulatory measures on international trade. For example, members might impose restrictions on imported goods to protect public morals and health and prevent deceptive practices. Measures like import prohibitions, dual retail schemes, and labelling schemes can significantly impact a member's economic interest. If these economic interests significantly improve a member's poverty status, social concerns could counter efforts to eradicate poverty or reduce inequality. Similarly, environmental concerns can affect poverty eradication or inequality reduction in WTO member countries. Existing WTO cases highlight conflicts between international trade and environmental issues, such as animal protection, biodiversity, and the conservation of natural resources.

Balancing these conflicting sustainable development interests involves two key requirements. First, a panel must determine whether a member's regulatory measure aimed at eradicating poverty or reducing economic inequality takes precedence over conflicting sustainability elements, considering all relevant circumstances. A member cannot justify its regulatory measure solely based on the importance of poverty eradication or inequality reduction, as all sustainability elements are equally important in legal value. The number of conflicting sustainability elements matters; the more conflicting interests there are, the more likely the measure will be deemed irrational. For instance, if a measure undermines three or more sustainable development interests, it cannot be considered proportionate to eradicating poverty or reducing inequality. If a measure protects more sustainable development interests than it harms, a panel should uphold it. Otherwise, the measure is not proportionate to achieving the relevant UN SDGs and is inconsistent with WTO rules.

Second, a panel's ruling cannot completely sacrifice a sustainability element and must minimise its undermining. This requirement implies that a panel cannot entirely prohibit a member from using its regulatory measures to eradicate poverty or reduce inequality. Additionally, the panel should not focus on whether the measure is more trade-restrictive than necessary to achieve its objective or if it discriminates against imported products. Economic concerns irrelevant to sustainable development should not be the panel's focus, as sustainability elements take precedence over purely economic elements in the sustainability test.

---

[48] See Ibid., para. 2.7.

5.2 Food Security and Sustainable Agriculture

5.2.1 SDG 2

SDG 2 is focused on ending hunger, achieving food security, improving nutrition, and promoting sustainable agriculture.[49] This goal encompasses several key targets. First, it aims to eradicate all forms of malnutrition by doubling the agricultural productivity and incomes of small-scale food producers.[50] Additionally, it calls for implementing sustainable food production systems and resilient agricultural practices. These practices are designed to enhance land and soil quality, boost productivity while preserving ecosystems, and strengthen the capacity to adapt to climate change, extreme weather, drought, flooding, and other disasters.[51]

Another crucial aspect of this goal is preserving biodiversity. This includes maintaining the genetic diversity of seeds, cultivated plants, farmed and domesticated animals, and their related wild species.[52] To achieve this, countries need to increase investments in rural infrastructure, agricultural research and extension services, promote technological development, and enrich plant and livestock gene banks.[53]

Finally, SDG 2 addresses trade restrictions and distortions in global agricultural markets.[54] It requires countries to implement measures that ensure the proper functioning of food commodity markets and their derivatives, with a primary emphasis on the timely provision of market information.[55]

5.2.2 *Relevance of the WTO Dispute Settlement*

A panel evaluating trade policies related to eradicating hunger could focus on two main areas: policies that liberalise international agricultural trade and those that restrict it. In theory, a panel could support the achievement of SDG 2 by either promoting or limiting the freedom of international agricultural trade.[56]

One approach that has gained recognition for its potential to combat hunger is the liberalisation of international agricultural trade.[57] By opening markets, trade can improve access to essential resources such as food, seeds, fertilisers, production

[49] See UNGA (2015), p. 15 and target 2.1.

[50] Ibid., target 2.2–2.3.

[51] Ibid., target 2.4.

[52] Ibid., target 2.5.

[53] Ibid., target 2.a.

[54] Ibid., target 2.b.

[55] Ibid., target 2.c.

[56] See Anderson (2017), pp. 1–2.

[57] Ibid., p. 8.

tools, and technologies, all critical for agricultural production and food security. Additionally, liberalised markets promote fair competition, enabling countries to benefit from exporting agricultural products, which can stimulate the growth of their agricultural sectors.[58]

However, this potential benefit must be weighed against the challenges faced by developing countries, particularly those heavily reliant on agriculture. These countries often depend on imported agricultural products to meet the growing domestic demand driven by expanding populations. This reliance is frequently exacerbated by the influx of cheap agricultural goods from developed countries, which can undermine local agricultural industries. When such imports dominate domestic markets, rural farmers struggle to sell their produce in urban centres, which intensifies rural poverty and places additional economic strain on urban areas. Urban centres, in turn, may experience higher food prices and increased unemployment rates due to the influx of cheap rural labour.

In light of these challenges, international agricultural trade can pose risks to food safety and security in developing countries. Adequate domestic food stocks are necessary for maintaining food security.[59] In some major agricultural exporters, regulators must control exports to ensure they do not undermine their own food security. However, if regulatory measures excessively restrict agricultural exports, they may violate WTO rules. This underscores the importance of WTO panels in evaluating the legality of such trade-restrictive policies.

Moreover, developing countries must navigate the volatility of international commodity prices.[60] The Nairobi Decision on Export Competition, adopted in 2015, mandates that developed countries eliminate their agricultural export subsidies. This measure is designed to help developing countries revitalise their agricultural sectors and enhance food security.[61] The WTO dispute settlement system plays a crucial role in enforcing this decision.

5.2.3 Existing WTO Rules and Jurisprudence

The *Indonesia-Import Licensing Regimes* case represents a significant episode in the saga of international agricultural trade disputes. Indonesia, the respondent in this dispute, implemented protectionist measures to safeguard its food self-sufficiency and argued that these measures could be justified under GATT Article XX.[62] The

[58] See Arda (2007), pp. 322–344.

[59] See De Schutter and Cordes (2011), pp. 6–7 and 14–15; See also WTO (2022), para. 10.

[60] See De Schutter and Cordes (2011), p. 4.

[61] WTO members agreed to abolish agricultural export subsidies and to set rules for other forms of farm export support in 2015. There is a transition period, though. See WTO (2015).

[62] See WTO Panel Report, Indonesia-Importation of Horticultural Products, Animals and Animal Products, WT/DS477/R (WT/DS478/R), adopted on 22 November 2017, pp. 28–42.

case illustrates the challenges faced by WTO members in justifying their agricultural trade regulations within the existing WTO framework.

The dispute, which began in the aftermath of the 1990s financial crisis and intensified during the 2008 food crisis, involved Indonesia, the US, and New Zealand. By 2014, the parties were engaged in consultations regarding Indonesian regulations. Indonesia defended its measures before the panel by invoking GATT Articles XX (a), (b), and (d), claiming that these measures were designed to protect public morals and human health and ensure compliance with Indonesian laws and regulations.[63] Additionally, Indonesia argued that under GATT Article XI: 2 (c), it had the right to restrict imports of agricultural products to address a temporary surplus of similar domestic products.[64] However, the panel deemed these measures protectionist[65] and found that Indonesia had failed to justify its WTO-inconsistent actions through the exception clauses.[66]

The panel's legal analysis centred on determining whether a member's measure was necessary to achieve its regulatory objective. In reviewing the Indonesian measures, the panel found that many were too far-fetched for Indonesia to make a prima facie defence.[67] For example, Indonesia argued that its harvest period requirement, referred to as Indonesian measure 4, was intended to protect the health of Indonesian consumers. Indonesia explained that the hot weather accelerated food spoilage, and thus, preventing an oversupply of horticultural products would protect consumers from rotten produce.[68] However, the panel found no mention of this purpose in the Indonesian regulation text.[69]

Another example is Indonesian measure 5, which concerns storage ownership and capacity requirements. Indonesia argued that this measure was intended to ensure the halal status of horticultural products to protect public morals. However, the panel disagreed, finding that measure 5 was irrelevant to maintaining the halal status of food. According to the panel, the horticultural products in question did not need to be stored under specific conditions because they were inherently halal under Indonesian laws.[70] Similarly, the panel disagreed with Indonesia's argument regarding the sale and distribution requirement for horticultural products, known as measure 6. The panel noted that these products should not be subject to halal requirements since they were not used in food but in industrial production.[71] Furthermore, the panel criticised

---

[63] See Ibid., p. 174.

[64] See Ibid., p. 67.

[65] See Ahn and Gnutzmann-Mkrtchyan (2019), pp. 197–218; See also Fane and Warr (2008), pp. 133–150.

[66] Jack (2016); Kugler et al. (2019), pp. 427–430.

[67] See WTO Panel Report, Indonesia-Importation of Horticultural Products, Animals and Animal Products, WT/DS477/R (WT/DS478/R), adopted on 22 November 2017, para. 7.515.

[68] See Ibid., para. 7.608.

[69] See Ibid., paras. 7.630–631.

[70] See Ibid., para. 7.654.

[71] See Ibid., paras. 7.712 and 7.715.

Indonesia's reference prices for chillies and shallots for consumption, referred to as measure 7. While Indonesia claimed this measure aimed to protect food safety, the panel found no indication of this purpose in the relevant Indonesian laws.[72]

Among the Indonesian measures, the panel considered measure 8, the six-month harvest requirement, to be the only one with the potential to protect human health.[73] Scientific evidence indicated that certain horticultural products harvested more than 6 months prior to importation could pose a food safety threat.[74] Indonesia argued that this measure significantly contributed to public health by ensuring that imported horticultural products retained their nutritional value until the date of importation, thereby providing adequate nutrition for Indonesian consumers. However, the panel disagreed, ruling that the measure was unnecessary to achieve Indonesia's objective. First, the measure could not guarantee that consumers would eat these products before their nutritional value diminished.[75] Second, the measure could be effectively replaced by existing Indonesian laws or regulations, such as health and SPS requirements outlined in Articles 21 and 22 of MOT 16/2013.[76]

This case highlights the limitations of using the necessity test to support SDG 2. The necessity test, rather than accommodating practical sustainability measures, tends to protect only those measures that are strictly necessary to achieve specific objectives. Consequently, members might be compelled to abandon many practical sustainability measures to comply with WTO rules. A clause like GATT Article XI:2 (c), which permits contingent trade protection measures, could offer a more suitable alternative. Unfortunately, both the panel and AB ruled that Article XI:2(c) was inoperative in this case due to Article 4.2 of the Agreement on Agriculture, which prohibits restrictive measures other than customs duties.

This decision effectively closes the door on the prohibition of importing agricultural products. From a sustainable development perspective, one might hope that the AB could overturn this ruling in a future decision. In my view, the panel's decision is questionable for several reasons. First, there is no indication that GATT Article XI:2 (c) is an agriculture-specific provision.[77] Indeed, GATT Article XI:2(c) addresses both agricultural and fisheries products. Paragraph (c) states:

> Import restrictions on any agricultural or fisheries product, imported in any form, necessary to the enforcement of governmental measures which operate: ....

---

[72] See Ibid., paras. 7.770–7.776.

[73] See Ibid., para. 7.793.

[74] See Ibid., para. 7.797.

[75] See Ibid., para. 7.801.

[76] See Ibid., para. 7.802. Article 21 of MOT 16/2013 states that every horticultural product imposed by a producer importer of horticultural products (PI) or a registered importer of horticultural products (RI) must first undergo verification or technical inquiry at its port of origin.

[77] See WTO Appellate Body Report, Indonesia-Importation of Horticultural Products, Animals and Animal Products, WT/DS477/AB/R (WT/DS478/AB/R), adopted on 22 November 2017, para. 5.65.

It is worth noting that GATT Article XI:2 lists 'agricultural products' and 'fisheries products' as two independent items, implying that agricultural products, as referred to in this clause, do not include fisheries products.[78] Accordingly, the panel should have interpreted this clause as not being specific to agriculture, thereby allowing Indonesia to use it.

Second, it is important to note that Article 4.2 of the Agreement on Agriculture does not explicitly state that GATT Article XI:2(c) is inoperative. According to Article X:3 of the Marrakesh Agreement, amendments to provisions of the multilateral trade agreements in Annexes 1A and 1C that would alter the rights and obligations of the members shall take effect for those members that have accepted them upon acceptance by two-thirds of the members, and thereafter for each other member upon its acceptance. Additionally, Article X:4 of the Marrakesh Agreement stipulates that amendments to provisions of the multilateral trade agreements in Annexes 1A and 1C that would not alter the rights and obligations of the members shall take effect for all members upon acceptance by two-thirds of the members. Accordingly, the panel and the AB should not have the discretion to declare any WTO rule inoperative.

Third, the panel's interpretation contradicts the objectives expressed in the preamble to the Agreement on Agriculture. The second recital of the preamble states:

> Their long-term objective as agreed at the Mid-Term Review of the Uruguay Round is to establish a fair and market-oriented agricultural trading system and that a reform process should be initiated through the negotiation of commitments on support and protection and through the establishment of strengthened and more operationally effective GATT rules and disciplines.

Therefore, the panel's interpretation that GATT Article XI:2(c) is rendered inoperative by Article 4.2 of the Agreement on Agriculture is inconsistent with this recital. Moreover, the sixth recital of the preamble to the Agreement on Agriculture requires that commitments under the reform programme be made equitably among all members, considering non-trade concerns such as food security and the need to protect the environment. However, the panel's decision effectively prohibited Indonesia from protecting its food security by abrogating GATT Article XI:2(c) through legal interpretation.

Findings outside the panel suggest that WTO law could, in some cases, tolerate more agricultural protectionist measures, as seen in the *Indonesia-Import Licensing Regimes* case. According to Ahn and Mkrtchyan, the impact of Indonesia's agricultural protectionist measures on agricultural exports was less significant than initially thought. Notably, many major exporters, including Australia (the largest exporter of animal products) and China (the largest exporter of agricultural products), did not join the dispute as complainants. Unlike the US and New Zealand, these countries benefited from Indonesia's measures. Additionally, smaller exporters such as Thailand, Vietnam, and Canada also reaped benefits from the Indonesian measures. The US and New Zealand, the fourth and fifth largest exporters respectively, were

[78] See Ibid., para. 5.75.

the only significant countries adversely affected, together accounting for less than 15% of the affected imports.[79]

The Indonesian measures did not significantly disrupt international agricultural markets, but they did distort competition, harming the interests of a few countries. From the perspective of trade liberalisation, preventing trade distortions is as crucial as maintaining the smooth operation of international markets since fair and equitable competition underpins a liberal market. Therefore, the panel had valid reasons to ensure that the US and New Zealand could compete on equal terms with other exporters. However, the panel should have reconsidered the importance of fairness and equity in international markets from the perspective of sustainable development when reviewing the Indonesian measures because sustainable development interests are paramount.

5.2.4 Key Considerations for Balancing Sustainability Elements

5.2.4.1 Structural Hunger

Structural hunger is a severe food security issue driven by the demands of international agricultural markets and irrational national trade policies.[80] A notable example is the Brazilian food crisis.[81] In this case, the liberalised international agricultural market, instead of addressing Brazil's domestic needs, encouraged Brazilian agricultural producers to allocate excessive land for cultivating specific crops to meet foreign importers' demands. Consequently, Brazil, a major agricultural nation, experienced a food supply shortage. This case demonstrates the necessity for countries' trade policies to prevent the oversupply of exported agricultural products to avoid structural hunger.

Structural hunger highlights two significant negative externalities of the liberalised international agricultural market. First, such a market can sometimes disrupt the supply of domestic food commodities. This issue is particularly evident in agricultural exporting developing countries, such as Brazil. When farmers prioritise production for export, they often reduce their supply to the domestic market, as arable land is limited and must be allocated accordingly.

Second, the liberalised international agricultural market can also increase unemployment in rural areas, leading to poverty. This problem arises when foreign agricultural producers dominate a country's domestic market.[82] Additionally, the liberalised market can exacerbate this issue by altering agricultural production

[79] See Ahn and Gnutzmann-Mkrtchyan (2019).

[80] See De Schutter and Cordes (2011), p. 2.

[81] Wise (2007), pp. 535–536 and 538.

[82] See Ibid., p. 544; See also De Schutter and Cordes (2011), pp. 17–18.

patterns. International trade often promotes a shift from traditional farming to large-scale commercial agriculture, which relies more on mechanised production rather than human labour. This transition inevitably reduces job opportunities in the agricultural sector. In developed countries, robust secondary and tertiary industries can absorb displaced farmers. However, this is often not the case in less developed agricultural countries. Moreover, the oversupply of exported agricultural products can trigger or intensify other social problems, potentially leading to social unrest. For instance, the over-exploitation of arable land may displace indigenous communities from territories that have historically been theirs.[83]

5.2.4.2 Corporate Social Responsibility

Corporate social responsibility (CSR) is becoming a crucial element in mitigating the negative externalities of international agricultural trade. With the removal of export subsidies, the behaviour of multinational agricultural companies now plays a pivotal role in addressing hunger. Today, major international agribusinesses dominate the production, sale, and technological innovation of agricultural products. Companies like Cargill, Bunge, Archer Daniels Midland, Glencore, Monsanto, and Dreyfus control the food supply chains and wield significant bargaining power over their suppliers.[84]

The dominance of these giant companies can be seen as a legacy of the export subsidies that once fuelled their growth. To advance the achievement of SDG 2, any review of regulations on international agricultural trade need to consider the impact of this monopoly power. Indeed, many protectionist measures can be attributed to the influence of these dominant firms. As De Schutter and Cordes pointed out, the protectionist measures implemented during the 2008 food crisis were a response to the manipulation of agricultural prices by international agribusinesses.[85]

From a sustainable development perspective, international agribusinesses have a responsibility to conduct their operations in alignment with SDG 2. If these companies exploit their monopoly power, countries may need to implement trade protection measures to enforce CSR. In this context, a panel should acknowledge the legitimacy of a member's measures that are designed to uphold these responsibilities.

5.2.4.3 Food Security

Food security requires a country to have the capacity to produce, acquire, and process agricultural products, as well as to build the necessary infrastructure to

[83] See Wise (2007), pp. 536–537.

[84] See De Schutter and Cordes (2011), p. 10.

[85] See Ibid., p. 10.

ensure a stable food supply.[86] This is a key sustainability factor that a panel must consider when evaluating a member's regulatory measures concerning international agricultural trade.[87]

It is important to note that while liberalised international agricultural markets benefit countries with a strong comparative advantage in agriculture, such as the Cairns Group members, they can be detrimental to those most vulnerable to hunger—net food-importing countries.[88] As previously discussed, dependency on imported agricultural products poses significant risks. For net food-importing countries, the removal of export subsidies from developed countries can exacerbate these risks rather than mitigate them. These countries often have no alternative but to import agricultural products to feed their populations. Therefore, excluding dumped agricultural products can lead to higher food costs and worsen food shortages.

Given this situation, it is crucial for a country to achieve a certain level of food self-sufficiency.[89] Strengthening a country's local food system and ensuring food security is a long-term process. During this transition, the WTO dispute settlement mechanism plays a vital role in supporting net food-importing countries as they work to build a more resilient food system.[90]

5.2.4.4 Climate Change

Climate change has a profound impact on achieving SDG 2, which aims to end hunger and ensure food security. There are two primary reasons for this. First, climate change significantly reduces agricultural productivity by affecting crop yields and soil health.[91] Second, agricultural products are crucial for biofuel production, which plays a key role in adapting to and mitigating climate change.[92] However, the increased use of biofuels comes at a cost: it diminishes the global food supply and drives up food prices. This creates a challenge for food security, particularly in countries with vulnerable agricultural sectors.

Despite these challenges, banning biofuels is not a viable option due to their importance in renewable energy production. Thus, there is an inherent conflict between addressing hunger and tackling climate change.[93] This conflict is likely to

---

[86] See Peres and Daibert (2017), p. 57; See also Feunteun (2015), pp. 341–343; See also Rome Declaration on World Food Security and World Food Summit Plan of Action, World Food Summit, November 1996, para. 1; See also Diaz-Bonilla (2014), p. 1.

[87] See Häberli (2013), pp. 80 and 97; See also Howse and Teitel (2009), p. 48; See also Trebilcock and Howse (2005), p. 579; See also Mutua and Howse (2001), p. 81; See also Howse (2008), p. 948.

[88] See De Schutter (2011).

[89] See Ibid.

[90] See Moreu (2011).

[91] See Anderson (2017), p. 31.

[92] See Cloots (2011).

[93] See Ibid., pp. 131–132.

persist until a technical solution is found. In the meantime, WTO panels must carefully balance these two sustainability objectives when evaluating international agricultural trade regulations.

5.3 Human Health and Life

5.3.1 SDG 3

SDG 3 is dedicated to ensuring healthy lives and promoting well-being for people of all ages. It addresses a range of global health challenges through three essential aspects. First, the goal focuses on safeguarding the health and lives of pregnant women and children. It aims to reduce the global maternal mortality ratio to below 70 per 100,000 live births[94] and to eliminate preventable deaths among newborns and children under five.[95] This aspect also emphasises reducing premature mortality from non-communicable diseases by one-third,[96] establishing robust sexual and reproductive health care services,[97] and enhancing prevention and treatment for substance abuse.[98]

The second aspect targets the prevention of deaths from various diseases and accidents. This includes eradicating epidemics of AIDS, tuberculosis, malaria, and neglected tropical diseases, as well as combating hepatitis, water-borne diseases, and other communicable diseases.[99] Additionally, it calls for halving global deaths and injuries from road traffic accidents,[100] achieving universal health coverage,[101] and reducing deaths and illnesses caused by hazardous chemicals and pollution.[102]

Finally, SDG 3 emphasises strengthening public health systems. It highlights the importance of implementing international health agreements, such as the World Health Organisation Framework Convention on Tobacco Control[103] and advancing research and development of vaccines and medicines for both communicable and non-communicable diseases.[104] This aspect also calls for a significant increase in health financing and improvements in the recruitment, development, training, and

[94] See UNGA (2015), target 3.1.

[95] Ibid., target 3.2.

[96] Ibid., target 3.4.

[97] Ibid., target 3.7.

[98] Ibid., target 3.5.

[99] Ibid., target 3.3.

[100] Ibid., target 3.6.

[101] Ibid., target 3.8.

[102] Ibid., target 3.9.

[103] Ibid., target 3.a.

[104] Ibid., target 3.b.

retention of the health workforce in developing countries.[105] Additionally, it underscores the need to enhance the capacity of all countries, especially developing ones, to manage national and global health risks through effective early warning systems and risk reduction strategies.[106]

5.3.2 Relevance of the WTO Dispute Settlement

The achievement of targets 3.3, 3.9, and 3.a of SDG 3 is closely intertwined with the WTO dispute settlement system. Target 3.3 aims to end and prevent the spread of diseases by addressing epidemics such as AIDS, tuberculosis, malaria, and neglected tropical diseases, as well as combating hepatitis, water-borne diseases, and other communicable diseases. Targets 3.9 and 3.a build on this by providing additional details and specific requirements. Target 3.9 focuses on health issues related to environmental problems, such as hazardous chemicals and pollution of air, water, and soil. In contrast, Target 3.a zeroes in on regulating tobacco products.

The intersection between international trade and disease control underscores the importance of the WTO's role in this context. Scientists recognise that diseases and contaminants can spread through imported goods, which may include agricultural products like fruits and vegetables, meat products such as cattle, poultry, and fish, as well as cigarettes and other raw materials. These goods can carry communicable or non-communicable diseases, pose imminent or chronic health threats, or introduce agricultural pests.

To meet these targets effectively, countries must implement SPS measures to eliminate sources of disease transmission. In addition to SPS measures, countries can leverage domestic laws and regulations or impose quantitative restrictions on imported goods to address health risks and curb disease spread. For example, WTO members may use the Agreement on Agriculture (AoA) provisions to justify restrictions on certain agricultural imports or impose cigarette labelling requirements to raise consumer awareness about health risks. Given that WTO Agreements regulate these domestic regulatory measures, the WTO dispute resolution mechanism plays a vital role in assessing compliance with WTO rules. Failure to justify such measures properly could result in violations of GATT Article III (regarding internal regulations), GATT Article XI (concerning quantitative restrictions), or provisions of the SPS or TBT Agreements.

Moreover, WTO members can use WTO rules to ensure access to medicinal products. The TRIPS Agreement outlines rules for the sale and purchase of patented medicines, with WTO panels and the AB consistently upholding IPRs,[107] including

[105] Ibid., target 3.c.

[106] Ibid., target 3.d.

[107] For example, see WTO Panel Report, India-Patents (US), WT/DS50/R, adopted 5 September 1997 and WTO Appellate Body Report, India-Patents (US), WT/DS50/AB/R, adopted

exclusive marketing rights and protection terms.[108] However, the Doha Declaration on TRIPS and Public Health affirms members' rights to issue compulsory licenses.[109] Consequently, a WTO member could request other members to waive patent rights for medicinal products to ensure access. As highlighted in Sect. 2.3.4.3, WTO members decided to waive patent rights for COVID-19 vaccines at the WTO's 12th Ministerial Conference. In principle, members could also seek support from WTO panels for their compulsory license claims.

5.3.3 *Existing WTO Rules and Jurisprudence*

5.3.3.1 Unsolved Scientific Issues and the Precautionary Principle

Between 1997 and 2018, a significant number of WTO cases related to human health and life were adjudicated. Out of 12 major cases, defending parties won only four,[110] indicating the challenge members face in justifying their regulatory measures based on public health (See Table 5.1). This difficulty primarily stems from the inability of WTO members to demonstrate that their measures are necessary for achieving their regulatory objectives.

To justify a measure, a member must provide sufficient evidence to show that: (1) the measure aims to achieve a legitimate objective as outlined in the exception clauses; (2) the measure is necessary to meet that objective; and (3) no less trade-restrictive measure would provide the same level of protection. Since panels base their findings on scientific evidence, which is often uncertain, members face a high bar in proving the necessity of their measures. Consequently, panels impose stringent requirements for demonstrating that trade restrictions are warranted on public health grounds.

The inherent uncertainty of science exacerbates this challenge. In many cases, members struggle to provide sufficient evidence to justify their regulatory measures, especially when facing unresolved scientific questions. When a scientific issue remains unresolved, it becomes difficult for a panel to determine which party has presented adequate scientific evidence to support its claims. This is because any scientific finding, whether representing a mainstream or minority view, could potentially be correct. Such scenarios are not rare but rather common in WTO disputes involving human health.

Due to the inherent uncertainty of science, a panel must establish its own standards for what constitutes sufficient scientific evidence, essentially defining

19 December 1997; See also WTO Panel Report, India-Patents (EC), WT/DS79/R, adopted 24 August 1998.

[108] See WTO Panel Report, Canada-Patent Term, WT/DS170/R, adopted 5 May 2000 and WTO Appellate Body Report, Canada-Patent Term, WT/DS170/AB/R, adopted 18 September 2000.

[109] See WTO (2001).

[110] See Table 5.1.

Table 5.1 (The results of WTO decisions on trade-related human health issues)

WTO Case Law	The Panel	The Appellate Body
EC-Hormones I (1997)	×	×
Japan-Agricultural Products I (1998)	×	×
EC-Asbestos (2001)	✓	✓
Japan-Apples (2003)	×	×
Dominican Republic-Import and Sale of Cigarettes (2004)	×	×
Brazil-Retreaded Tyres (2007)	×	×
EC-Hormones II (2008)	×	✓
US-Poultry (China) (2010)	×	×
US-Clove Cigarettes (2011)	✓	✓
US-Animals (2015)	×	×
Indonesia-Chicken (2017)	×	×
Korea-Radionuclides (2018)	×	✓

✓: Justifying the member's regulatory measure(s)
×: Accusing the member's regulatory measure (s)

what is considered legally adequate. This approach is guided by three key principles. First, the panel must assess whether the defending party was in compliance with WTO obligations up until the moment of dispute.[111] Second, the panel should recognise that a measure is not necessarily unscientific merely because it is based on a divergent or minority scientific view.[112] Finally, the panel should acknowledge that a member has the right to implement provisional measures to prevent health risks when scientific evidence is insufficient to fully establish necessity. These principles underpin the precautionary principle, helping panels navigate the complexities of scientific uncertainty in WTO dispute settlements related to human health.

However, the precautionary principle frequently becomes a contentious point in WTO disputes. This is understandable because it allows a member to justify provisional measures based on limited or minority scientific findings, which may not align with the majority view. Consequently, applying this principle presents numerous challenges and issues that require careful consideration.

5.3.3.2 Article 5.7 of the SPS Agreement and Provisional Measures

Article 3.3 of the SPS Agreement allows WTO members to implement higher levels of protection than international standards when addressing health risks related to unresolved scientific questions, provided that these measures comply with due

[111] See WTO Panel Report, Japan-Agricultural Products II, WT/DS76/R, adopted on 19 March 1999, para. 8.31.

[112] See WTO Panel Report, Japan-Measures Affecting the Importation of Apples, WT/DS245/R, adopted 15 July 2003, para. 8.98.

process requirements under WTO Agreements. Specifically, Article 8 of the SPS Agreement and the provisions in Annex C outline the procedural rules that members must follow when adopting sanitary and phytosanitary measures. Since these procedural rules are straightforward and not dependent on complex scientific issues, WTO panels can more easily assess whether a member has adhered to them. For instance, the *US-Poultry* and *Indonesia-Chicken* panels both identified procedural errors made by members in the adoption or implementation of their measures.[113]

It is important to note that adopting a provisional measure based on the precautionary principle involves more than just meeting due process requirements. To fully understand this, one must consider both Article 3.3 and Article 5.7 of the SPS Agreement together. In practice, provisional measures often set a high standard, as outlined in Article 3.3, to address health risks related to unresolved scientific questions. Article 5.7 establishes specific conditions that must be met to justify the high standard of these provisional measures.

According to Article 5.7, a member must adhere to several conditions to justify a provisional measure.[114] First, the measure must be applied in situations where relevant scientific evidence is insufficient. Second, it must be based on available pertinent information. Third, the member must actively seek additional information necessary for a more objective risk assessment. Lastly, the measure must be reviewed within a reasonable period.[115] These requirements make it clear that while a member is allowed to implement provisional measures, it must consider all relevant scientific findings in designing and applying these measures and make efforts to address gaps in scientific evidence as promptly as possible.

Accordingly, WTO members are obligated to review their regulatory measures, explore alternative approaches based on different regulatory models, and contribute to research on unresolved scientific questions. If a member fails to advance such research, it cannot justify taking a provisional measure solely on the basis of potential risks associated with that scientific uncertainty. For instance, in *Japan-Agricultural Products II*, the panel determined that Japan, as the defending party, was ineligible to impose a provisional measure because it had not sought additional information for a more objective risk assessment over a period of more than 30 years.[116] This ruling indicated that Japan had both the responsibility and sufficient time to conduct research on the varietal testing requirements since the SPS Agreement came into force on January 1, 1995.

[113]See WTO Panel Report, United States-Certain Measures Affecting Imports of Poultry from China, WT/DS392/R, adopted on 29 September 2010; See also WTO Panel Report, Indonesia-Measures Concerning the Importation of Chicken Meat and Chicken Products, WT/DS484/R, adopted on 17 October 2017, paras. 7.529–7.530.

[114]See WTO Panel Report, United States-Measures Affecting the Importation of Animals, Meat and other Animal Products from Argentina, WT/DS447/R, 24 July 2015, paras. 7.255–7.257.

[115]See WTO Appellate Body Report, Japan-Measures Affecting Agricultural Products, WT/DS76/AB/R, adopted on 19 March 1999, para. 89.

[116]See WTO Panel Report, Japan-Apples, WT/DS245/R, adopted on 10 December 2003, para. 8.57.

Therefore, while implementing a provisional measure might appear straightforward and subject to procedural rules that a member can adhere to in good faith, it is intricately linked to addressing unresolved scientific questions. This responsibility aims to prevent misuse of the precautionary principle but also restricts many members' ability to apply provisional measures. In reality, most WTO members lack the capacity to contribute effectively to research on these unresolved scientific issues, making it challenging to meet the conditions outlined in Article 5.7 of the SPS Agreement. As a result, Article 5.7 imposes a significant burden of proof on members to justify their provisional measures, thereby substantially limiting the practical application of the precautionary principle.

5.3.3.3 The Impact of Trade Restrictiveness on the Protection of Human Health

According to WTO jurisprudence, a measure's necessity in achieving its regulatory objective is assessed based on both its contribution to legitimate goals and its trade restrictiveness. A more trade-restrictive measure must demonstrate a greater contribution to achieving its intended objective. The AB in *Brazil-Retreaded Tyres* illustrates this principle, stating:

> When a measure produces restrictive effects on international trade as severe as those resulting from an import ban, it appears to us that it would be difficult for a panel to find that measure necessary unless it is satisfied that the measure is apt to make a material contribution to the achievement of its objective.[117]

Consequently, if a measure significantly restricts international trade, a member must provide more substantial scientific evidence to demonstrate its contribution to achieving legitimate objectives. This implies that members face varying burdens of proof when justifying different domestic regulatory measures. This principle is exemplified in *Korea-Various Measures on Beef*, where the panel specifically noted that a labelling measure, which was less trade-restrictive than a dual retail scheme that entirely banned imports, should be used instead.

The rulings in *Brazil—Retreaded Tyres* and *Korea-Various Measures on Beef* also have significant implications for public health protection. They establish that members face varying burdens of proof when justifying domestic regulatory measures based on the nature and severity of the risks involved. As previously discussed, these rulings indicate that more trade-restrictive measures must demonstrate a greater contribution to achieving legitimate objectives. Moreover, the more significant the measure's contribution to its regulatory goals must be, the more scientific evidence members need to provide. Consequently, when members implement highly trade-restrictive measures to address unknown or severe health risks, they bear a

[117] See WTO Appellate Body Report, Brazil-Retreaded Tyres, WT/DS332/AB/R, adopted on 17 December 2007, para. 7.211.

heavier burden of proof to show that these measures are necessary for protecting public health.

Two WTO cases illustrate the impact of trade restrictiveness on the burden of proof: while the panel in *US-Clove Cigarettes* prioritised public health over trade interests, the *Korea-Radionuclides* panel favoured trade interests. Both cases involved import bans, yet the panel reports highlight the differing burdens of proof imposed on the members. In *US-Clove Cigarettes*, which dealt with preventing non-communicable diseases caused by smoking, the panel acknowledged that the US technical regulation amounted to an import ban. Despite this, the panel gave significant weight to public health concerns and firmly rejected the two dozen alternative measures proposed by Indonesia, the defending party.[118] In addressing Indonesia's request for special and differential treatment under Article 12 of the TBT Agreement, the panel remarked that:

> Considering that the measure is a ban on cigarettes with characterising flavours for reasons of public health, the panel fails to see how it could be possible, under WTO rules, to exclude from the ban cigarettes with characterising flavours from developing countries. Indeed, a requirement to exclude a product that is harmful to human health from a ban, solely on the grounds that the product is produced and exported by a developing country, would limit Members' ability to regulate for public health purpose.[119]

As a result, the measure in question outweighed Indonesia's economic interests, and the US was not required to prove that the proposed alternative measures would fail to achieve its desired level of protection.

However, the *Korea-Radionuclides* panel placed less emphasis on public health concerns compared to the *US-Clove Cigarettes* case, as the measure aimed to address an unknown risk posed by radioactive materials. Two critical findings influenced the panel's rulings.

The first important finding concerned the trade restrictiveness of the Korean measure. Korea argued that, following the principle of 'as low as reasonably achievable,' the measure aimed not only to comply with the international standard of 1 mSv/year of radionuclides in food products but also to minimise Korean consumers' exposure to radionuclides as much as possible. While the panel recognised Korea's right to determine its appropriate level of protection, it disagreed that the measure was necessary to achieve this objective according to international standards. The panel reasoned that the exposure of Korean consumers to radionuclides should be significantly lower than the international standard, given that Korean consumers could also purchase similar products from other sources.[120]

---

[118] See WTO Panel Report, US-Clove Cigarettes, WT/DS406/R, adopted on 24 April 2012, paras. 7.421–7.423.

[119] See Ibid., para. 7.647.

[120] See WTO Panel Report, Korea-Radionuclides, WT/DS495/R, adopted on 26 April 2019, para. 7.181.

Additionally, the panel was unwilling to prioritise preventing an unknown risk, such as that posed by radioactive substances, at the expense of trade interests.[121]

The second important finding concerns the definition of 'like products.' The panel identified like products in its analysis of Article 2.3 of the SPS Agreement (regarding 'non-discrimination') and Annex C(1)(a) of the SPS Agreement (regarding 'no less favourable treatment than similar domestic products'). Regarding Article 2.3, the panel ruled that products originating from Japan and other countries were considered similar because scientific evidence showed that both categories were exposed to radionuclides due to nuclear experimentation and leak incidents. In other words, all products exported to Korea might pose the same health risk. To illustrate this similarity, the panel specifically referred to European mushrooms, which contained even higher levels of radionuclides than other products.[122]

Interestingly, the panel made a contradictory finding regarding similar products under Annex C(1)(a) of the SPS Agreement. It rejected Japan's claim and explained that:

> Although a mushroom and a fish could both be contaminated by the same substance and hence pose a similar or identical health risk, this is insufficient to say that they are like products.[123]

This contradictory finding clearly illustrates how the panel became entangled in the complexities of scientific evidence. Overall, the panel's analysis is questionable. From a sustainable development perspective, the panel arguably needs to consider more factors when reviewing the Korean measure. Regarding the risks associated with radioactive substances, it is premature for the panel to assert that Korean consumers will not be exposed to excess radionuclides. Conceivably, some Korean consumers might choose to buy only fish products of Japanese origin, which could undermine their health. Therefore, the panel's reasoning is insufficient to conclude that Japan's alternative measure can achieve the level of protection chosen by Korea and that the Korean measure is more trade-restrictive than necessary to achieve this level of protection.

The *Korea-Radionuclides* case highlights that applying a consistent burden of proof, regardless of the levels of scientific research associated with different health risks, is unreasonable. This approach, rather than enhancing public health protection, hinders members from safeguarding their populations against unknown and severe health risks. Specifically, members find it easier to justify measures for preventing non-communicable diseases compared to those aimed at preventing communicable diseases, which can be more harmful to public health. A more effective approach for panels would be to follow detailed guidelines on health risks when reviewing members' regulatory measures rather than treating all health risks as a single, unified category. Panels should recognise the relevance of the severity of health risks in

[121] See Ibid., para. 7.95.

[122] See Ibid., paras. 7.313–7.322.

[123] See Ibid., paras. 7.388–7.392.

determining the burden of proof that members must meet to show that their measures are necessary to protect human health. For instance, if a member faces an unknown risk with the potential to cause severe harm, that member should be allowed to implement provisional measures based on the precautionary principle.

While the AB Report reversed the panel's analysis, it did not alter the analytical framework, which remains incompatible with the need to protect human health. From a sustainable development perspective, WTO panels should prioritise human health—a key component of sustainable development—over purely economic interests unrelated to sustainability. Therefore, WTO panels must acknowledge that a member can justify its regulatory measures on human health grounds based on the precautionary principle.

5.3.3.4 Article 30 of the TRIPS Agreement

In *Canada-Patent Protection of Pharmaceutical Products*, the European Communities (EC), the complainant, argued that Sections 55.2(1) and 55.2(2) of the Canadian Patent Act violated patentees' exclusive rights and the term of protection provided by the TRIPS Agreement. The Canadian law allowed generic medicine producers to stockpile patented pharmaceutical products without the patentees' consent (the stockpiling exception) and to manufacture similar products during the 6 months before the 20-year patent term expired (the regulatory review exception).[124] Canada attempted to justify its measures under Article 30 of the TRIPS Agreement,[125] which permits WTO members to provide limited exceptions to patentees' exclusive rights, provided these exceptions do not unreasonably conflict with the normal exploitation of the patent and do not prejudice the patentees' legitimate interests.[126]

The panel determined that the stockpiling exception provided by Section 55.2 (2) of the Canadian Patent Act was clearly incompatible with Article 30 of the TRIPS Agreement, as it constituted a substantial curtailment of patentees' exclusive rights to use and make patented products under Article 28.1 of the TRIPS Agreement.[127] Members can only make slight diminutions of patent rights,[128] and the panel found that the stockpiling exception exceeded this limitation.

To legitimise its regulatory review exception, Canada presented three arguments. First, Canada contended that the protection of IPRs should not become a barrier to legitimate trade.[129] In this context, Canada argued that the TRIPS Agreement

[124] See WTO Panel Report, Canada-Patent Protection of Pharmaceutical Products, WT/DS114/R, adopted 17 March 2000, para. 3.1.

[125] Ibid., para. 4.9.

[126] Article 30, TRIPS Agreement.

[127] See WTO Panel Report, Canada-Patent Protection of Pharmaceutical Products, WT/DS114/R, adopted 17 March 2000, para. 7.36.

[128] Ibid., para. 7.30.

[129] Ibid., para. 4.13.

permitted members to introduce limited exceptions to patentees' exclusive rights to promote full competition in regulated-product markets after the expiration of the patent protection term. Additionally, Canada asserted that this limited exception could help realise the cost-saving benefits that competition in these markets would bring to society, especially in the healthcare products market.[130]

Second, Canada argued that the measure in question was legitimate as it served social welfare and public health interests. To support this claim, Canada cited Articles 7 and 30 of the TRIPS Agreement, which provide WTO members with general and flexible authority to adopt measures that balance a patentee's interests with other societal interests.[131]

Third, Canada argued that the measure did not violate WTO rules because it prohibited producers from selling generic medicines before the patent expired. Canada contended that this restriction mitigated the measure's impact on patent rights. Additionally, Canada claimed that the measure represented a Bolar exception, which was essential for expediting the lengthy regulatory review processes for generic products.[132] In this context, Canada suggested that the measure prevented EC pharmaceutical manufacturers from unjustifiably extending their patent protection periods to secure an unwarranted monopoly.[133]

Based on these arguments, Canada contended that its patent law complied with the TRIPS Agreement. The panel's analysis examined both the three conditions for exceptions outlined in Article 30 and the non-discrimination requirement under Article 27.1 of the TRIPS Agreement. The panel determined that Canada's regulatory review exception constituted a permissible limited exception under Article 30. Unlike the stockpiling exception, which impaired patentees' rights, the regulatory review exception solely permitted production for regulatory approval purposes without affecting patent rights.[134] Additionally, the panel found that the regulatory review exception adhered to the non-discrimination requirement of Article 27.1. Consequently, the panel ruled that the regulatory review exception was consistent with WTO rules.

The panel's rulings have significant implications for patent rights, affirming that WTO members may utilise the Bolar exception for regulatory purposes beyond commercial use. However, the panel did not address the sustainable development factors related to these measures in its legal analysis. As Canada pointed out, compulsory patent licenses facilitate the availability of cost-effective generic products, alleviating pressure on public healthcare systems[135] and addressing concerns about rising drug costs.[136] These benefits have been acknowledged by the World

[130]Ibid., para. 4.12.

[131]Ibid., para. 4.13.

[132]Ibid., para. 4.12.

[133]Ibid., para. 4.14.

[134]Ibid., para. 7.45.

[135]Ibid., para. 4.14.

[136]Ibid., para. 4.21.

Health Organisation.[137] In this case, the panel focused solely on whether the measures complied with the TRIPS Agreement's limited exception and non-discrimination requirements. Since the 'limited exception' requirement mandates that such exceptions be confined to rare and narrowly defined circumstances, it imposes stringent limitations on members' policy options, making it difficult for them to convince WTO panels to uphold their claims for compulsory licenses.

5.3.4 Key Considerations for Balancing Sustainability Elements

Existing WTO jurisprudence indicates that panels often confront a tension between public health and trade interests when resolving disputes related to human health.[138] From the perspective of sustainable development, addressing this conflict should be relatively straightforward for WTO panels. Since human health is a key aspect of sustainable development and takes precedence over trade interests, WTO panels should prioritise measures aimed at protecting human health. In this context, there is no need to balance health concerns against trade interests. The primary consideration for a panel is to determine whether a member's measure effectively contributes to safeguarding human health. Therefore, panels may need to reassess their approach to evaluating scientific evidence in this regard.

As previously discussed, it is problematic for WTO panels to apply a uniform standard of review across cases involving different health risks. Scholars have highlighted this issue and proposed various solutions. A commonly suggested approach is for panels to fully defer to members' risk assessments, provided there are no procedural issues with the scientific reports.[139] However, this approach is not ideal for safeguarding human health. Relying solely on procedural correctness fails to adequately address the role of scientific evidence in health protection and is as inadequate as the panel's current de novo review standard.[140]

In cases where scientific research on health risks is extensive and widely accepted, panels should consider expert opinions to ensure that measures effectively promote human health protection. The optimal standard of review should, therefore, strike a balance between respecting the role of scientific evidence and safeguarding against unknown, severe health risks.

---

[137] Ibid., para. 4.14.

[138] This conclusion is based on existing WTO jurisprudence. I do not deny that public health can conflict with the economic interests associated with sustainable development. However, this conflict currently exists primarily at the domestic level.

[139] See Button (2004), p. 218.

[140] See Gruszczynski (2010), p. 143.

I propose that WTO panels adopt varying standards of review depending on the type of health risk involved. Disputes related to human health typically involve three categories of health risks: (1) traditional health risks, (2) chronic health risks, and (3) imminent and severe health risks. The level of scientific research available for each category varies significantly. Traditional health risks are generally associated with well-known diseases, pathogens, and hazardous substances. Since these causes are well understood, members can usually provide sufficient scientific evidence to support their regulatory measures. In contrast, chronic health risks are challenging to assess in the short term, and thus, members may struggle to provide conclusive scientific evidence. Similarly, for imminent and severe health risks, which are often poorly understood or not yet fully researched, members may find it difficult to substantiate their measures with existing scientific data. It would be beneficial for WTO panels to tailor the burden of proof based on the state of scientific research for each type of health risk. This approach would ensure that the review process is more aligned with the complexities and uncertainties inherent in different health risk categories.

5.4 Climate Change, the Ocean, and Clean Energy

5.4.1 SDG 13, SDG 14, and SDG 7

SDG 13 seeks to tackle climate change and its effects through several key strategies.[141] The goal first calls for the integration of climate change considerations into national policies, strategies, and planning.[142] This includes a specific focus on enhancing climate change planning and management in the least-developed countries and small island developing states.[143] Additionally, the goal aims to build capacity for adapting to climate-related hazards and natural disasters across all nations.[144] It also underscores the need for developed countries to meet their commitment under the United Nations Framework Convention on Climate Change by jointly mobilising $100 billion annually by 2020 to support these vulnerable nations in combating climate change.[145] Finally, SDG 13 stresses the importance of public engagement in climate action, requiring countries to advance education, raise awareness, and strengthen both human and institutional capacities for effective climate change mitigation and adaptation.[146]

---

[141] See UNGA (2015), Goal 13.

[142] Ibid., target 13.2.

[143] Ibid., target 13.b.

[144] Ibid., target 13.1.

[145] Ibid., target 13.a.

[146] Ibid., target 13.3.

SDG 14 aims to ensure the conservation and sustainable use of oceans, seas, and marine resources. Countries committed to this goal are tasked with significantly reducing all forms of marine pollution and implementing effective, science-based management plans to prevent adverse impacts on marine and coastal ecosystems.[147] This commitment includes addressing ocean acidification and associated ecological challenges,[148] regulating harvesting practices, and ending overfishing and illegal, unreported fishing activities.[149] Moreover, SDG 14 highlights the need to eliminate subsidies that contribute to harmful fishing practices detrimental to marine ecosystems.[150] A key target is to conserve at least 10 per cent of coastal and marine areas while eliminating these damaging fisheries subsidies.[151] This goal also emphasises empowering vulnerable nations and communities to address climate change. It seeks to enhance the capacity of small island developing and least-developed countries to manage their marine resources effectively and increase their economic benefits.[152] This includes ensuring that small-scale artisanal fishers have access to marine resources and markets.[153] To support these efforts, developed countries are expected to assist vulnerable nations by boosting research capacities, sharing scientific knowledge, and transferring marine technology.[154]

SDG 7 focuses on ensuring the affordable and sustainable supply of modern energy for all nations. Consequently, countries are committed to ensuring universal access to affordable, reliable, and modern energy services[155] and substantially increasing the share of renewable energy in the global energy supply chains.[156] Additionally, this goal aims to double the global rate of improvement in energy efficiency.[157] Lastly, SDG 7 places emphasis on the role of research and innovation in promoting renewable energy use, including expanding infrastructure and upgrading technology to ensure the supply of modern and sustainable energy services worldwide.[158] It, therefore, calls for global cooperation in developing clean energy research and mobilising the required financial resources.[159]

---

147 Ibid., targets 14.1–14.2.

148 Ibid., target 14.3.

149 Ibid., target 14.4.

150 Ibid., target 14.6.

151 Ibid., target 14.5.

152 Ibid., targets 14.c and 14.7.

153 Ibid., target 14.b.

154 Ibid., target 14.a.

155 Ibid., target 7.1.

156 Ibid., target 7.2.

157 Ibid., target 7.3.

158 Ibid., target 7.b.

159 Ibid., target 7.a.

5.4.2 Relevance of the WTO Dispute Settlement

SDG 13, SDG 14, and SDG 7 are interconnected, as clean energy helps reduce GHG emissions that cause climate change and ocean acidification. The relevance of WTO agreements to energy is significant, though often underestimated.[160] As Cottier noted, major OPEC countries (except for Iran, Iraq, Algeria, and Libya) have joined the WTO,[161] making their energy trade subject to WTO agreements.[162] Additionally, WTO dispute settlements influence the implementation of national regulatory measures in energy-importing countries that aim to achieve these UN SDGs. These measures include carbon taxes, domestic green incentive policies, and contingent trade protective measures.[163]

The carbon tax is an effective method of promoting green production and reducing global GHG emissions.[164] Implementing such policies, like the EU's Carbon Border Adjustment Mechanism (CBAM), must comply with GATT Article III (2), which prohibits the imposition of internal taxes or charges on imported products in excess of those imposed on comparable domestic products.[165]

Green incentive policies are another strategy that countries often use to combat climate change.[166] These policies leverage procurement to promote the development of domestic green industries, including energy developers and producers of green production installations.[167] Specifically, such a policy ensures that state-owned companies purchase electricity from domestic energy developers who use green production installations produced locally.[168] This approach has multiple benefits: it incentivises the growth of local environmental industries and promotes the use of clean energy.[169]

Nevertheless, the implementation of this policy is subject to Article 2 of the TRIMs Agreement and its illustrative list, which prohibit any domestic content requirement on the grounds that such a requirement constitutes 'less favourable treatment' for foreign products compared to domestic similar products, as defined in

---

[160] See Cottier (2014), p. 42.

[161] Ibid., p. 42.

[162] Ibid., p. 42. These WTO agreements include the GATT, GATS, Agreement on TRIMs, Agreement on TRIPs, GPA, ASCM, Agreement on SG, and the Agreement on TBT.

[163] See Kent and Jha (2014), p. 248.

[164] See Branger and Quirion (2014), pp. 53–71; See also Vranes (2016), p. 77.

[165] Vranes (2016), p. 80.

[166] See Ibid., p. 248. Some scholar calls these green incentive policies market-pull policies.

[167] See Bahar (2013).

[168] See Kent and Jha (2014), p. 249; Weber (2015), p. 161; WTO Panel Report, Canada-Certain Measures Affecting the Renewable Energy Generation Sector (Canada-Renewable Energy), WT/DS412/R (WT/DS426/R), 19 December 2012, paras. 7.64–7.68; See also WTO Panel Report, India-Certain Measures Relating to Solar Cells and Solar Modules, WT/DS456/R, 24 February 2016, paras. 2.1 and 7.1–7.3.

[169] See Ofgem e-serve (2013); Ontario Ministry of Energy (2012), p. 7; Kent and Jha (2014), p. 249; See also Griffin (2013), p. 205.

GATT Article III (4).[170] Additionally, members who adopt green incentive policies often attempt to justify their regulatory measures under GATT Article III (8).[171] This clause allows for procurements made for governmental purposes, provided they are not intended for commercial resale or to produce goods for commercial sale.

Furthermore, WTO rules provide for contingency protection mechanisms, including antidumping, countervailing, and safeguard measures. WTO members can use these tools to shield domestic environmental industries from unfair competition by imported goods. Compared to green incentive policies, WTO panels generally adopt a more lenient stance towards contingency measures. Members can, in principle, implement any contingent trade protective measure that complies with WTO procedural requirements. Following the recent *US-Safeguard Measure on PV Products,*[172] which established groundbreaking jurisprudence supporting US safeguards on photovoltaic products,[173] an increasing number of countries have started to impose safeguards on imports of environmental goods. This case demonstrates that WTO members must show compliance with GATT Article XIX and Article 2.1 of the Safeguards Agreement to justify their safeguard measures.

5.4.3 Existing WTO Rules and Jurisprudence

5.4.3.1 Green Incentive Policy and GATT Article III (8) (a)

The *Canada-Renewable Energy* and *India-Solar Cells* cases both address issues related to members' green incentive policies and their domestic content requirements.[174] In both instances, the member that implemented such policies sought to invoke GATT Article III (8)(a) to exempt itself from the obligation to provide equal treatment to imported products and domestic similar products as required by GATT Article III (4). However, WTO rules prohibited the use of local content requirements in these cases.[175]

GATT Article III (8)(a) outlines four cumulative conditions under which a measure violating GATT Article III (4) due to domestic content requirements may be exceptionally justified. Specifically, the measure must: (1) be a law or regulation governing procurement; (2) pertain to procurement by governmental agencies; (3) be

[170] See Weber (2015), p. 161.

[171] See WTO Panel Report, Canada-Certain Measures Affecting the Renewable Energy Generation Sector (Canada-Renewable Energy), WT/DS412/R (WT/DS426/R), 19 December 2012, p. 64; See also WTO Panel Report, India-Certain Measures Relating to Solar Cells and Solar Modules, WT/DS456/R, 24 February 2016, p. 59.

[172] See Fang (2022), p. 266.

[173] See Ibid.

[174] See Kent and Jha (2014), p. 269; See also Espa and Holzer (2018), p. 417.

[175] See Espa and Holzer (2018), p. 417.

intended for purchasing products for governmental purposes; and (4) not be aimed at commercial resale.

These requirements are not explicitly stated in GATT Article III (8)(a) but were established by the panel in *Canada-Renewable Energy Generation Sector* and further developed by the panel in *India-Certain Measures Relating to Solar Cells and Solar Modules.*[176] Importing countries can generally meet the first two conditions by carefully designing and integrating their national climate policies into domestic laws and regulations.[177] The government only needs to ensure these climate policies enforce legally binding measures. Meeting the last two requirements is more challenging for importing countries.[178] On the one hand, green production installations subject to domestic content requirements are often not procured by governmental agencies.[179] On the other hand, when a governmental agency purchases electricity, it must resell it to distributors and consumers.

The AB's interpretation of the third requirement presents a significant challenge for members seeking to justify a green incentive policy under GATT Article III (8)(a). This difficulty is illustrated by the *India-Certain Measures Relating to Solar Cells and Solar Modules* case. In this case, India, the defending party, contended that even though the Indian government did not take title or custody of the solar cells and modules, its purchase of electricity generated from these cells and modules amounted to effective procurement of the products.[180] However, the panel rejected this argument and determined that India failed to justify its measure under GATT Article III (8)(a).[181]

Regarding the last requirement, the panels provided an extremely vague interpretation of 'view to commercial resale.' In both cases, the panels and the AB did not define the term 'commercial sale.' In *Canada-Renewable Energy Generation Sector*, the panel concluded that the governmental agency sold renewable electricity to retailers with a view to commercial sale because these retailers profited from their electricity distribution activities.[182] However, it is nearly impossible for state-owned companies to resell electricity without charging for it. Therefore, although a

[176] See Charnovitz and Fischer (2015), pp. 177–210 and 178; WTO Panel Report, Canada-Certain Measures Affecting the Renewable Energy Generation Sector (Canada-Renewable Energy), WT/DS412/R (WT/DS426/R), 19 December 2012, para. 7.122; See also WTO Panel Report, India-Certain Measures Relating to Solar Cells and Solar Modules, WT/DS456/R, 24 February 2016, pp. 70–84.

[177] See WTO Panel Report, India-Certain Measures Relating to Solar Cells and Solar Modules, WT/DS456/R, 24 February 2016, paras. 7.301 and 7.319.

[178] See Charnovitz and Fischer (2015), pp. 189–190.

[179] See WTO Appellate Body Report, Canada-Certain Measures Affecting the Renewable Energy Generation Section (Canada-Renewable Energy), WT/DS412/AB/R, adopted on 24 May 2013, para. 5.75.

[180] See WTO Panel Report, India-Certain Measures Relating to Solar Cells and Solar Modules, WT/DS456/R, 24 February 2016, para. 7.114.

[181] See Ibid., paras. 7.114–7.115 and 7.120.

[182] See Ibid., p. 74.

government may minimise profits as much as possible, it still does not meet the last requirement of Article III (8)(a).

5.4.3.2 Safeguard Measures and the Unforeseen Development Requirements

GATT Article XIX and the Agreement on Safeguards allow WTO members to protect domestic producers of similar products by imposing safeguard measures on rapidly increasing imports, provided they meet specific requirements.[183] Article 2.1 of the Safeguards Agreement stipulates that for safeguards to be imposed, a product must be imported in such increased quantities, either in absolute terms or relative to domestic production, that it causes serious injury to the domestic industry. Additionally, GATT Article XIX requires that the increase in imports result from unforeseen developments and the effect of the obligations incurred under the GATT.

It is worth noting that Article 2.1 of the Safeguards Agreement, which was negotiated after the GATT 1994, does not include the requirement for unforeseen developments. There is a general consensus among academics that the conditions for imposing safeguards under Article 2 of the Safeguards Agreement do not require that increased imports result from unforeseen developments.[184] However, in *Argentina-Footwear (EC)* and *Korea-Dairy*, the AB held that any safeguard measure imposed after the entry into force of the WTO Agreement must comply with the provisions of both the Safeguards Agreement and Article XIX of the GATT 1994.[185] This means that a WTO member's competent authority must demonstrate the existence of unforeseen developments in its investigation report.

Currently, safeguards are used to protect domestic photovoltaic (PV) industries. These measures play a vital role in sustainable development by breaking the monopoly of multinational companies in the PV industry, thereby creating a level playing field that is essential for promoting the healthy growth of the environmental sector. This is crucial for achieving the SDGs of combating climate change. The imposition of safeguards can: (1) diversify the supply chains of PV products in importing countries, ensuring their clean energy security; and (2) reduce environmental pollution throughout the life cycle of PV products.[186]

Nevertheless, it is extremely challenging for WTO members to demonstrate compliance with the unforeseen development requirements embodied in GATT Article XIX. Investigating authorities must demonstrate in their investigation report:

---

[183] See WTO Appellate Body Report, Argentina-Footwear (EC), WT/DS121/AB/R, adopted on 12 January 2000, para. 84; WTO Appellate Body Report, Korea-Dairy, WT/DS98/AB/R, adopted on 12 January 2000, paras. 76–77; See also Fang (2022), p. 245.

[184] Mavroidis et al. (2010), p. 500.

[185] See WTO Appellate Body Report, Argentina-Footwear (EC), WT/DS121/AB/R, adopted on 12 January 2000, para. 84; WTO Appellate Body Report, Korea-Dairy, WT/DS98/AB/R, adopted on 12 January 2000, paras. 76–77.

[186] I have developed this argument in detail elsewhere. Please see Zhao (2023), pp. 193–197.

(1) the existence of unforeseen developments, (2) that imports increased as a result of these unforeseen developments, and (3) that imports increased due to the effect of the obligations incurred.[187] Historically, members' safeguards have rarely passed panel examinations of these requirements until the recent *US-Safeguard Measure on PV Products* case.

In this case, the panel found that the USITC report (i.e., the investigation report of the US authority) met all unforeseen development requirements.[188] While the panel report offers groundbreaking jurisprudence on the legality of safeguards, it is insufficient to significantly improve the state of importing countries' domestic PV industries. Considering that these rulings are highly fact-specific,[189] WTO members may not be able to defend their safeguards in the same manner as the US did in this case. Consequently, it is challenging to expect future panel reports to achieve the same effect.

In addition, the panel report has significant limitations. It does not address the legality of green subsidies, despite numerous studies confirming their contribution to combating climate change.[190] To effectively slow global warming, governments must use green incentive policies, such as Feed-in Tariff (FIT) programs (often referred to as renewable energy subsidies),[191] to complement contingent protection measures.[192] Most importantly, the panel did not integrate sustainable development elements into its consideration of this case. Therefore, there is a high degree of contingency between the panel's rulings and their contribution to combating climate change.

5.4.4 Key Considerations for Balancing Sustainability Elements

5.4.4.1 Development Level

It is argued that economic growth contributes to combating climate change for three key reasons. First, economic growth can drive the development of environmentally friendly industries, thereby reducing GHG emissions from traditional, polluting industries. Second, economic growth fosters the development of environmental technologies that help reduce GHG emissions. Third, economic growth increases

[187] WTO Panel Report, United States-Safeguard Measure on Imports of Crystalline Silicon Photovoltaic Products, WT/DS562/R, circulated 2 September 2021, para. 7.11.

[188] Ibid., para. 7.62.

[189] Fang (2022), p. 262.

[190] Kent and Jha (2014), p. 249; Weber (2015), p. 161.

[191] Espa and Holzer (2018), p. 435; Borlini and Montanaro (2018), p. 86.

[192] Bougette and Charlier (2018), p. 182.

the number of people concerned about environmental protection.[193] Since international trade contributes to countries' economic growth and access to environmental technologies, many trade regulators believe that a high level of development is a precondition for protecting the environment and combating climate change.

In this context, the most well-known theory is the Environmental Kuznets Curve (EKC). This theory suggests that various indicators of environmental degradation tend to worsen with economic growth until average income reaches a certain threshold, after which environmental conditions begin to improve.[194] Essentially, the EKC advocates a 'pollution first, then governance' development pattern, implying that countries initially experience environmental degradation as they grow economically but later invest in environmental protection as they become wealthier.

From the perspective of sustainable development, trade regulators must reconsider the role of development levels in shaping trade-related environmental policies. The EKC theory is flawed as it overstates economic growth's contribution to environmental improvement,[195] misleading people to believe that international trade inherently aids in combating climate change. In reality, for international trade to be environmentally friendly, it must meet several specific conditions.

Furthermore, the EKC theory is incompatible with the urgent requirements of SDG 13, which calls for all countries to take immediate action to combat climate change. According to this theory, only developed countries can establish environmentally friendly industries to reduce GHG emissions. In contrast, developing countries are expected to industrialise and accumulate wealth at the expense of environmental degradation, thereby undermining global efforts to combat climate change. This development pattern not only fails to support climate change mitigation and adaptation but also exacerbates the problem. Additionally, the adverse effects of climate change are likely to further damage the economies of developing countries.[196] Therefore, it is crucial for these countries to participate in global efforts to reduce GHG emissions actively.

5.4.4.2 International Coordination

The premise for developing countries to move away from the 'pollution first, then governance' model is that they must obtain the necessary economic and technological support for sustainable development through alternative means. International coordination is a key approach to providing developing and least-developed countries with the financial resources and environmental technologies they need.

---

[193] See Cottier and Shariff (2013), p. 417.

[194] To know more details about Kuznets Curve theory, please see Kaika and Zervas (2013), pp. 1392–1402; See also Özcan and Öztürk (2019).

[195] See Kaika and Zervas (2013), p. 1403; See also Shafik (1994), pp. 757–758.

[196] See Stern (2006); See also Kent and Jha (2014), p. 247.

Currently, many international environmental treaties stipulate that developed countries are responsible for supplying this essential support.[197]

Nevertheless, achieving consensus on the exact amount of economic and technological support can be challenging. Countries may lack incentives to assist others in addressing environmental issues that do not directly impact them.

Moreover, even if developed countries provide sufficient financial and technological aid to developing nations, international donations may not effectively reduce global GHG emissions. This is because air pollution problems often become more severe in middle-income economies, which tend to be more energy-intensive and industrialised.[198] These middle-income developing countries are often relatively wealthier than other developing nations and may even compete with developed countries in specific economic or industrial sectors. As a result, developed countries frequently exclude these more polluted developing nations from receiving aid. Consequently, it becomes difficult for these countries to establish green industries and effectively combat and mitigate the impacts of climate change.

5.4.4.3 Export and Transfer of Pollution

When reviewing the impact of a member's trade policies on climate change, panels must consider the export and transfer of pollution. While the positive contribution of international trade and investment to the economic aspect of sustainable development is often recognised, it is also evident that multinational companies' substantial investments help developing countries build financial capacity to support their environmental policies.

However, the outsourcing of pollution is a significant aspect of international trade and investment. In addition to adopting clean energy and green production technologies, a country can invest in building factories abroad, thereby transferring its domestic pollution to these foreign locations. Liberalised international trade has facilitated this shift in industrial production from developed to developing countries. While developed countries have significantly improved their environmental quality by reducing industrial output, pollution—such as water, air, solid waste, and noise pollution—has worsened in developing countries, particularly in Asia, which has become the new global industrial centre.

Therefore, any form of international trade has the potential to exacerbate climate change rather than help countries reduce global GHG emissions. Additionally, the shift in industrial production can lead to structural unemployment and other social issues. Given these considerations, a panel should assess the outsourcing of pollution when evaluating whether a trade policy contributes to combating climate change. If a trade policy significantly transfers pollution to developing countries, the panel must determine whether these countries have the capacity to mitigate the environmental

[197] See, for example, UNGA (2015), target 13.3.

[198] See Shafik (1994), p. 770.

impacts of increased industrial activity. In particular, the panel should carefully evaluate the reasonableness of a trade policy under two scenarios: first, if the environment of the importing country is highly vulnerable, and second, if the importing country lacks the capacity to offset the environmental damage caused by industrial production. In both scenarios, it would be prudent for the panel to consider supporting restrictions on international trade to prevent the outsourcing of pollution.

5.4.4.4 Domestic Laws and Regulations on the Corporate and Social Responsibility

Companies play a crucial role in mitigating the adverse effects of international trade on climate change as private actors.[199] Therefore, domestic laws and regulations on corporate and social responsibility in exporting countries should address concerns about the outsourcing or transfer of pollution. Such laws must compel multinational companies to adopt environmentally sustainable practices in their investment and trade activities. These regulations, to be effective, should include two key obligations. First, multinational companies should be prohibited from relocating their factories to other countries solely to reduce pollution within their home country. They must ensure that the production methods of their overseas facilities do not harm the environment, animals, or human health. Second, multinational companies should not evade their responsibilities under national laws. This means they must implement sustainable practices in their domestic operations before expanding internationally. By adhering to these obligations, developing and least-developed countries can benefit from liberalised global markets without being negatively impacted by polluting industries.

5.4.4.5 Environment

Unsurprisingly, the environment should be central to the panel's analysis.[200] Often, WTO disputes involving environmental issues centre on the tension between importing countries' environmental protection efforts and exporting countries' economic interests. As seen in cases like *Canada-Renewable Energy Generation Sector* and *India-Certain Measures Relating to Solar Cells and Solar Modules*, measures implemented by importing countries that restrict international trade are typically aimed at achieving specific environmental objectives. This is crucial because,

---

[199] See Lin and Streck (2009), pp. 70–101; Newell (2010) pp. 253–269; Saggi (2007), pp. 191–235; Mathews et al. (2010), pp. 3263–3265; See also Kent and Jha (2014), pp. 247–248.

[200] To know climate change's impact on the environment, please see the following publications: Kent and Jha (2014), pp. 245–247; UNFCCC, Copenhagen Accord, COP Dec.2/CP. 15, UNFCCCOR, UN Doc. FCCC/CP/2009/11/Add.1, para. 1; Lynch (2007).

inherently, international trade contributes to GHG emissions and exacerbates climate change.

Today, with SDG 13 explicitly requiring countries to reduce GHG emissions, WTO members can more easily justify that national climate policies contribute to sustainable development. This was confirmed by the panel in both *Canada-Renewable Energy Generation Sector* and *India-Certain Measures Relating to Solar Cells and Solar Modules*. In *Canada-Renewable Energy Generation Sector*, the panel and the complainants acknowledged that Canada's Feed-in Tariff program aimed to encourage investment in local renewable energy equipment production in Ontario.[201] The panel found that this measure was clearly intended to reduce GHG emissions and foster the growth of local environmental industries.[202] Similarly, in *India-Certain Measures Relating to Solar Cells and Solar Modules*, both the EU and the US recognised that India's green incentive policy was designed to promote sustainable development.[203] The panel, in both cases, confirmed that green incentive policies could significantly advance the development of environmental industries and the use of clean energy.[204]

Moreover, when evaluating the contribution of green incentive policies to combating climate change, a panel should consider not just GHG emissions during production but also throughout the entire life cycle of products. This includes production processes, usage, and waste and recycling management. For instance, research indicates that the production and disposal (e.g., landfill) of crystalline silicon are highly polluting.[205]

5.4.4.6 Market Openness

A panel must also consider market openness when addressing conflicts between environmental and economic interests related to sustainable development. Existing case law demonstrates that while green incentive policies may restrict international trade to some extent, they do not entirely prohibit it. In both *Canada-Renewable Energy Generation Sector* and *India-Certain Measures Relating to Solar Cells and Solar Modules*, the panels found that the domestic content requirement measures in question did not fully prevent the importation of green production installations and related components.[206] In fact, green incentive policies that foster the growth of

[201] See WTO Panel Report, Canada-Certain Measures Affecting the Renewable Energy Generation Sector (Canada-Renewable Energy), WT/DS412/R (WT/DS426/R), 19 December 2012, para. 7.109.

[202] See Ibid., para. 7.216.

[203] See Ibid., para. 7.18.

[204] See Ibid., para. 7.110; See also WTO Panel Report, India-Certain Measures Relating to Solar Cells and Solar Modules, WT/DS456/R, 24 February 2016, para. 7.10.

[205] See Huang et al. (2017), pp. 132–141; See also Mulvaney (2014).

[206] See WTO Panel Report, Canada-Certain Measures Affecting the Renewable Energy Generation Sector (Canada-Renewable Energy), WT/DS412/R (WT/DS426/R), 19 December 2012, para.

domestic environmental industries can simultaneously encourage the importation of technological patents, equipment, parts, and raw materials. Consequently, multinational companies can benefit from these policies, as the government's green incentive programmes create the market for these products.[207] Therefore, it should not be assumed that green incentive policies and their domestic content requirements substantially undermine international trade.

5.4.4.7 Technological Innovation

Clean energy sources like solar and wind power still face significant challenges. Research shows that many renewable energy sources are intermittent and reliant on specific geographic and weather conditions. As a result, integrating these sources into the grid can be costly.[208] Consequently, the contribution of renewable energy sources to sustainable development can be limited in certain scenarios. To address these deficiencies, continued investment in technological innovation for renewable energy is essential. When reviewing cases, a panel must recognise the substantial impact its decisions can have on innovation in environmental technologies. Panels should be mindful not to impede the progress of renewable energy innovation. Therefore, their rulings should avoid unreasonably obstructing investment in the development and commercialisation of renewable energy technologies.

5.5 Conclusion

WTO members must realise SDG 1, SDG 9, and SDG 10 to fulfil their commitments to eradicating poverty and reducing inequality.[209] The WTO dispute settlement mechanism can play a crucial role in supporting this commitment. On the one hand, WTO rules can prevent protectionism and ensure fair and equitable participation in global trade.[210] On the other hand, they can protect essential and labour-intensive industries, thereby providing sufficient employment opportunities to enhance people's well-being.[211]

7.110; See also WTO Panel Report, India-Certain Measures Relating to Solar Cells and Solar Modules, WT/DS456/R, 24 February 2016, pp. 31–35.

[207] See Charnovitz and Fischer (2015), p. 178.

[208] See UNEP (2008), p. 8.

[209] See Sect. 5.1.1.

[210] See Sect. 5.1.2.

[211] Ibid.

However, existing WTO case law reveals several shortcomings.[212] The WTO Agreements lack exception clauses that would allow members to deviate from trade obligations in order to achieve SDG 1 and SDG 10.[213] Consequently, countries have limited policy space to implement these goals and can only safeguard their domestic industries under the guise of public morals or the prevention of deceptive practices.[214] Unfortunately, current WTO provisions are insufficient for ensuring that members effectively commit to eradicating poverty and reducing inequalities between developed and developing countries.

First, the applicability of these provisions is narrowly defined.[215] Second, WTO panels are not mandated to incorporate various sustainable development elements when assessing domestic measures, limiting their ability to enforce commitments to poverty eradication and inequality reduction effectively.[216] To address these gaps, WTO panels should consider a comprehensive range of factors, including economic growth, employment, wages, trade income, public morals, public health, animal welfare, biodiversity, and natural resource conservation, in their analyses.[217]

WTO members must pursue SDG 2 to fulfil their commitment to food security and sustainable agriculture.[218] WTO panels can influence this commitment by either facilitating or restricting international agricultural trade.[219] However, current WTO jurisprudence suggests that it is challenging for members to justify their agricultural trade regulations under existing WTO rules.[220] The *Indonesia-Import Licensing Regimes* case illustrates the limitations of using the necessity test to advance SDG 2, revealing that members often have to forgo practical sustainability measures to comply with WTO rules. Moreover, the panel and AB have ruled that members cannot employ contingent trade protective measures to shield their agricultural sectors, resulting in very limited policy space for achieving food security and sustainable agriculture.[221] To effectively expand policy space, WTO panels should incorporate key sustainability elements into their assessments. These include addressing structural hunger, promoting corporate social responsibility, ensuring food self-sufficiency, and mitigating the impacts of climate change.[222]

To fulfil their commitment to human health and life, WTO members must achieve SDG 3, which aims to ensure healthy lives and promote well-being for all at all

[212] WTO members proposed the joint statement initiative on electronic commerce at the WTO Ministerial Conference in 2017. However, there has not yet e-commerce related disputes.

[213] See Sect. 5.1.3.1.

[214] See Sects. 5.1.3.2 and 5.1.3.3.

[215] Ibid.

[216] Ibid.

[217] See Sect. 5.1.4.

[218] See Sect. 5.2.1.

[219] See Sect. 5.2.2.

[220] See Sect. 5.2.3.

[221] Ibid.

[222] See Sect. 5.2.4.

ages.[223] WTO rules permit members to implement SPS measures to address sources of transmission, including diseases, contaminants, and other harmful substances in imported goods.[224] These rules also allow members to adopt domestic measures to prevent the spread of diseases and health risks,[225] as well as to ensure access to essential medicinal products.[226]

However, the policy space for WTO members to apply these rules is significantly constrained for several reasons. Demonstrating that their measures are necessary to achieve regulatory objectives is challenging due to the complex and often unresolved scientific issues involved.[227] As a result, members frequently rely on the precautionary principle to support their claims. Yet, WTO jurisprudence reveals numerous difficulties in applying this principle effectively.[228] Furthermore, WTO panels generally show reluctance to endorse trade-restrictive measures.[229] From a sustainable development perspective, WTO panels should prioritise measures that protect human health over purely trade-related concerns.[230] To enable effective domestic measures, panels should consider lowering the standards of review, particularly when scientific issues are less well-studied.[231]

To fulfil their commitments to addressing climate change, protecting the oceans, and advancing clean energy, WTO members must achieve SDG 13, SDG 14, and SDG 7. These goals mandate that members combat climate change, sustainably manage the oceans, seas, and marine resources, and promote the use of clean energy.[232] WTO members can employ tools such as carbon taxes, domestic green incentive policies, and contingent protection measures to reduce greenhouse gas emissions and foster the growth of domestic environmental industries.[233]

However, demonstrating that these measures comply with WTO rules can be challenging.[234] Regarding green incentive policies, WTO panels have previously found such measures in violation of WTO rules due to their domestic content requirements.[235] In the case of contingent trade protective measures, some WTO members have implemented safeguard measures to protect local PV industries. Notably, in *US-Safeguard Measure on PV Products*, the panel upheld the consistency of these safeguards with WTO rules for the first time. However, the panel did

[223] See Sect. 5.3.1.

[224] See Sect. 5.3.2.

[225] Ibid.

[226] Ibid.

[227] See Sect. 5.3.3.1.

[228] See Sect. 5.3.3.2.

[229] See Sect. 5.3.3.3.

[230] See Sect. 5.3.4.

[231] Ibid.

[232] See Sect. 5.4.1.

[233] See Sect. 5.4.2.

[234] WTO panels have not yet heard a case on carbon taxes.

[235] See Sect. 5.4.3.1.

not consider sustainable development elements in its ruling. This oversight highlights the gap between the panel's decisions and their actual impact on combating climate change.[236]

To ensure that WTO members fulfil their commitments to addressing climate change, protecting the ocean, and advancing clean energy, WTO panels should incorporate the following elements into their sustainability assessments: (1) development level, (2) international coordination, (3) the export and transfer of pollution, (4) domestic corporate and social responsibility regulations, (5) environmental considerations, (6) market openness, and (7) technological innovation.[237]

Jurisprudence

WTO Panel Report, United States-Measures Affecting Imports of Woven Wool Shirts and Blouses from India, WT/DS33/R, adopted on 23 May 1997

WTO Panel Report, India-Patents (US), WT/DS50/R, adopted 5 September 1997

WTO Appellate Body Report, India-Patents (US), WT/DS50/AB/R, adopted 19 December 1997

WTO Panel Report, India-Patents (EC), WT/DS79/R, adopted 24 August 1998

WTO Panel Report, Japan-Agricultural Products II, WT/DS76/R, adopted on 19 March 1999

WTO Appellate Body Report, Korea-Dairy, WT/DS98/AB/R, adopted on 12 January 2000

WTO Appellate Body Report, Argentina-Footwear (EC), WT/DS121/AB/R, adopted on 12 January 2000

WTO Panel Report, Canada-Patent Protection of Pharmaceutical Products, WT/DS114/R, adopted 17 March 2000

WTO Panel Report, Canada-Patent Term, WT/DS170/R, adopted 5 May 2000

WTO Panel Report, Korea-Measures Affecting Imports of Fresh Chilled and Frozen Beef, WT/DS161/R (WT/DS169/R), 31 July 2000

WTO Appellate Body Report, Canada-Patent Term, WT/DS170/AB/R, adopted 18 September 2000

WTO Appellate Body Reports, Korea-Various Measures on Beef, WT/DS161/AB/R (WT/DS169/AB/R), adopted on 10 January 2001

WTO Panel Report, Japan-Measures Affecting the Importation of Apples, WT/DS245/R, adopted 15 July 2003

WTO Panel Report, Japan-Apples, WT/DS245/R, adopted on 10 December 2003

WTO Panel Report, US-Gambling, WT/DS285/R, adopted on 20 April 2005

WTO Appellate Body Report, US-Gambling, WT/DS285/AB/R, adopted on 20 April 2005

[236]See Sect. 5.4.3.2.

[237]See Sect. 5.4.4.

WTO Appellate Body Report, Brazil-Retreaded Tyres, WT/DS332/AB/R, adopted on 17 December 2007
WTO Panel Report, United States-Certain Measures Affecting Imports of Poultry from China, WT/DS392/R, adopted on 29 September 2010
WTO Panel Report, United States-Certain Country of Origin Labelling (COOL) Requirements, WT/DS384/R (WT/DS386/R), adopted on 23 July 2012
WTO Appellate Body Reports, US-COOL, WT/DS384/AB/R (WT/DS386/AB/R), adopted on 23 July 2012
WTO Panel Report, Canada-Certain Measures Affecting the Renewable Energy Generation Sector (Canada-Renewable Energy), WT/DS412/R (WT/DS426/R), 19 December 2012
WTO Appellate Body Report, Canada-Certain Measures Affecting the Renewable Energy Generation Section (Canada-Renewable Energy), WT/DS412/AB/R, adopted on 24 May 2013
WTO Panel Report, United States-Measures Affecting the Importation of Animals, Meat and other Animal Products from Argentina, WT/DS447/R, 24 July 2015
WTO Panel Report, India-Certain Measures Relating to Solar Cells and Solar Modules, WT/DS456/R, 24 February 2016
WTO Panel Report, Indonesia-Measures Concerning the Importation of Chicken Meat and Chicken Products, WT/DS484/R, adopted on 17 October 2017
WTO Panel Report, Indonesia-Importation of Horticultural Products, Animals and Animal Products, WT/DS477/R (WT/DS478/R), adopted on 22 November 2017
WTO Panel Report, Korea-Radionuclides, WT/DS495/R, adopted on 26 April 2019
WTO Panel Report, United States-Safeguard Measure on Imports of Crystalline Silicon Photovoltaic Products, WT/DS562/R, circulated 2 September 2021

References

Bibliography

Ahn D, Gnutzmann-Mkrtchyan A (2019) Indonesia-import licensing regimes: GATT rules for agricultural trade? World Trade Rev 18(2):197–218
Anderson K (2017) Finishing global farm trade reform. University of Adelaide Press, Adelaide
Arda M (2007) Food retailing, supermarkets, and food security: highlights from Latin America. In: Guha-Khasnobis B et al (eds) Food security: indicators, measurement, and the impact of trade openness. OUP, New York, pp 322–344
Bahar H et al. (2013) Domestic incentive measures for renewable energy with possible trade implications. OECD Trade and Environment Working Paper. 2013/01. Available via https://www.oecd-ilibrary.org/trade/domestic-incentive-measures-for-renewable-energy-with-possible-trade-implications_5k44srlksr6f-en. Accessed 21 July 2024
Borlini LS, Montanaro F (2018) Climate change and trade: challenges and lingering questions on the relationship between renewable energy subsidies and WTO disciplines. China-EU Law J 6: 81–101
Bougette P, Charlier C (2018) Antidumping and feed-in tariffs as good buddies? Modeling the EU-China solar panel dispute. Energy J 39(6):171–188

Branger F, Quirion P (2014) Climate policy and the 'carbon haven' effect. Wiley Interdiscip Rev Clim Chang 5(1):53–71
Button C (2004) The power to protect: trade, health and uncertainty in the WTO. Hart, Oxford
Charnovitz S, Fischer C (2015) Canada-renewable energy: implications for WTO law on green and not-so-green subsidies. World Trade Rev 14(2):177–210
Cloots AS (2011) Biofuels and the right to food: an uneasy partnership. In: Schutter D, Olivier C, Kaitlin Y (eds) Accounting for hunger: the right to food in the era of globalisation. Hart, New York, pp 95–134
Cottier T (2014) Renewable energy and WTO law: more policy space or enhanced disciplines? Renew Energy Law Policy Rev 5(1):40–51
Cottier T, Shariff N (2013) International trade and climate change. In: Van Calster G, Prévost D (eds) Research handbook on environment, health and the WTO. Edward Elgar, Cheltenham, pp 413–447
De Schutter O (2011) International trade in agricultural and the right to food. In: De Schutter O, Cordes KY (eds) Accounting for hunger: the right to food in the era of globalisation. Hart, New York, pp 137–192
De Schutter O, Cordes KY (2011) Accounting for hunger: an introduction to the issues. In: De Schutter O, Cordes KY (eds) Accounting for hunger: the right to food in the era of globalisation. Hart, New York, pp 1–26
Diaz-Bonilla E (2014) Agricultural trade and food security: some thoughts about a continuous debate. E15 Initiative Expert Group. International Centre for Trade and Sustainable Development (ICTSD) and World Economic Forum Working Paper
Espa I, Holzer K (2018) Negotiating 21st century rules on energy: what is at stake for the European Union, the United States and the BRICS? J World Invest Trade 19(3):415–443
Fane G, Warr P (2008) Agricultural protection in Indonesia. Bull Indones Econ Stud 44(1):133–150
Fang MM (2022) Shedding any new light? The WTO's latest ruling in the US-China solar Battle. Asian J WTO Int Health Law Policy 17(1):239–278
Feunteun T (2015) Cartels and the right to food: an analysis of states' duties and options. J Int Econ Law 18(2):341–382
Grant H (2015) Social crime prevention in the developing world: exploring the role of police in crime prevention. Springer, Cham
Griffin S (2013) The World Trade Organization: a barrier to green energy. Transnatl Law Contemp Probl 22:205–226
Gruszczynski L (2010) Regulating health and environmental risks under WTO law. OUP, New York
Häberli C (2013) God, the WTO-and hunger. In: Schefer KN (ed) Poverty and the international economic legal system: duties to the world's poor. CUP, New York, pp 79–106
Howse R (2008) Human rights, international economic law and constitutional justice: a reply. Eur J Int Law 19(5):945–953
Howse R, Levy PI (2013) The TBT panels: US-cloves, US-tuna, US-COOL. World Trade Rev 12(2):327–375
Howse R, Teitel RG (2009) Beyond the divide: the international covenant on economic, social and political rights and the World Trade Organization. In: Joseph S (ed) The World Trade Organization and human rights. Interdisciplinary perspectives. Edward Elgar, Cheltenham, pp 39–68
Huang B et al (2017) Environmental influence assessment of China's multi-crystalline silicon (multi-Si) photovoltaic modules considering recycling process. Sol Energy 143:132–141
Hudec RE (1993) Enforcing international trade law-the evolution of the modern GATT legal system. Butterworth Legal Publishers, Salem (New Hampshire)
Jack C (2016) U.S. scores sweeping win at WTO in agricultural import licensing dispute. Inside US Trade 34(50)
Kaika D, Zervas E (2013) The environmental Kuznets curve (EKC) theory-part a: concept, causes and the CO2 emissions case. Energy Policy 62:1392–1402

Kanter J, Rivlin G (2007) WTO gives Antigua right to violate U.S. copyrights in gambling dispute. The New York Times. New York. 21 December 2007. Available via https://www.nytimes.com/2007/12/21/business/worldbusiness/21iht-wto.html. Accessed 21 July 2024
Kent A, Jha V (2014) Keeping up with the changing climate: the WTO's evaluative approach in response to the trade and climate conundrum. J World Invest Trade 15:254–271
Kugler K et al (2019) Overview of WTO jurisprudence in 2017. In: Bungenberg M et al (eds) European yearbook of international economic law 2019. Springer, Cham, pp 389–434
Lin J, Streck C (2009) Mobilising finance for climate change mitigation: private sector involvement in international carbon finance mechanisms. Melbourne J Int Law 10(1):70–101
Lynch C (2007) U.N. secretary general calls global warming a priority. Washington Post. Washington. 2 March 2007. Available via https://www.washingtonpost.com/wp-dyn/content/article/2007/03/01/AR2007030101484.html. Accessed 21 July 2024
Mathews J et al (2010) Mobilizing private finance to drive an energy industrial revolution. Energy Policy 38(7):3263–3265
Mavroidis PC et al (2010) The law and economics of contingent protection in the WTO. Edward Elgar, Cheltenham
Moreu LF (2011) Food aid: how it should be done. In: Schutter D, Olivier C, Kaitlin Y (eds) Accounting for hunger: the right to food in the era of globalisation. Hart, New York, pp 239–264
Mulvaney D (2014) Solar energy isn't always as green as you think. IEEE Spectrum. 13 November 2014. Available via https://spectrum.ieee.org/green-tech/solar/solar-energy-isnt-always-as-green-as-you-think. Accessed 21 July 2024
Mutua M, Howse R (2001) Protecting human rights in a global economy: challenges for the World Trade Organization. In: Stokke H, Tostensen A (eds) Human rights in development yearbook 1999/2000. Martinus Nijhoff Publishers, Leiden, pp 51–82
Newell R (2010) The role of markets and policies in delivering innovation for climate change mitigation. Oxf Rev Econ Policy 26(2):253–269
Özcan B, Öztürk I (eds) (2019) Environmental Kuznets curve (EKC): a manual. Academic, London
Peres ALS, Daibert L d S (2017) Negotiating agriculture in the World Trade Organization: food security as a non-trade concern. Direito Alimentar E Combate à Fome 14(1):54–67
Saggi K (2007) Trade, foreign direct investment and international technology transfer: a survey. World Bank Res Obs 17(2):191–235
Shafik N (1994) Economic development and environmental quality: an econometric analysis. Oxf Econ Pap 46:757–773
Stern N (2006) The stern review on the economics of climate change. HM Treasury. 30 October 2006. Available via http://webarchive.nationalarchives.gov.uk/+/http://www.hm-treasury.gov.uk/stern_review_report.htm. Accessed 21 July 2024
Trebilcock MJ, Howse R (2005) The regulation of international trade. Routledge, New York
Vranes E (2016) Carbon taxes, PPMs and the GATT. In: Delimatsis P (ed) Research handbook on climate change and trade law. Edward Elgar, Cheltenham, pp 77–108
Weber RH (2015) Renewable energy: subsidies and taxes as competition distortion. In: Kreiser L et al (eds) Environmental pricing. Edward Elgar, Cheltenham, pp 161–176
Wise J (2007) Hunger and thieves: anticipating the impact of WTO subsidies reform on land and survival in Brazil. Am Indian Law Rev 31(2):531–551
Wohl I (2009) The Antigua-United States online gambling dispute. Journal of International Commerce and Economics. United States International Trade Commission. July 2009. Available via https://www.usitc.gov/staff_publications/jice/antigua_united_states_online_gambling_dispute. Accessed 21 July 2024
Zhao X (2023) The WTO panel report on US-safeguard measure on PV products: a decisive victory for the fight against climate change? In: Bäumler J et al (eds) European yearbook of international economic law 2022. Springer, Cham, pp 175–204

Documents

Ofgem e-serve (2013) Feed-in Tariffs Update. June 2013. Available via https://www.ofgem.gov.uk/sites/default/files/docs/2013/06/feed-in-tariff-update-quarterly-report-issue-12_0.pdf. Accessed 21 July 2024

Ontario Ministry of Energy (2012) Feed-in Tariff Program: Two-year Review Report. Available via https://www.ofgem.gov.uk/publications/feed-tariff-fit-annual-report-2011-2012. Accessed 21 July 2024

UNEP (2008) Reforming Energy Subsidies: Opportunities to Contribute to the Climate Change Agenda

UNGA (2015) Transforming Our World: the 2030 Agenda for Sustainable Development. UNGA Res/70/1. 21 October 2015

WTO (2001) Declaration on the TRIPS Agreement and Public Health. WT/ MIN (01) DEC/2. 20 November 2001). Available via https://www.wto.org/english/thewto_e/minist_e/min01_e/mindecl_trips_e.htm. Accessed 21 July 2024

WTO (2015) Export Competition. WT/MIN (15)/45, WT/L/980. Ministerial Decision of 19 December 2015. 21 December 2015

WTO (2022) Ministerial Declaration on the Emergency Response to Food Insecurity. WT/ MIN (22)/ 28 WT/ L/ 1139. 22 June 2022

Part III
Integrating Sustainability into WTO Law

Chapter 6
The Constitutionalisation of the WTO: An Ideal Methodology

Abstract WTO members must incorporate TSDCs and new analytical frameworks into WTO rules to enable panels to apply sustainability tests. Constitutionalising the WTO appears to be a promising approach for integrating these necessary elements into its legal framework. However, while desirable, the constitutionalisation of the WTO faces significant challenges associated with global constitutional theories. Given this, several doubts arise regarding the constitutionalisation of the WTO. The first doubt concerns its utility and whether the WTO should possess constitutional features, particularly in terms of its structure. The second doubt pertains to the institutional structure of global governance and the relationship among international governmental organisations. Can there be a hierarchical or fragmented global constitutional system that allows international organisations to govern global affairs? For the WTO, this raises the question of whether the current global governance system permits it to address non-trade issues and, if so, how. The third doubt involves the harmony of WTO constitutional norms, specifically whether WTO rules can reconcile economic freedoms with human rights. The final doubt relates to the ideology of economic constitutionalism. Constitutionalism can be used to defend various liberal ideologies and protect different legal values. What legal values should the constitutionalisation of the WTO aim to protect? In this Chapter, I will discuss the role of constitutionalisation in integrating the UN SDGs into WTO law and address these critical issues associated with this global constitutional theory.

X. Zhao, *Integrating the UN SDGs into WTO Law*, EYIEL Monographs - Studies in European and International Economic Law 41,
https://doi.org/10.1007/978-3-031-73876-0_6

6.1 Constitutionalising the WTO: The Way Towards Sustainability

6.1.1 Constitutionalisation and Sustainable Development: The Twins in the WTO

Constitutionalism was originally a domestic law concept, evolving through a combination of self-contained development and the fusion of different constitutional theories.[1] Scholars in various countries, such as John Locke, Abbé Sieyès,[2] Montesquieu,[3] Jean-Jacques Rousseau,[4] and Thomas Paine,[5] developed constitutional ideas within the context of their own social and legal backgrounds, including their legal systems, the Enlightenment, the Industrial Revolution, and globalisation. These thoughts, shaped by entrenched national identities, reflect the characteristics of nation-states but collectively form the common elements of constitutionalism.[6]

In the earliest constitutional documents,[7] such as the Magna Carta (1215), the Declaration of Independence of the United States (1776), and the Declaration of the Rights of Man and of the Citizen (1789),[8] we can see that liberty, equality, natural (human) rights, the rule of law, the limitation of power, the institutionalisation of power, the separation of powers, and social contract theory[9] were recognised as the core elements of constitutionalism. These commonalities demonstrate a tendency to acknowledge universal values and even suggest the possibility of drafting a global constitution.

Many German-speaking scholars, including notable figures such as Kant from Germany and Kelsen and Verdross from Austria, along with others sharing similar educational backgrounds, have endeavoured to create and develop global

---

[1] For a recent overview see Albert (2020).

[2] See, for example, Fioravanti (2007), pp. 87–103.

[3] Montesquieu (2013).

[4] Rousseau (2013).

[5] Paine (1791).

[6] One can try to trace back these concepts to English antecedents going back to Magna Carta (1215), as it is for example done by Schwartz (1980).

[7] The term "constitutionalism" is very much due to the idea what the ideal constitution of a State should look like in particular by limiting the absolute power of the monarch or government. In view of the importance of the respect of these principles enshrined in the constitution, it normally entails a request for a strict supremacy of these constitutional principles over any other regulations. This leads automatically to the question of the role of courts or independent bodies controlling this hierarchy. The French Declaration of Human Rights (*Déclaration des droits de l'homme et du citoyen*) in 1789 particularly shows this idea.

[8] One should not forget the influence in the Caribbean, Central America and South America.

[9] In the views of social contract philosophers, one must interpret a constitution as a contractual relationship between a State and its citizens, upon which citizens fulfil their equality and natural rights due to their status as constituent members.

constitutionalism. This movement first emerged with Kantianism in 1795[10] and particularly gained momentum after the end of the Second World War and the Cold War. However, scholars like Lauterpacht, who succeeded thinkers such as Grotius,[11] Oppenheim,[12] and Pufendorf,[13] favoured realism over global constitutionalism. They believed that international relations played a more crucial role in developing international law, viewing the efforts of constitutionalists as somewhat utopian.

Despite this, constitutional theory has shifted from a national to an international (and even supranational) concept,[14] especially since Verdross developed a theory of international constitutionalism based on the UN Charter and jus cogens norms.[15] Verdross's international constitutional theory provides a theoretical foundation for how international relations and institutions should be grounded in the idea of the social contract to ensure the enforcement of jus cogens norms[16] in international law.[17]

The Verdrossian theory has been primarily developed by successive German-speaking scholars and their pupils,[18] including Hermann Mosler,[19] Christian Tomuschat,[20] and Bruno Simma,[21] as well as more recently by Anne Peters, Bardo Fassbender,[22] Armin von Bogdandy, and Nico Krisch. Other scholars influenced by their writings, such as Erika de Wet, Jan Klabbers, and Geir Ulfstein,[23] have also contributed to the development of this theory.

International constitutional theory aims to establish a norms-based international law, contrasting with the power-driven order depicted by Hegel.[24] International

---

[10] Kant (1795).

[11] See, for example, Borschberg (2006); See also Scheltens (1983), pp. 43–60.

[12] See Oppenheim (1921).

[13] See Andrade and Sahd (2009), pp. 143–163.

[14] See for a good overview: Tsagourias (2007) or Schwöbel-Patel (2011).

[15] Kleinlein (2012), pp. 385–416; See also O'Donoghue (2012), p. 813.

[16] Jus cogens norms include State sovereignty, human rights, common values recognised by the UN Charter, and other domestic constitutional norms. See Verdross (1919), p. 291; Verdross (1949), pp. 435–440; See also O'Donoghue (2012), p. 808.

[17] See Ziegler and Zhao (2021), pp. 416–435; Weatherall (2015), p. 17; See for a recent overview Suami et al. (2018); See also Atilgan (2018).

[18] Some go so far to claim the constitutional approach to international law is a German concept: Wahl (2010), p. 225.

[19] Mostly towards the end of his life but thereby influencing many young scholars as Director of the Max Planck Institute for Comparative Public Law and International Law in Heidelberg, e.g., Mosler (1980).

[20] See, for example, Tomuschat (1995), pp. 1–20.

[21] See, for example, Simma (1998), pp. 266–277.

[22] See O'Donoghue (2012), pp. 814–821.

[23] See Klabbers et al. (2009). Some later also questioned these developments: by Follesdal and Ulfstein (2018).

[24] Fassbender (1998), pp. 529–619; See also O'Donoghue (2012), p. 804.

economic law scholars, such as John H. Jackson, Ernst-Ulrich Petersmann, and Gunther Teubner, have used this theory to interpret the role of international organisations in enhancing the international rule of law and protecting individuals' constitutional rights, including non-economic and social rights. As Ziegler notes, the German experience in the 1930s significantly influences scholars who believe that international judicial institutions like the WTO are essential for safeguarding human rights and democracy.[25] These scholars often consider international law issues from the perspective of the international community and are dedicated to using constitutional theory to reshape current international law and modernise existing international organisations.

Among international law scholars, certain experts in international economic law, such as John H. Jackson, Ernst-Ulrich Petersmann, Gunther Teubner, and Robert Cass, have explored how the WTO can be used to institutionalise international trade rules and ensure the rule of law in international trade. Their central argument is to establish a constitutional system centred around the WTO, imbuing international trade with constitutional features at both the structural and normative levels.

Notably, Jackson has highlighted the necessity of constitutionalism to address human rights, environmental, and social rights within the WTO framework. As early as 1996, Jackson emphasised the importance of addressing non-trade issues, including these rights, within the WTO. He argued that the constitutionalisation of the WTO is crucial for tackling these non-trade concerns. In his influential article, "The WTO 'Constitution' and Proposed Reforms: Seven 'Mantras' Revisited," Jackson outlines several constitutional issues arising from the WTO's institutional deficiencies.[26]

Although Jackson did not use the term 'sustainable development,' the issues he referred to as 'link issues' are closely aligned with contemporary sustainability concerns.[27] These link issues include the relationship between environmental regulations and trade policy, labour standards and trade policy, human rights and trade policy (including economic sanctions), monetary policy and trade policy, and arms control or non-proliferation issues and their intersection with trade policy.[28]

Jackson's study demonstrates that constitutionalisation and sustainable development are intrinsically linked within the WTO framework. The evolution of WTO law to include comprehensive trade-related sustainable development commitments—encompassing economic, environmental, and social dimensions—requires the WTO's constitutionalisation. This process involves integrating these interconnected rights into the existing WTO Agreements.

The connection between the WTO's constitutionalisation and the incorporation of sustainable development into WTO law was evident as early as 1994, during the WTO's formative years. Jackson argues that enhancing the WTO's constitution,

---

[25] See Ziegler and Zhao (2021), p. 5.

[26] Jackson (2001), p. 78.

[27] See Jackson (1996), pp. 39–40.

[28] Ibid., p. 41.

specifically the Marrakesh Agreement, to further entrench constitutional principles at both structural and normative levels is crucial for addressing sustainability challenges. Thus, integrating sustainability into WTO law is essential for resolving these enduring development issues.

6.1.2 The Functions of Constitutionalism in Integrating the UN SDGs into WTO Law

The WTO's constitutional features facilitate the enforcement of a comprehensive set of WTO commitments, creating favourable conditions for integrating sustainability into WTO law. Theoretically, constitutionalism serves three primary functions in embedding SDGs into WTO law.

First, constitutionalisation strengthens the rule of law, which is crucial for ensuring the enforcement of rules.[29] Scholars such as John Jackson,[30] Robert Hudec,[31] and Kenneth Dam[32] have extensively examined the role of constitutionalism in developing the WTO dispute settlement system. This system has been instrumental in reducing protectionism, rent-seeking, nationalism, and arbitrariness. By addressing these elements that undermine the rule of law, constitutionalisation significantly enhances the enforcement of sustainability rules.

The need to further constitutionalise the WTO may encourage its members to continue improving the rule of law within the organisation. As discussed in Chap. 5, potential improvements could include refining the analytical frameworks of WTO panels and the standards of review concerning scientific evidence.

Second, the constitutionalisation of the WTO aids in integrating sustainability into its values and norms, ensuring coherence among them. Constitutional norms are designed to pursue and protect various, often conflicting, rights. Thus, the WTO's constitutionalisation will enable the creation of an inclusive rulebook that encompasses a broad range of interests related to sustainable development and addresses conflicts between them. This process will obligate WTO members to incorporate sustainability commitments and elements into WTO Agreements, enabling the dispute settlement system to contribute to achieving the UN SDGs. Modernising the existing WTO Agreements to include environmental and social rights is essential for this transformation.[33] Enriching WTO values and norms will fill gaps in current

[29] On Jackson's role from a US perspective see Trachtman (1999), pp. 175–181 and more critically Kennedy (1995), pp. 671–716.

[30] See Jackson (1969).

[31] See Hudec (1990).

[32] See Dam (1970).

[33] Kumm suggested that global constitutional norms should include certain moral principles as domestic constitutional norms do. See Kumm (2018), pp. 168 and 171–172; See also Schill (2017), pp. 652–662. Stephan Schill claimed that the universal values of IEL consist of national

rules, providing a legal foundation for sustainable development commitments and supporting domestic regulatory measures aimed at sustainability.

Finally, the constitutionalisation of the WTO contributes to coordinating international organisations to ensure global multilevel governance of UN SDGs. Enhancing this global coordination will help enforce UN SDGs at both the national and international levels. It will also promote global efforts within international organisations, particularly those in the UN system, to achieve sustainable development by harmonising various domestic regulatory measures and optimising the management of financial and technical resources.

The role of constitutionalism in enhancing multilevel governance for sustainable development is not merely theoretical. The success of European integration has demonstrated that a robust system of multi-layered governance based on constitutionalism can effectively realise public goods, including economic growth, environmental protection, and various social rights.[34] Consequently, constitutionalizing the WTO would equip its members with a permanent, coordinated framework among international organisations. This framework would enable them to address global crises related to sustainable development with an immediate and comprehensive socio-economic response.

Despite these laudable functions, using constitutionalism to extend the regulatory effect of economic international organisations like the WTO in a way that protects fundamental human rights and pursues sustainable development remains controversial. Advocates of this approach primarily come from a relatively small group of scholars influenced by the German experience in the first half of the twentieth century. In the US, few successors to Jackson continue to endorse the constitutional functions of international economic law and organisations, with Joel Trachtman being a notable exception.[35] Prior to addressing the most critical doubts regarding the WTO's constitutionalisation in Sect. 6.2, I will examine two key challenges in realising global constitutionalism that underpin these doubts.

constitutional principles, the UN Charter, and sustainable development principles and international law; Petersmann always claims in his many writing on this approach that it is necessary to establish a constitutional mind-set within IEL so as to create a citizen-centred conception of international law favourable to rights of citizens and their democratic demand for transnational public goods (PGs). See Petersmann (2014), pp. 639–651; Petersmann (2017b), p. 39; Petersmann (2006), p. 640; Other authors in the United States and Germany advocated creating societal constitutionalism so as to bring more social rights in IEL. They are trying to use constitutionalism as a method for creating a more inclusive rulebook of IEL, which includes human rights, the rule of law, and democracy, and social values for achieving sustainable development. See Peters (2018), pp. 277–350; Anderson (2013), pp. 881–906.

[34] See Ziegler (1996).

[35] See, for example, Trachtman (2006), p. 27 ff.

6.1.3 Difficulties in Realising Global Constitutionalism

Many scholars have explored the usefulness of global constitutionalism as a method in international economic law.[36] However, the most pressing political question is whether constitutionalism is feasible at the international level or if it is too idealistic compared to traditional realist theories that uphold nearly unlimited sovereignty for nation-states.[37]

That question encompasses two sub-questions: (1) how to reach a consensus on the content of international constitutional norms, and (2) whether states will enforce these international constitutional norms at the national level in a state-centred world order.[38]

Reaching a consensus on international constitutional norms is nearly impossible in the current international political landscape. The prolonged deadlock of WTO negotiations since 1995 epitomises this challenge. A citizen-centred international community seems necessary to modernise existing international economic law and ensure the implementation of international constitutional norms. However, such a community, envisioned in the preamble to the UN Charter, has yet to be realised. Additionally, consensus on these norms can erode over time, as evidenced by Brexit and its impact on the purpose and objectives of the European Union.

The history of constitutionalisation in Japan offers a lesser-known but valuable insight into the consensus-building of international constitutional norms. The Japanese experience reveals the dynamics between autocratic rulers and their people and between democracies and non-democratic countries in the process of constitutionalisation. It highlights two significant points: Japanese rulers and their people did not share a common understanding of the constitution's implications, and the intervention of the US, a powerful democracy, was instrumental in drafting a robust constitution for the Empire of Japan, a non-democratic country.[39]

Citizens hold the primary constituent power to formulate a constitution. However, in non-democratic countries, the bottom-up process of constitutionalisation seen in democracies is entirely reversed and thus the primary constituent power does not belong to the citizens. This was evident in the process of promulgating the Meiji Constitution in the Empire of Japan.[40]

---

[36] Cass (2005); Dunoff (2006), pp. 647–675; Besson (2009); Collins (2009), pp. 251–287; Schwöbel (2010a, b), pp. 611–635; Schwöbel (2010a, b), pp. 529–553; Van Mulligen.(2011), pp. 277–304; Rosenfeld (2014), pp. 177–199; Schneiderman (2018), pp. 585–613; Vanoverbeke (2018), pp.203–244; Burnay (2018), pp. 225–244; Tushnet (2019), pp. 29–39; Stopler (2019), pp. 94–122.

[37] See, most recently, Zidar (2019). It should be noted that this debate applies also to constitutional thinking in many domestic systems. See, for example, for the United States Waldron (2009), pp. 267–282 or Rothbard (1978), p. 48.

[38] See, for example, Dunoff (2006), pp. 647–675.

[39] See Kumm (2018), pp. 209–211.

[40] Ibid.

Similarly, realising constitutionalism at the international level is challenging. In the current international community, both democratic and non-democratic countries participate in drafting international law. When members of the international community attempt to establish a series of international constitutional norms, both democracies and non-democratic countries, with their differing views on the implications of these norms, become constituent members. This diversity makes reaching a consensus on international constitutional norms difficult. Consequently, using constitutionalism as a method to create an inclusive rulebook for international economic law, encompassing the full range of constitutional norms, is problematic. As Jean d'Aspremont notes, the significant obstacle to cooperation between democracies and non-democracies is self-evident, even though the possibility of finding a diplomatic solution remains open.[41]

Nonetheless, WTO members cannot sacrifice the substantive values of a constitution for the sake of the WTO's constitutionalisation. These values serve as the baseline for recognising a robust constitution.[42] If such values are excluded from the WTO rulebook as a compromise, the objective of the WTO's constitutionalisation would be undermined. Therefore, trade policies that, for example, blatantly disregard human rights, systematically violate the rule of law, or endorse the enslavement of people must not be tolerated under WTO rules.

Obviously, this objective is too ambitious, given the current international political landscape. Nonetheless, the SDGs are more acceptable to WTO members than other constitutional norms due to their inherent ambiguities. All WTO members have signed the UN 2030 Agenda for Sustainable Development, and many have their own national development policies aimed at sustainable development. Therefore, it is conceivable that sustainable development could be recognised as a common value within the WTO's constitution.

Enforcing international constitutional norms at the domestic level is more difficult than building consensus. As Kun notes, addressing a global legal issue always hinges on solutions at the local level.[43] These national-level solutions are crucial for implementing international consensus, especially when states do not share a common legal tradition or political stance. Nonetheless, in most cases, there is no common solution at the national level. As Besson points out, there is no shared understanding of many terminologies used in international law practice that serve as a consensus of the international community. This results in a dilemma: even if the international community universally accepts international constitutional norms, they could be implemented differently in each state due to varying interpretations.[44]

An additional debate often concerns the protection of agreed international rules, especially treaty provisions, at the domestic level. To what extent should these rules hold supremacy over domestic laws? Should domestic courts be entitled to rule on

[41] See d'Aspremont (2008).

[42] See Besson (2009), pp. 385–386.

[43] For example, Fan (2016), pp. 244–292.

[44] See Besson (2009), p. 394.

individual claims based on these international treaties? These questions highlight the typical challenges regarding the role of international law at the domestic level, which becomes particularly controversial when it involves individual human rights, including those of investors, consumers, and traders.

The creation of international courts can be a remedy, but such courts remain rare and vulnerable. Current discussions on the future of international (regional) human rights courts, investor-state dispute settlement mechanisms, and the WTO DSU exemplify this aspect of global constitutionalism. Until more robust international judicial institutions are established, existing bodies like the WTO DSB must collaborate with national courts to enforce their decisions at the domestic level.

Failing to address these two critical issues—establishing common norms and determining how to protect them—results in an impasse. Constitutionalism cannot effectively be used to create and defend global public goods, such as a healthy environment, sustainability, and peace, nor can it integrate social rights-based values, such as respect for human dignity, the absence of violence and poverty, equality, and access to food, into the regulation of international trade. Given this context, the acceptance of constitutionalism as a methodology for transforming WTO law into a social rights-based framework that recognises international constitutional norms is doubtful. Consequently, an increasing number of scholars advocate for achieving this objective through alternative methods.

Cass and Schwöbel have highlighted the influence of politics on the use of constitutionalism in international economic law and have sought to minimise political interference. Cass proposed shifting the constitutionalisation of the WTO into a purely legal process, arguing that the WTO AB can generate constitutional norms. She described her approach as a judicial norm-generation method aligned with traditional US constitutional law theory.[45] According to Cass, this approach enables the WTO to adopt constitutional norms without making explicit normative claims in its decision-making process[46] and allows it to realign traditional sovereign relationships among constituent entities and between itself and its sub-parts through a deliberative process.[47] Schwöbel advocated for treating the constitutionalisation of the WTO as an ongoing process that does not rely on pre-existing political values. He placed the discursive political determination of constitutionalism at the core of his solution.[48] On the other hand, Patel emphasised political economy's role in achieving consensus on international constitutional norms among states.[49]

Schneiderman's insights also serve as a notable example in this context. Like many scholars, he argued that applying constitutionalism as a method in international economic law is excessive. He believed that international investment law itself should serve as the constitution for international investment. However,

[45] See Cass (2005), pp. 177–178.

[46] Ibid., p. 22.

[47] Ibid., p. 19.

[48] Schwöbel (2010a, b), p. 539.

[49] See Schwöbel-Patel (2018), p. 104.

Schneiderman considered the ideal balanced constitution, as defined by Tocqueville, unrealistic at the international level.[50] Consequently, he suggested that the application of constitutionalism in international investment law—and potentially in WTO law—should be confined to its current scope.[51] In the next section, I will examine in detail the concerns surrounding the constitutionalisation of the WTO.

6.2 Doubts on the Constitutionalisation of the WTO

6.2.1 Doubt on the Utility of the WTO's Constitutionalisation

The foremost and primary concern regarding the WTO's constitutionalisation is its utility. From the early stages of negotiations to establish the WTO, there has been intense debate over whether the organisation should incorporate constitutional features, particularly in its structural framework. The positions of negotiators on this issue have been starkly divided. Some delegates championed the benefits of constitutionalising the WTO, while others argued that such features should be excluded from the Marrakesh Agreement Establishing the World Trade Organisation, deeming them incompatible with the nature of international organisations.

Today, we know that WTO members ultimately accepted many proposals regarding the WTO's institutional structure—such as the establishment of the decision-making body, the administrative body, and the dispute settlement body—based on Jackson's constitutional theories.[52] These proposals were notably advanced by Canada and the US during the negotiations.[53] However, several significant proposals, such as adopting a flexible approach to amending the WTO Agreements, were not included in the final text of the Marrakesh Agreement.[54] The divided stance on the utility of constitutionalising the WTO has shaped its structure and continues to influence ongoing WTO negotiations.

Over the past decades, scholars have extensively debated the utility of the WTO's constitutionalisation, drawing on the experiences of WTO members during negotiations. Two main perspectives emerge from the existing literature.

The first view argues that constitutionalisation is essential to modernise the old, power-driven GATT regime and establish a genuine multilateral trading system

[50] See Schneiderman (2018), pp. 585–613.

[51] See Ibid.

[52] The WTO consists of the Ministerial Conference, Dispute Settlement Body, General Council, and other branches and specialized working groups. Please see 'WTO Organisation Chart'. https://www.wto.org/english/thewto_e/whatis_e/tif_e/org2_e.htm. Accessed 21 July 2024.

[53] See Steger (2016), pp. 339–341.

[54] Ibid.

governed by the rule of law.[55] Proponents believe that the Marrakesh Agreement should provide an institutional framework akin to a democratic constitutional structure to mitigate the influence of superpowers and ensure adherence to the rule of law. They advocate for a clear division of power within the WTO, comprising three main bodies: the Ministerial Conference (MC), the General Council (GC), and the DSB.

The MC is the highest decision-making body, where member states have an equal voice and decisions are made by consensus. The GC, as the principal administrative body, manages day-to-day operations and implements policies set by the MC. The DSB is responsible for resolving disputes related to the WTO Agreements and has the authority to establish dispute settlement panels and the AB. The representatives within the DSB decide on whether to accept or reject the reports of these panels and the AB by consensus. Notably, since the party winning the dispute also participates in the voting, decisions from panels and the AB are rarely rejected. This arrangement is designed to maintain the independence of the DSB.

The separation of powers among these three divisions is clear, with minimal overlap or influence among them. This structure, characterised by independent and balanced branches, embodies several attributes of a constitutional system, including separation of powers, democratic decision-making, an independent judiciary, checks and balances, and transparency.[56] While the WTO may not encompass all features of a traditional constitutional polity, its independent and balanced institutions significantly institutionalise member power and mitigate the risks of arbitrariness inherent in a member-driven organisation.

The essence of the first perspective lies in a value judgment regarding the international trade order. It advocates for the constitutionalisation of the WTO to enhance the international rule of law in trade, thereby establishing an international trade system governed by legal principles rather than by power dynamics. This law-based trade order would be characterised by common values such as multilateralism, limitations on state power, the rule of law, social rights, sustainable development, and the protection of individual constitutional rights beyond national borders. It would also emphasise a citizen-centred approach to international law. Many scholars argue that these values are crucial for ensuring sustainable development, peace, and transnational cooperation. Consequently, they assert that WTO law must safeguard international trade from threats posed by unilateralism, plurilateralism, nationalism, a power-driven world order, state capitalism, and state-centred international law, all of which undermine the international rule of law.

The second perspective, rooted in realism, argues that the constitutionalisation of the WTO is unfeasible. Realists who hold this view do not necessarily dispute the value of constitutionalising the WTO but question the practicality of replicating

[55] The GATT Agreement is an alternative to the failed International Trade Organisation (ITO), which was regarded as a power-driven system dominated by superpowers, and the Agreement also has less institutional setup, especially for ensuring the rule of law, than the present WTO Agreement. For reviewing the position of countries on the ITO and GATT, please read Hudec (1987); See also Hudec (1990).

[56] See Horwitz (2013); Kozel (2014), p. 957; See also Papaspyrou (2018).

domestic constitutional structures at the international level.[57] Some critics argue that the idea of WTO constitutionalisation is utopian and incompatible with current realities. For example, Jeffrey L. Dunoff points out that there has never been, and is unlikely to be in the foreseeable future, a political process aimed at creating a constitutional document for the WTO.[58]

Some realists acknowledge the efforts to draft constitutional documents at the international level but remain sceptical about their effectiveness. Michel Rosenfeld, for instance, suggests that even if a particular transnational legal order does not fully embody constitutional characteristics, it is still more useful to evaluate its potential and limitations through the lens of constitutionalism than through other frameworks, such as administrative law or international law.[59]

Other realists contend that constitutionalism is fundamentally a domestic law concept and should not be applied to international organisations like the WTO. Proponents of this view often downplay or outright reject the value of constitutionalising the WTO, instead advocating for nationalist, unilateralist, or isolationist positions. They argue that national interests should guide trade policies, and integrating international constitutional norms into national frameworks is seen as counterproductive. A closer examination reveals that this perspective struggles to adapt to the evolving global landscape. While policymakers might occasionally invoke this theory to support their agendas, it is increasingly difficult to sustain the argument that international constitutional norms are always contrary to national interests.

Examining these perspectives makes it apparent that some scholars may exaggerate the differences in attitudes toward the utility of the WTO's constitutionalisation. While realists express scepticism about the feasibility of constitutionalising the WTO, they do not outright reject the value of many constitutional norms. Therefore, there remains significant potential for WTO members to initiate negotiations to incorporate interests related to sustainable development—viewed as constitutional norms within the framework of global constitutionalism—into WTO Agreements.

6.2.2 Doubt on the Institutional Structure of Global Governance: A Hierarchical System vs. a Fragmented System

6.2.2.1 Hierarchical Global Governance System Led by the United Nations

Many scholars argue that the institutional structure of global governance functions as a hierarchical system led by the UN, often described as a form of 'world

[57] See Schneiderman (2018), pp. 585–613.

[58] Dunoff (2006), p. 650.

[59] Rosenfeld (2014), p. 199.

government.' Among these scholars, including those from NYU School of Law such as Benedict Kingsbury, Nico Krisch, and Richard B. Stewart, there is an interpretation of the current global governance regime through the lens of global administrative theory.[60] This theory draws a parallel between the existing global governance institutions and domestic administrative systems, positing that a set of global administrative bodies oversees international affairs similarly to how domestic administrative bodies operate. These global entities are tasked with ensuring high standards of transparency, participation, reasoned decision-making, and legality, while also providing robust mechanisms for reviewing rules and decisions, thereby offering a sound international adjudication system.[61]

Therefore, this group of scholars explains that the hierarchical system of global governance represents a model of trans-governmental regulation, which relies on cooperation among international governmental organisations.[62] For instance, Kingsbury, Krisch, and Stewart argue that environmental administrative bodies should collaborate with non-environmental institutions such as the WTO, the World Bank, and the OECD to effectively manage environmental issues.[63] In this framework, the establishment of a hierarchical global governance system and the WTO's jurisdiction over non-trade matters are, in principle, not mutually exclusive.

However, the so-called hierarchical system of global governance often compromises the pursuit of high standards for international constitutional norms in order to maintain states' consent to this hierarchy. Essentially, this hierarchy undermines the concept of global constitutionalism. Critics argue that existing global administrative bodies frequently overlook or fail to enforce crucial international constitutional norms, such as human rights and sustainable development. Instead, these organisations tend to impose only superficial and vague common rules that member states must adhere to in order to participate in the international community.[64] More often than not, these rules focus on procedural aspects rather than substantive norms, as procedural rules are more readily accepted by states on a universal basis.

The consequence of such a hierarchical system is that the global order is governed more by relativism than by strict adherence to the law. This relativism implies that international law must tolerate numerous violations of fundamental constitutional rights, which are typically protected at the domestic level, unless they breach

[60] Kingsbury et al. (2005), pp. 15–161.

[61] These bodies include formal intergovernmental regulatory bodies, informal intergovernmental regulatory networks and coordination arrangements, and national regulatory bodies operating with reference to an international intergovernmental regime, hybrid public-private regulatory bodies, and some private regulatory bodies exercising transnational governance functions of particular public significance. See Kingsbury et al. (2005), pp. 17 and 20.

[62] Ibid., p. 16.

[63] Ibid., p. 19.

[64] Ibid., p. 29. '...The focus of the field of global administrative law is not, therefore, the specific content of substantive rules, but rather the operation of existing or possible principles, procedural rules, review mechanism, and other mechanisms relating to transparency, participation, reasoned decision-making, and assurance of legality in global governance.'

procedural rights.[65] This compromise dilutes the essence of global constitutionalism by significantly restricting the pursuit and protection of fundamental rights, including those related to the economic, social, and environmental dimensions of sustainable development.

The hierarchical system of global governance is thus a double-edged sword. On the one hand, it acknowledges the importance of institutionalising global governance and enhances the international rule of law, particularly in procedural aspects. By transferring common administrative laws from domestic governments to intergovernmental organisations, global administrative bodies can effectively adapt and enforce these rules with the support of national governments and courts.[66] On the other hand, this hierarchical structure often fails to adequately address and enforce international constitutional norms, limiting the effectiveness of many major international organisations. For example, during the early stages of the Doha Round negotiations, many WTO members criticised the organisation for overstepping national regulatory sovereignty and prematurely extending its jurisdiction to non-trade issues, without first establishing a robust legal framework.

Consequently, addressing the gaps in enforcing international constitutional norms is crucial.[67] However, scholars such as Kingsbury, Krisch, and Stewart argue that there is no ideal solution to this ambitious goal.[68] They suggest that the greatest contribution of the hierarchical system of global governance lies in its ability to offer a global mechanism that integrates hierarchical organisations, oversight, and legal accountability. This system is supported by market and peer pressures, financial controls, and public reputational dynamics.[69]

Nevertheless, such a mechanism may undermine rather than enhance the foundation of the international rule of law. First, the hierarchical system of global governance is inherently power-driven, relying heavily on the authority of superpowers. The very presence of these superpowers can undermine the principle of a world order governed by law, making the hierarchical system fundamentally at odds with efforts to strengthen the international rule of law. Second, improving the existing hierarchical system of global governance faces significant challenges, including political interference, high costs, and inefficient enforcement at the national level.[70] These issues create a cycle where the hierarchical system, while contributing to the institutionalisation and stability of global governance, simultaneously erodes the international rule of law that underpins the protection of fundamental rights as envisioned by international constitutional norms.

[65] See for example Callaway (2015).

[66] Ibid., p. 26.

[67] See Kingsbury (2005), pp. 40 and 42–44.

[68] Ibid., pp. 56–59.

[69] Ibid., p. 58.

[70] Ibid., p. 58.

6.2.2.2 Power-Driven, Fragmented Governance System

The second group of scholars argues that the institutional structure of global governance is fundamentally power-driven and fragmented. They recognise that the existing system's core attributes—specialised institutions and a decentralised regime—pose significant barriers to establishing a unified hierarchical government. According to this perspective, overcoming these challenges is not only extremely difficult but may be virtually impossible.

One major issue is the high cost of creating a hierarchical regime, as illustrated by the financial constraints faced by the UN. Additionally, the presence of diverse and often conflicting values among states makes it impractical to establish a single system that can govern all international affairs comprehensively. Even if such an institution were successfully created, it might still struggle with inefficiencies and lack the enforcement power needed to operate as effectively as a national government.

Some scholars argue that enforcing international constitutional norms requires a hierarchical structure among states, drawing on the domestic law concept where enforcement relies on state institutions such as the army, police, prosecuting authorities, and prisons. These institutions are responsible for implementing laws and punishing offenders.[71] In this view, if a global constitution were to exist, it would need to be promulgated and enforced by a world government with powers equivalent to those of a sovereign state's central government. Since such a global government does not exist, these scholars contend that global governance remains effectively anarchic. They assert that the current institutional structure of global governance is inherently power-driven and fragmented, which fosters unregulated global competition and prevents the cohesive implementation of international constitutional norms.

A power-driven, fragmented system describes the existing global governance regime, where the lack of a cohesive regulatory framework impedes the establishment of a rules-based, hierarchical global governance structure and undermines the enforcement of international constitutional norms. However, this description may overstate the difficulties and wrongly imply that nations should abandon efforts to institutionalise global governance and enhance the international rule of law.[72] Such a radical, pessimistic view risks providing a discursive strategy for lawbreakers to advocate a power-based approach to international law and justify their violations. Clearly, this outcome is unacceptable. Fortunately, global governance does not have to be constrained by such an unregulated system.

[71] See He and Sun (2020), p. 9.

[72] See Ibid., pp. 238–240 and 212.

6.2.2.3 Multi-Layered and Fragmented, Global Constitutional System

The third group of scholars, including Teubner, Petersmann, and Somek, advocates for a multi-layered and fragmented global constitutional system. According to their view, this system should enable international governmental organisations, such as the WTO, to operate as autonomous constitutional regimes.[73] Unlike hierarchical models, this approach does not depend on a central authority to enforce international constitutional norms.[74] Instead, each international organisation can develop its own constitutional features at both structural and normative levels, akin to a national constitutional system. Consequently, this global constitutional system is characterised by its decentralisation, consisting of various independent international constitutions.

The multi-layered and fragmented structure of global governance is crucial for maintaining the effectiveness of autonomous constitutional regimes in upholding international constitutional norms. A hierarchical approach would hinder the specialised functions of various international organisations. As Teubner points out, auto-constitutional regimes generate specialised primary rules in distinct legal domains alongside procedural norms for law-making, recognition, and legal sanctions.[75] Teubner's characterisation aligns with the reality of global governance, which is composed of a network of separate and specialised international organisations.

Despite some level of cooperation, these organisations often focus on specific, sometimes conflicting areas—such as the trade interests of multinational companies versus the constitutional rights of individuals, a notable international tension.[76] The need for expertise and tailored services in different fields necessitates specialisation among these bodies, which, in turn, drives a division of labour and exacerbates conflicts between their roles in international governance. Rather than facilitating the enforcement of international constitutional norms, the interactions or interventions among these organisations can undermine the effectiveness of an autonomous constitutional regime. Therefore, decentralising the global constitutional system is essential to preserving the self-contained nature of individual international organisations and ensuring their effective operation.

However, a fragmented global constitutional system should not be equated with an unregulated, power-driven system devoid of legal order. In fact, such a system can uphold the international rule of law through a multilevel approach to the enforcement of international constitutional norms. This approach requires cooperation among local, regional, and international actors to implement these norms

---

[73] See Petersmann (2017a).

[74] Teubner was of the view that there is no unitary global constitution overlying all areas of society but rather the constitutionalization of a multiplicity of autonomous subsystems of world society. See Teubner (2010), p. 329.

[75] See Ibid., p. 333.

[76] See Ibid., p. 330.

effectively at various levels.[77] Essentially, it facilitates order among diverse international constitutions by ensuring that all participants in global governance—ranging from international governmental organisations to NGOs and local governments—recognise, respect, and uphold a common set of values. This mode of global governance promotes a form of 'managerial anarchy,[78]' which mitigates the disorder often associated with the anarchical nature of transnational governance. By ensuring that governors at different levels adhere to shared values and rules, this system fosters a structured and cohesive international order.[79]

This multi-layered and fragmented global constitutional system offers significant flexibility and potential for realising a global constitutional order. Unlike a hierarchical approach, which would impose a rigid hierarchy among international governmental organisations, this system integrates constitutional features into each component of global governance. In the context of the WTO, constitutionalisation refers to incorporating these features into the WTO itself rather than imposing a hierarchical framework across all international organisations.

Such an approach can be seen as a practical form of constitutionalisation within international law and politics. It allows for meaningful integration of constitutional principles into specific branches of international law, such as WTO law, without necessitating a top-down hierarchy among international institutions. The WTO exemplifies this concept through its multi-layered governance approach. For instance, the WTO and various national or regional regulatory bodies apply the same international standards set by the Codex Alimentarius Commission to address issues related to permissible food ingredients.[80]

This method of institutionalising global governance effectively balances the protection of international constitutional norms with the need to avoid a rigid hierarchical structure. It represents a pragmatic compromise between a purely hierarchical system and a fragmented, power-driven approach. Such constitutionalisation is, therefore, feasible and relevant, even within the complexities of contemporary international politics.

6.2.3 Doubt on the Harmony of WTO Constitutional Norms: Economic Freedoms v. Human Rights?

There is notable criticism within the constitutionalist movement—particularly from those advocating for robust human rights protections through international law—regarding the application of constitutional principles to economic activities.[81]

---

[77] See Walker (2001), p. 33.

[78] See Somek (2014), pp. 228–232.

[79] See Ibid., p. 230.

[80] See Ibid., p. 231.

[81] For a business-friendly analysis see CATO (2011).

Essentially, this criticism represents a broader debate over whether property rights and economic freedoms should be considered as valuable as other human rights, especially those commonly classified as economic, social, and cultural rights. Critics question whether economic freedoms can be harmonised with human rights.[82]

Proponents like Petersmann, Hilf, and Cottier argue that individual rights to trade and invest are fundamental human rights deserving of the same protections and limitations as other human rights.[83] However, traditional human rights advocates often dispute this perspective.[84] They worry that prioritising economic freedoms could undermine efforts to protect traditional human rights, particularly those of workers, indigenous peoples, and other vulnerable groups.[85] They also express concerns that the rule of law, direct effect, judicial enforcement, and transparency in economic matters might negatively impact the safeguarding of these traditional human rights.[86]

In summary, some scholars and advocacy groups object to treating the rights of traders and investors—often represented by multinational corporations—as equal to the economic and social rights of workers and ordinary individuals. They argue that regulators should prioritise the human rights of these individuals, particularly given the widespread issues of famine, poverty, and environmental pollution. These critics caution against a lack of hierarchy between economic freedoms—such as trade, investment, and intellectual property rights—and fundamental human rights, including the right to work, access to clean water, adequate food, and a healthy environment.[87]

The debate over the conflict between economic and human rights often hinges on different value judgments depending on one's perspective. This debate becomes particularly intense when viewed through the lens of sustainable development. In this context, economic freedom—often synonymous with globalisation—is sometimes criticised as being at odds with environmental protection and social progress. Indeed, there is a noticeable tension between economic development and efforts to combat climate change. Globalisation can negatively impact individual health and national security, including economic security, as foreign direct investment and commercial activities are frequently associated with pollution, labour exploitation, and even economic colonisation.

Nevertheless, these activities also generate wealth and create jobs for large populations, enhancing the well-being of residents in host countries of investment

---

[82] See for example, Nicol (2010), pp. 47–81.

[83] See for example, Cottier (2009), pp. 317–333.

[84] See Hilf (2005), p. 397; See also Ziegler and Boie (2012), pp. 272–299.

[85] See, for example, Hilf (1997), pp. 321–356.

[86] See the famous debate between Ernst-Ulrich Petersmann and Philippe Alston in the European Journal of International Law: Petersmann (2002b), pp. 621–650; Howse (2002), pp. 651–659; Alston (2002), pp. 815–844; Petersmann (2002a), pp. 845–851; and a second time Petersmann (2008a), pp. 769–798; Carozza (2008), pp. 931–944; Howse (2008), p. 945; Petersmann (2008b), pp. 955–960. See also ILA (2008).

[87] See, for example, Alston (2002), pp. 815–844.

and importing countries of goods. Multinational corporations facilitate financial flows and the movement of people across borders through international business ventures, investments in foreign programmes, and the establishment of factories and other business entities abroad. Furthermore, economic interdependence between nations, such as that between the US and China, can help mitigate political tensions and promote peace. Therefore, economic interests and environmental and social concerns are not necessarily at odds. In fact, economic interests often align with the economic dimension of sustainable development, which supports the realisation of human rights related to the environmental and social dimensions of sustainability.

The relationship between economic and human rights necessitates the WTO to find a balance to achieve sustainable development. As a specialised international governmental organisation, the WTO should uphold the economic values it traditionally protects while also prioritising the broader values of sustainable development, including economic, environmental, and social rights. These rights, often referred to by constitutionalists as international constitutional norms, must be harmonised within the WTO's framework. To achieve this balance, the WTO needs to integrate these constitutional norms, considering all relevant factors in specific cases. Constitutionalism offers a robust framework for establishing a hierarchy among ordinary and constitutional norms and resolving conflicts between them. Drawing on the extensive experience gained from national constitutional frameworks can provide valuable insights for ensuring the coherence of the WTO's constitutional norms.

6.2.4 *Doubt on the Ideology of Economic Constitutionalism: Which Type of Liberalism?*

A specific aspect of this debate arises from the economic theories employed by constitutionalists to examine the economic foundations of international economic law's constitutional norms. Scholars such as Petersmann often reference economists like Wilhelm Röpke (1899–1966), Ludwig von Mises (1881–1973), Michael A. Heilperin (1909–1971), Friedrich August von Hayek (1899–1992), Lionel Robbins (1898–1984), Gottfried Haberler (1900–1995), and Jan Tumlir (1926–1985). Critics label these theories as neo-liberal, ordo-liberal, or part of the Geneva School of Neoliberalism.

The central issues in this debate involve the role of government in society and the extent to which economic rights (such as those related to investment, trade, and intellectual property) should be shielded from state intervention. This controversy pertains to the degree of market intervention, ranging from complete laissez-faire (or Manchester liberalism) to total state control.

Most scholars I have identified as proponents of constitutional ideas in international economic law argue that while some degree of state intervention is justified, it must adhere to general principles governing limitations on human rights—such as transparency, predictability, and rule-based regulation. Specifically, state intervention should serve a legitimate public interest (such as promoting economic stability or public goods like competition), be non-discriminatory, and be proportionate. These scholars emphasise that ordoliberalists were more focused on establishing an international economic order that would enable welfare states and curb abuses of both public and private power compared to nineteenth-century Manchester Liberals like John Bright and Richard Cobden or twentieth-century neoliberals such as Milton Friedman and the Second Chicago School.

At the heart of the debate is whether twentieth-century liberal ideas—often simplified as neoliberalism—adequately balance the protection of public goods and human rights (such as those associated with a welfare state) while limiting state intervention. Proponents of ordoliberalism, such as Walter Eucken and Franz Böhm, and even social ordoliberals like Alexander Rüstow and Wilhelm Röpke, attempted to integrate social cohesion and ethical principles into their frameworks. However, their efforts were not always deemed successful or sufficient. These scholars were responding to three major challenges: World War I, the Great Depression, and decolonisation. Additionally, opposition to communism (and socialism) also played a significant role, particularly in the work of Friedrich A. Hayek.[88]

To address these issues, they proposed that international institutions establish a regime to ensure property rights and maintain stability in the international division of labour.[89] Critics, however, argue that these neoliberal ideas overlook the economic, social, and cultural rights of vulnerable groups and the impoverished. Some contend that these ideas merely protect the property and economic rights of investors and multinational corporations to preserve the existing status quo.[90]

Ziegler challenges this critique. He argues that it is an unfair assessment, noting that the early neoliberal thinkers of the interwar period were responding to a 1930s world characterised by arbitrary property confiscation often based on national, racial, or religious identity—a practice that led to marginalisation, dehumanisation, and destruction. According to Ziegler, their emphasis on property rights was intended to safeguard human dignity rather than to oppose taxation, even if it is high and progressive, provided it is applied equally and impartially.[91] Ziegler contends that economic rights, such as property rights, are not absolute. They can be limited by principles like the rule of law, public interest, non-discrimination, and proportionality. At the same time, he maintains that these rights still deserve protection.

In my opinion, sustainable development aims to achieve not only environmental protection and social rights but also economic prosperity. In some situations, economic prosperity is even more crucial because it underpins the entire concept of sustainable development. Therefore, the WTO's ideology must protect trade interests and continue to promote global economic prosperity. It is clear that

---

[88] See for example, Hayek (2007) and much later Hayek (1988).

[89] See James (2018), p. 59.

[90] See the most recent debate between Slobodian (2018) versus Petersmann (2018), pp. 915–921.

[91] Ziegler and Zhao (2021), p. 17.

neoliberalism or embedded liberalism alone may not enable international governmental organisations, particularly the WTO, to achieve sustainable development. If the WTO aims to balance economic rights with non-economic rights, it must acknowledge both the side effects of liberalism on environmental and social rights and its contributions to fulfilling these rights. In other words, while the WTO should continue to pursue economic interests and promote global economic prosperity, it must also mitigate the negative impacts of unchecked capitalism.

To this end, WTO agreements need to provide a more inclusive framework that encompasses not only economic rights but also environmental and social rights—integrating what constitutionalists refer to as international constitutional norms into WTO law. Incorporating these international constitutional norms into WTO law would enable the WTO to help countries establish and uphold environmental and social standards at a transnational level, thereby complementing and sustaining economic globalisation.[92]

6.3 Conclusion

Constitutionalisation and sustainable development are closely intertwined within the WTO framework.[93] WTO members must constitutionalise the organisation to integrate SDGs into WTO rules. Constitutionalising the WTO is essential for achieving sustainable development for three key reasons. First, constitutionalisation enhances the rule of law, which is necessary for enforcing TSDCs.[94] Second, it facilitates the incorporation of sustainability into WTO law, ensuring synergy between trade and sustainable development.[95] Finally, constitutionalisation helps coordinate international organisations, promoting global multilevel governance of trade-related SDGs.[96]

Nevertheless, constitutionalizing the WTO faces two significant challenges: (1) reaching a consensus on the content of international constitutional norms and (2) enforcing these norms at the domestic level.[97] These challenges raise several doubts about the constitutionalization of the WTO, including its utility, the necessary institutional structure for global governance, its ability to reconcile economic freedoms with human rights, and the type of liberalism it promotes.[98]

By reviewing existing opinions on the utility of the WTO's constitutionalisation, I found that some scholars have overstated the differences in attitudes toward this

---

[92] See Mishra (1999), pp. 41 and 50–51.

[93] See Sect. 6.1.1.

[94] See Sect. 6.1.2.

[95] Ibid.

[96] Ibid.

[97] See Sect. 6.1.3.

[98] See Sect. 6.2.

issue.[99] Although realists raise doubts about the viability of the WTO's constitutionalisation, they do not outright reject the value of constitutional norms.[100] Consequently, there remains significant potential for WTO members to initiate negotiations on the constitutionalisation of the WTO.

Regarding the institutional structure of global governance, scholars generally hold three different perspectives: (1) a hierarchical global governance system led by the UN, (2) a power-driven, fragmented governance system, and (3) a multi-layered and fragmented global constitutional system.[101] I argue that the power-driven, fragmented governance system is unacceptable because it allows law-breakers to use discursive strategies to promote a power-based international law and justify their wrongful actions.[102] Between the hierarchical global governance system and the multi-layered and fragmented global constitutional system, I believe the latter is more effective and preferable.[103]

I analysed the gaps in the existing hierarchical global governance system through the lens of global administration theory. Although this hierarchy aims to establish global governance order, it remains content with the absence of international constitutional norms.[104] Such a system cannot fulfil the substantive fundamental rights guaranteed by national constitutions, as existing international governmental bodies do not prioritise protecting those rights. In this sense, the hierarchical system does not fully reject a power-driven world order.[105] It can only ensure procedural rights, which may enhance the international rule of law to some extent. This limitation reflects the lack of international constitutional norms in UN-led global governance[106] and reveals that the hierarchical global governance system cannot ensure a rules-based international legal order.[107]

Fortunately, the absence of a hierarchical global governance system does not impede the constitutionalisation of the WTO for two reasons. First, hierarchy among international governmental organisations, instead of enhancing the enforcement of international constitutional norms, may actually undermine the international rule of law.[108] Second, decentralised global governance does not hinder the realisation and dissemination of common constitutional norms among domestic and international

[99] See Sect. 6.2.1.

[100] Ibid.

[101] See Sect. 6.2.2.

[102] See Sect. 6.2.2.2.

[103] See Sects. 6.2.2.1 and 6.2.2.3.

[104] See Sect. 6.2.2.1.

[105] Ibid.

[106] Ernst-Ulrich Petersmann invariably emphases that global constitutionalism must ensure the protection of substantive rights of constitution at the international level. See Petersmann (2012), p. 195.

[107] See Sect. 6.2.2.1.

[108] See Sect. 6.2.2.3.

governmental bodies.[109] Therefore, I believe that a multi-layered and fragmented global constitutional system is both desirable and viable.[110]

Regarding the concern about reconciling economic freedoms with human rights, I believe that it is indeed possible. Economic interests and environmental and social interests are not mutually exclusive.[111] In fact, economic interests often align with the economic dimension of sustainable development, which can support environmental protection and the fulfilment of social rights.[112]

Finally, I find that the WTO must adopt a liberalised trade approach that balances economic, environmental, and social interests.[113] I argue that neither neoliberalism nor embedded liberalism alone can guide international governmental organisations, particularly the WTO, toward achieving sustainable development, as they fail to address the negative side effects of capitalist pursuits.[114] Therefore, WTO agreements should encompass an inclusive framework that integrates economic, environmental, and social rights, incorporating what constitutionalists refer to as international constitutional norms into WTO law.[115]

References

Bibliography

Albert R (ed) (2020) Revolutionary constitutionalism - law, legitimacy, power. Hart, Oxford

Alston P (2002) Resisting the merger and Acquisition of Human Rights by trade law: a reply to Petersmann. Eur J Int Law 13(4):815–844

Anderson GW (2013) Societal constitutionalism, social movements, and constitutionalism from below. Indiana J Global Legal Stud 20(2):881–906

Andrade L, Sahd SL (2009) The Samuel Pufendorf's social contract. Revista de Filosofia: Aurora 21(28):143–163

Atilgan A (2018) Global constitutionalism: a socio-legal perspective. Springer, Berlin

Besson S (2009) Whose constitution(s)? International law, constitutionalism, and democracy. In: Dunoff JL, Trachtman JP (eds) Ruling the world? Constitutionalism, international law, and global governance. Cambridge University Press, New York, pp 381–408

Borschberg P (2006) Grotius, the social contract and political resistance. A study of the unpublished theses LVI. International Law and Justice Working Paper. 2006/7. https://iilj.org/wp-content/uploads/2016/08/Borschberg-Grotius-the-Social-Contract-and-Political-Resistance-A-Study-of-the-Unpublished-Theses-LVI-2006-2.pdf. Accessed 21 July 2024

[109] Ibid.

[110] Ibid.

[111] See Sect. 6.2.3.

[112] Ibid.

[113] See Sect. 6.2.4.

[114] Ibid.

[115] Ibid.

Burnay M (2018) Chinese perspectives on the rule of law. In: Suami T et al (eds) Global constitutionalism from European and east Asian perspectives. Cambridge University Press, Cambridge, pp 225–244
Callaway HG (2015) What's wrong with relativism, anyway? A political-ethical approach to extreme ethical skepticism. https://www.researchgate.net/publication/275041797_What's_Wrong_with_Relativism_anyway. Accessed 21 July 2024
Carozza PG (2008) Human dignity and judicial interpretation of human rights: a reply. Eur J Int Law 19(5):931–944
Cass DZ (2005) The Constitutionalization of the World Trade Organization legitimacy, democracy, and Community in the International Trading System. Oxford University Press, New York
Collins R (2009) Constitutionalism as Liberal-juridical consciousness: echoes from international Law's past. Leiden J Int Law 22(2):251–287
Cottier T (2009) The constitutionalism of international economic law. In: Meessen KM et al (eds) Economic law as an economic good, its rule function and its tool function in the competition of systems. Sellier European Law, Munich, pp 317–336
d'Aspremont J (2008) L'état non démocratique en droit international Étude critique du droit international positif et de la pratique contemporaine. Editions A. Pedone, Paris
Dam KW (1970) The GATT: law and international economic organization. University of Chicago Press, Chicago
Dunoff JL (2006) Constitutional conceits: the WTO's 'constitution' and the discipline of international law. Eur J Int Law 17(3):647–675
Fan K (2016) 'Glocalization' of international arbitration-rethinking tradition: modernity and east-west binaries through examples of China and Japan. Univ Pa East Asia Law Rev 11(2):244–292
Fassbender B (1998) The United Nations charter as constitution of the international community. Columb J Transnational Law 36(3):529–619
Fioravanti M (2007) Sieyès et le jury constitutionnaire: Perspectives historico-juridiques. Annales historiques de la Révolution française 349:87–103
Follesdal A, Ulfstein G (eds) (2018) The Judicialization of international law—a mixed blessing? Oxford University Press, New York
Hayek FA (1988) The fatal conceit: the errors of socialism. University of Chicago Press, Chicago
Hayek FA (2007) The road to serfdom, first published 1944. University of Chicago Press, Chicago
He Z, Sun L (2020) A Chinese theory of international law. Springer, Singapore
Hilf M (1997) The role of National Courts in international trade relations. Mich J Int Law 18(2): 321–356
Hilf M (2005) Die WTO - Eine Gefahr für die Verwirklichung von Menschenrechten? Archiv des Völkerrechts 43(4):397–465
Horwitz P (2013) First amendment institutions. Harvard University Press, Cambridge
Howse R (2002) Human rights in the WTO: whose rights, what humanity? Comment on Petersmann. Eur J Int Law 13(3):651–659
Howse R (2008) Human rights, international economic law and constitutional justice: a reply. Eur J Int Law 19(5):945–953
Hudec RE (1987) Developing countries in the GATT legal system. Gower, London
Hudec RE (1990) The GATT legal system and world trade diplomacy. Butterworth Legal Publishers, Salem (New Hampshire)
Jackson JH (1969) World trade and the law of GATT. Bobbs-Merrill, Indianapolis
Jackson JH (1996) Appraising the launch and functioning of the WTO. German Yearb Int Law 39: 20–41
Jackson JH (2001) The WTO 'constitution' and proposed reforms: seven 'mantras' revisited. J Int Econ Law 4(1):67–78
James H (2018) From Vienna to the WTO: book review, -globalists: the end of empire and the birth of neoliberalism. By Quinn. Finance Dev 55(3):59
Kant I (1795) Perpetual peace: a philosophical sketch. F. Nicolovius, Königsberg

Kennedy D (1995) The international style in postwar law and policy: John Jackson and the field of international economic law. Am Univ J Int Law Policy 10(2):671–716

Kingsbury B et al (2005) The emergence of global administrative law. Law Contemp Probl 68(3/4): 15–61

Klabbers J et al (2009) The Constitutionalization of international law. Oxford University Press, New York

Kleinlein T (2012) Alfred Verdross as a founding father of international constitutionalism? Goettingen J Int Law 4(2):385–416

Kozel RJ (2014) Institutional autonomy and constitutional structure. Mich Law Rev 112(6): 957–978

Kumm M (2018) On the history and theory of global constitutionalism. In: Suami T et al (eds) Global constitutionalism from European and east Asian perspectives. Cambridge University Press, Cambridge, pp 168–200

Mishra R (1999) Globalization and the welfare state. Edward Elgar, Cheltenham

Montesquieu C-L d S (2013) De l'esprit des loix. Flammarion, Paris

Mosler H (1980) The international society as a legal community. Sijthoff and Noordhoff, Alphen aan den Rijn

Nicol D (2010) The constitution protection of capitalism. Hart, Oxford

O'Donoghue A (2012) Alfred Verdross and the contemporary Constitutionalization debate. Oxf J Leg Stud 32(4):799–822

Oppenheim LFL (1921) The future of international law. Oxford Clarendon Press, Oxford

Paine T (1791) Rights of man. J. Johnson, London

Papaspyrou N (2018) Constitutional argument and institutional structure in the United States. Hart, Oxford

Peters A (2018) Global constitutionalism-the social dimension. In: Suami T et al (eds) Global constitutionalism from European and east Asian perspectives. Cambridge University Press, Cambridge, pp 277–350

Petersmann E-U (2002a) Taking human dignity, poverty and empowerment of individuals more seriously: rejoinder to Alston. Eur J Int Law 13(4):845–851

Petersmann E-U (2002b) Time for a United Nations 'global compact' for integrating human rights into the law of worldwide organizations: lessons from European integration. Eur J Int Law 13(3):621–650

Petersmann E-U (2006) Human rights, constitutionalism and the World Trade Organization: challenges for World Trade Organization jurisprudence and civil society. Leiden J Int Law 19(3):633–667

Petersmann E-U (2008a) Human rights, international economic law and constitutional justice. Eur J Int Law 19(4):769–798

Petersmann E-U (2008b) Human rights, international economic law and constitutional justice: a rejoinder. Eur J Int Law 19(5):955–960

Petersmann E-U (2012) Cosmopolitan 'aggregate public goods' must be protected by cosmopolitan access rights and judicial remedies. In: Petersmann E-U (ed) Multilevel governance of interdependent public goods: theories, rules and institutions for the central policy challenge in the 21st century. EUI working papers. ESCAS 2012/23. European University Institute, Fiesole

Petersmann E-U (2014) Need for a new philosophy of international economic law and adjudication. J Int Econ Law 17(3):639–669

Petersmann E-U (2017a) Multilevel constitutionalism for multilevel governance of public goods: methodology problems in international law. Hart, Oxford

Petersmann E-U (2017b) Reforming multilevel governance of transnational public goods through republican constitutionalism?—legal methodology problems in international law. Asian J WTO Int Health Law Policy 12(33):33–73

Petersmann E-U (2018) Book review, – globalists: the end of empire and the birth of neoliberalism. By Quinn Slobodian. J Int Econ Law 21(4):915–921

Rosenfeld M (2014) Is global constitutionalism meaningful or desirable. Eur J Int Law 25(1): 177–199
Rothbard MN (1978) For a new liberty: the libertarian manifesto. Collier Books, New York
Rousseau J-J (2013) Du contrat social ou Principes du droit politique. J'ai Lu, Paris
Scheltens DF (1983) Grotius' doctrine of the social contract. Netherlands Int Law Rev 30(1):43–60
Schill SW (2017) Reforming investor-state dispute settlement: a (comparative and international) constitutional law framework. J Int Econ Law 20(3):649–672
Schneiderman D (2018) Against constitutional excess: Tocquevillian reflections on international investment law. Univ Chicago Law Rev 85(2):585–613
Schwartz B (1980) The roots of the bill of rights. Chelsea House, New York
Schwöbel CEJ (2010a) Organic global constitutionalism. Leiden J Int Law 23(3):529–553
Schwöbel CEJ (2010b) Situating the debate on global constitutionalism. Int J Constitutional Law 8(3):611–635
Schwöbel-Patel C (2011) Global constitutionalism in international legal perspective. Martinus Nijhoff Publishers, Leiden
Schwöbel-Patel C (2018) Global constitutionalism and east Asian perspectives in the context of political economy. In: Suami T et al (eds) Global constitutionalism from European and east Asian perspectives. Cambridge University Press, Cambridge, pp 100–122
Simma B (1998) The "international community": facing the challenge of globalization. Eur J Int Law 9(2):266–277
Slobodian Q (2018) Globalists: the end of empire and the birth of neoliberalism. Harvard University Press, Cambridge
Somek A (2014) The cosmopolitan constitution. Oxford University Press, New York
Steger DP (2016) John H. Jackson—WTO institution builder. J Int Econ Law 19(2):339–341
Stopler G (2019) Semi-liberal constitutionalism. Global Constitutionalism 8(1):94–122
Suami T et al (eds) (2018) Global constitutionalism from European and east Asian perspectives. Cambridge University Press, Cambridge
Teubner G (2010) Fragmented foundations: societal constitutionalism beyond the nation state. In: Dobner P, Loughlin M (eds) The twilight of constitutionalism? Oxford University Press, Oxford, pp 327–342
Tomuschat C (1995) Die internationale Gemeinschaft. Archiv des Völkerrechts 33(1/2):1–20
Trachtman JP (1999) John Jackson and the founding of the World Trade Organisation. Mich J Int Law 20(2):175–181
Trachtman JP (2006) The international economic law revolution and the right to regulate. Cameron, London
Tsagourias NK (ed) (2007) Transnational constitutionalism. Cambridge University Press, Cambridge
Tushnet M (2019) The globalisation of constitutional law as a weakly neo-Liberal project. Global Constitutionalism 8(1):29–39
Van Mulligan JG (2011) Global constitutionalism and the objective purport of the international legal order. Leiden J Int Law 24(2):277–304
Vanoverbeke D (2018) Are we talking the same language? The Sociohistorical context of global constitutionalism in East Asia as seen from Japan's experiences. In: Suami T et al (eds) Global constitutionalism from European and east Asian perspectives. Cambridge University Press, Cambridge, pp 203–224
Verdross A (1919) Reichsrecht und Internationales Recht. Eine Lanze für Art 3 des Regierungsentwurfs der Deutschen Verfassung. Deutsche Juristenzeitung 24:291 et seq
Verdross A (1949) On the concept of international law. Am J Int Law 43(3):435–440
Wahl R (2010) In Defence of constitution. In: Dobener P, Loughlin M (eds) The twilight of constitutionalism? Oxford University Press, Oxford, pp 220–244
Waldron J (2009) Constitutionalism—a skeptical view. In: Christiano T, Christman J (eds) Contemporary debates in political philosophy. Blackwell, Oxford, pp 267–282

Walker N (2001) The EU and the WTO: constitutionalism in a new key. In: de Burca G, Scott J (eds) The EU and the WTO: legal and constitutional issues. Hart, Portland, pp 31–58
Weatherall T (2015) Jus Cogens: international law and social contract. Cambridge University Press, Cambridge
Zidar A (2019) The world community between hegemony and constitutionalism. Eleven, The Hague
Ziegler AR (1996) Trade and environmental law in the European Community. Oxford Clarendon Press, Oxford
Ziegler AR, Boie B (2012) The relationship between international trade law and international human rights law. In: Erika DW, Vidmar J (eds) Hierarchy in international law: the place of human rights. Oxford University Press, Oxford, pp 272–299
Ziegler AR, Zhao X (2021) Global constitutionalism as a method in international economic law. In: Rossana D, Nicholas T (eds) Research methods in international law. Edward Elgar, Cheltenham, pp 416–435

Documents

CATO, The Right to Property in Global Human Rights Law. CATO Institute Policy Report (2011) Available via https://www.cato.org/policy-report/may/june-2011/right-property-global-human-rights-law. Accessed 21 July 2024
ILA (2008) Resolution 5/2008 on Human Rights and International Trade Law. Available via http://www.ejil.org/pdfs/19/5/1706.pdf. Accessed 21 July 2024

Chapter 7
Progress in Embedding Sustainability into WTO Rules

Abstract Many WTO members are actively negotiating new rules to enhance the sustainability of their commitments within the organisation. These negotiations are pivotal in driving the constitutionalisation of the WTO and reforming its rules. Given the stagnation of multilateral negotiations, WTO members are engaging in plurilateral negotiations to advance these rules. I categorise these negotiations into three main areas: liberalising international trade, advancing social rights, and committing to environmental sustainability. This chapter examines the progress and outcomes of these negotiations to determine which TSDCs are being incorporated into WTO law and whether they are adequate to establish a legal basis for the sustainability test.

7.1 The Liberalisation of International Trade

7.1.1 The Services Domestic Regulation Agreement

The current commitments of WTO members largely ensure free trade in goods. However, their commitments in other trade areas remain relatively limited. One such area is the services trade market, where there is a significant need for increased trade freedom. Currently, 67 out of 164 WTO members have signed the Services Domestic Regulation Agreement, which has somewhat enhanced the liberalisation of the services market.[1]

It is worth noting that the signatories to the Services Domestic Regulation Agreement will not expand market access beyond their previous commitments.[2] Instead, the agreement will enhance the liberalisation of trade in services by reducing trade barriers in sectors already open to foreign service suppliers.[3] The agreement, to achieve this, primarily regulates the licensing regimes of the signatory countries.

---

[1] See WTO (2021c).

[2] Ibid., paras. 1–3.

[3] Ibid., paras. 1–3.

X. Zhao, *Integrating the UN SDGs into WTO Law*, EYIEL Monographs - Studies in European and International Economic Law 41,
https://doi.org/10.1007/978-3-031-73876-0_7

The agreement is divided into two parts: general services and financial services. Despite this division, the licensing rules for both parts are essentially the same and include several key requirements. First, the agreement mandates that signatories' competent authorities allow a reasonable period for submitting applications.[4] Second, it requires signatories to optimise the legal procedures for applications.[5] Relevant laws and regulations should enable competent authorities to accept electronic documents and copies.[6] Specifically, signatories need to simplify the application process and ensure its certainty and transparency.[7] Third, signatories commit to ensuring the reasonableness and transparency of authorisation fees.[8] Fourth, the agreement imposes the same requirements on the examination process for authorising the supply of a service, stipulating that the cost, time, and procedural rules of the assessment process should be reasonable, transparent, and convenient for applicants.[9] Fifth, the agreement requires signatories to actively communicate in order to develop mutually recognised licensing regimes and professional qualifications.[10] Finally, it demands that signatories ensure the independence and transparency of the licensing process through various provisions.[11] This requirement also applies to the development of licensing rules within signatory countries.[12]

By increasing the transparency of the licensing system and regulating procedural rules and fee standards, the Services Domestic Regulation Agreement aims to prevent corruption and bureaucratic inefficiencies in the application process. This will reduce the time and monetary costs for applicants, significantly facilitating cross-border trade in services. The agreement is particularly beneficial for developing countries and micro, small, and medium enterprises (MSMEs). As noted in the joint brief by the OECD and WTO on this plurilateral agreement, the Services Domestic Regulation Agreement will facilitate services trade by cutting red tape and reducing trade costs.[13]

[4] Ibid., para. 5.

[5] Ibid., para. 4.

[6] Ibid., para. 6.

[7] Ibid., para. 7.

[8] Ibid., para. 9.

[9] Ibid., para. 10.

[10] Ibid., para. 11.

[11] Ibid., paras. 12–20.

[12] Ibid., paras. 21–22.

[13] See WTO and OECD (2021), p. 1.

7.1.2 E-Commerce

The joint statement initiative on electronic commerce was initially proposed by WTO members at the 2017 WTO Ministerial Conference.[14] Two years later, at the World Economic Forum in Davos, 76 WTO members reaffirmed their commitment by agreeing to launch plurilateral negotiations on electronic commerce.[15] As of now, 91 WTO members are participating in these negotiations. On July 26, 2024, they reached a stabilised text.[16]

The negotiations cover six major areas. First, the participating countries aim to facilitate electronic commerce by ensuring all necessary conditions for electronic transactions are met.[17] This includes establishing robust electronic transaction frameworks and providing reliable electronic payment services.[18] Additionally, countries recognise the importance of using electronic authentication, e-signatures, electronic contracts, and electronic invoicing.[19] Signatories are also encouraged to optimise trade policies, reduce paperwork, streamline customs procedures, and enhance logistical support.[20] These objectives have been incorporated into the final text of the statement.

Second, the negotiating countries aim to address the complexities of electronic trade openness, focusing on four key topics:

1. Non-discriminatory treatment of digital products and limiting liability.
2. Cross-border transfer of information via electronic means or cross-border data flows.
3. Customs duties on electronic transmissions.
4. Access to the internet and data.[21]

There are significant differences in views between developing and developed countries on these issues. Developed countries, including EU member states, Canada, New Zealand, and Japan, have advocated for maintaining the moratorium on customs duties for electronic transmissions and promoting the free flow of data.[22] In contrast, some developing countries support imposing tariffs on electronic trade[23] and restricting data flows.[24] At the WTO's 12th Ministerial Conference, members

[14] See WTO (2017d).

[15] See WTO (2019c).

[16] See WTO (2024a).

[17] See WTO (2019j), para. 2.2(4); See WTO (2019e), para. 2.

[18] See WTO (2019h), Article 4.

[19] See WTO (2019j), paras. 2.1–2.2; See WTO (2019e), para. 2.1; See WTO (2019i).

[20] See WTO (2019e), para. 2.1; See also WTO (2021e), paras. 5–6.

[21] See WTO (2019j), para. 2.9; See WTO (2019e), para. 3.1; See also WTO (2019h), Article 6; See WTO (2019f).

[22] See WTO (2019k), paras. 1–2; See WTO (2019h), Article 5.

[23] See WTO (2020).

[24] See WTO (2019b), paras. 4.2–4.3.

agreed to extend the moratorium on customs duties for electronic transmissions until December 31, 2023.[25] In the final text of the joint statement, participants committed to not imposing customs duties on electronic transmissions between them.[26] However, the issue of free data flow was not included in the final text.

Third, negotiating countries are dedicated to ensuring the security of electronic trade, addressing concerns such as consumer protection, personal data security, and the use of source code and cryptography.[27] Fourth, they have agreed to address several important cross-cutting issues, including transparency in domestic regulations, cooperation mechanisms, cybersecurity,[28] capacity building, and technical assistance. Fifth, there is a commitment to updating trade rules applicable to telecommunications services.[29] Finally, the participating countries aim to achieve greater consensus on market access issues.[30] All these concerns are reflected in the final text of the statement.

To summarise, the participating countries have made significant progress in the negotiations. They have reached consensus on several key issues, including spam prevention, electronic signatures and authentication, e-contracts, online consumer protection, and open government data.[31] Unfortunately, they were unable to agree on the issue of the free flow of data.

7.1.3 Investment Facilitation for Development

Investment facilitation for development is a prominent plurilateral negotiation actively advanced by WTO members.[32] Over 120 WTO members involved in these discussions are committed to overcoming trade barriers that hinder and restrict investment processes between countries.[33] In November 2023, they finalised the agreement, which establishes investment regulation standards to enhance the transparency and predictability of domestic regulations.[34] The agreement also aims to streamline and expedite the administrative procedures for investment approval, reflecting the approach adopted in the services domestic regulation agreement.[35] Additionally, it includes provisions related to responsible business conduct,

[25] See WTO (2022b).
[26] See WTO (2024a), para. 11.3.
[27] See WTO (2019j), paras. 2.3 and 2.8; See WTO (2019d).
[28] See WTO (2019g), section IX.
[29] See WTO (2019j), para. 3; See WTO (2019e); See also WTO (2021e).
[30] See WTO (2019j), paras. 2.9 and 4.
[31] See WTO (2019e), para. 4.1.
[32] See WTO (2017b).
[33] See WTO (2019a).
[34] See WTO (2023), Articles 6–12.
[35] See WTO (2017b), para. 4; WTO (2023), Articles 13–21.

encouraging the adoption of corporate social responsibility standards among participating countries.[36]

7.1.4 *Micro Small and Medium Enterprises*

WTO members began discussing rules applicable to Micro, Small, and Medium Enterprises (MSMEs) in 2017.[37] Currently, 91 member countries are engaged in these negotiations[38] to remove trade barriers that impede MSMEs and enable them to drive economic growth more effectively.[39] The discussions are still in the early stages, focusing on several longstanding issues related to trade liberalisation. Key areas of negotiation include enhancing the transparency and predictability of regulatory measures, reducing the time and costs associated with administrative procedures, and improving trade facilitation measures, logistical and infrastructural support, and international cooperation.[40] Additionally, the negotiating countries, recognising the vulnerability of MSMEs, are committed to exploring mechanisms to provide financial aid in response to emergencies, such as the COVID-19 pandemic.[41]

7.2 The Pursuit of Social Rights

7.2.1 *Gender Equality*

Gender equality is currently the only social rights issue being negotiated by WTO member countries. In 2017, WTO members initiated discussions on trade and women's economic empowerment and established an informal working group on trade and gender in 2020.[42] Within this framework, 127 WTO members are negotiating strategies to enhance women's participation in international trade. This plurilateral negotiation on gender equality is still in its early stages, and substantial work remains to finalise the agreement's text.[43] Nevertheless, the negotiating countries have reached consensus on four key issues.

---

[36] See Jose and Oeschger (2022); WTO (2023), Article 37.

[37] See WTO (2017c).

[38] See WTO (2021a).

[39] Ibid., p. 1.

[40] Ibid., para. 1.

[41] See WTO (2017a), para. 5.

[42] See WTO (2021f), para. 1.

[43] Some countries have incorporated commitments to gender equality into their bilateral and regional trade agreements. Please see WTO Database on gender provisions in RTAs (2023).

First, the negotiating countries should continue to review, develop, and enhance the collection and analysis of national and/or regional gender-disaggregated data that is as comparable as possible, to support informed, gender-responsive policies.[44] Second, they should leverage research initiatives to guide trade policy instruments and programmes that promote women's economic empowerment and increase their participation and leadership in international trade.[45] Third, they should incorporate a gender perspective and address women's economic empowerment issues within the WTO's activities.[46] Fourth, they should encourage and highlight collaboration on trade and gender among international and regional organisations to embed a gender equality perspective into Aid for Trade.[47] In the coming years, the negotiating countries will establish specific trade rules to achieve these goals, while considering the impact of the COVID-19 pandemic on women.[48]

7.3 Environmental Commitments

7.3.1 Plastics Pollution

Plastic pollution has a severe negative impact on the environment, biodiversity, and health. To address these issues in the context of trade, 82 WTO members are negotiating a plurilateral agreement. They issued joint statements at both the WTO's 12th and 13th Ministerial Conferences.[49] According to the publicly available documents, the members have reached agreement on the following points.

First, they acknowledge the commitments made by countries under other international conventions, such as the Basel Convention on the Control of Transboundary Movements of Hazardous Wastes and their Disposal, along with its Plastic Waste Amendments.[50] They are dedicated to strengthening cooperation between the WTO and other international environmental regulatory bodies, particularly the Secretariats of Multilateral Environmental Agreements, and refining trade rules aligned with existing international standards.[51]

Second, the negotiating countries have committed to enhancing their collaborative efforts to identify joint actions that support global initiatives to reduce plastic pollution.[52] The future agreement is expected to include provisions that will:

---

[44] See WTO (2021d), p. 2.

[45] Ibid., p. 2.

[46] Ibid., p. 2.

[47] Ibid., p. 2.

[48] Ibid., p. 2.

[49] See WTO (2021b, 2024b).

[50] See WTO (2021b), p. 2; WTO (2024b), p. 3.

[51] Ibid.

[52] Ibid.

(1) Enhance understanding of global trade in plastics.[53]
(2) Establish mechanisms for sharing best practices to transition towards a more circular, resource-efficient, and environmentally sustainable plastic trade.[54]
(3) Address trade-related capacity building and provide technical assistance to developing members.[55]
(4) Integrate considerations of plastic pollution and environmentally sustainable plastics trade into Aid for Trade initiatives with sustainability objectives.[56]

Third, they have agreed to persist in engaging with and supporting actions across other international processes.[57] This includes enhancing cooperation with other international organisations,[58] identifying effective trade policies or measures to support the implementation of actions both within and outside the WTO,[59] and improving the collection of data on trade flows and supply chains.[60] Additionally, they are committed to promoting discussions to identify best practices and share experiences related to reducing plastic pollution associated with trade.[61]

The final text of the agreement is expected to establish robust trade rules in these areas, which could significantly contribute to reducing plastic pollution from trade. However, the relatively small number of WTO members currently involved in this plurilateral negotiation may limit the agreement's overall impact. As highlighted in the joint ministerial statement, the negotiating countries must persist in their efforts to attract additional WTO members to strengthen the agreement's effectiveness in combating plastic pollution.[62]

7.3.2 Fossil Fuel Subsidy Reform

Transitioning from carbon-based energy production to clean energy sources is crucial for reducing GHG emissions and effectively addressing global warming and climate change. Currently, carbon-based electricity generation from oil, gas, and nuclear power—despite its unresolved long-term storage and waste challenges—remains dominant worldwide.[63]

[53] Ibid.

[54] Ibid.

[55] Ibid.

[56] See WTO (2024b), p. 3.

[57] Ibid.

[58] Ibid. The cooperation includes definitions, scope, standards, design, and labelling for plastics.

[59] Ibid.

[60] Ibid.

[61] Ibid.

[62] Ibid.

[63] Ibid.

Forty-seven WTO members are engaged in a plurilateral negotiation on fossil fuel subsidy reform. According to the proposed ministerial statement, the negotiating countries aim to tackle fossil fuel subsidy issues through three main strategies. First, they seek to enhance transparency regarding fossil fuel subsidies and promote their reform. Second, they aim to ensure that support measures introduced in response to the global energy crisis are targeted, transparent, and temporary. Third, they intend to identify the most problematic fossil fuel subsidies from both trade and environmental perspectives and explore pathways to address them.[64]

Additionally, the negotiating countries have committed to considering the specific needs and conditions of developing countries to minimise any adverse impacts on their development while protecting vulnerable populations and affected communities.[65] They have also pledged to share their experiences with other WTO members and encourage broader participation in efforts to phase out fossil fuel subsidies.[66] It is anticipated that the negotiating countries will continue to advance their work and achieve further progress before the 14th Ministerial Conference.

7.3.3 *Trade and Environmental Sustainability Structured Discussions (TESSD)*

WTO members established the Trade and Environment Sustainability Structured Discussion (TESSD) in 2020[67] to complement and support the efforts of the Committee on Trade and Environment and other relevant WTO committees and bodies.[68] In the Ministerial Statement on Trade and Environmental Sustainability, the participating countries outline six key tasks for the TESSD forum on trade and sustainability.

First, the participating countries will enhance their efforts in areas of common interest and identify concrete actions.[69] Secondly, they will initiate focused discussions on how trade-related climate measures and policies can effectively support climate and environmental goals while remaining consistent with WTO rules and principles.[70] Currently, TESSD members are actively involved in various WTO plurilateral negotiations related to sustainability. For instance, some countries have already submitted proposals addressing specific sustainability issues, such as

64 See WTO (2021g), p. 2; WTO (2024c), p. 2.

65 See WTO (2021g), p. 2.

66 Ibid., p. 2.

67 See WTO (2021i).

68 Ibid., p. 2.

69 Ibid., p. 2.

70 Ibid., p. 2.

New Zealand's proposals on fossil fuel subsidy reform and agreements concerning climate change, trade, and sustainability.[71]

Another example is Japan's proposal for the WTO's role in achieving carbon neutrality.[72] Japan has suggested eliminating tariffs on goods that contribute to emission reductions and establishing rules to promote the dissemination of such products and technologies.[73] Iceland and the EU have put forward similar proposals. Iceland has called for removing environmentally harmful subsidies and eliminating barriers to trade in environmental goods and services.[74] The EU has advocated for liberalising trade in green and climate-friendly goods and services, as well as increased transparency in domestic measures like fossil fuel subsidies and carbon border adjustment mechanisms.[75] Additionally, Switzerland has outlined eight key issues in its proposal, including liberalising environmental goods and services, reforming fossil fuel subsidies, greening aid for trade, promoting a circular economy, addressing plastic pollution, climate adaptation, and biodiversity.[76]

Third, the participating countries will explore opportunities and potential approaches for promoting and facilitating trade in environmental goods and services to meet environmental and climate goals. This includes addressing issues related to supply chains, as well as technical and regulatory elements.[77] Fourth, they will develop trade policies that support a more resource-efficient circular economy, promote sustainable supply chains, tackle challenges arising from sustainability standards and related measures, and facilitate access to environmental goods and services, including low-emission and other climate-friendly technologies.[78]

Consensus among the participating countries on these two points, acknowledging the significance of green supply chains, is essential for achieving emission reduction targets. Although WTO members previously sought to negotiate an Environmental Goods Agreement to eliminate trade barriers like tariffs on green products, this negotiation has been suspended.[79] Should WTO members fail to resume these talks, the commitment of TESSD members to these issues will become even more pivotal.

Fifth, the participating countries will collaborate to enhance capacity building and technical assistance on trade and environmental sustainability, including through Aid for Trade.[80] This effort will play a crucial role in advancing the TESSD's plurilateral negotiations. Many countries, particularly developing ones, lack the technology and financial resources necessary to implement effective green trade

[71] See WTO (2021j); See also WTO (2022a).

[72] See WTO (2021h).

[73] Ibid., paras. 3–5.

[74] See WTO (2021k).

[75] See WTO (2021l).

[76] See WTO (2021m), para. 1.5.

[77] See WTO (2021i), p. 2.

[78] Ibid., p. 2.

[79] See Cottier and Payosova (2016), p. 23; See also Wu (2016), pp. 279–281.

[80] Ibid., p. 2.

policies, a challenge that will become even more urgent in the post-COVID-19 era. Additionally, the participating countries will need to encourage other WTO members to join their initiatives by offering increased financial and technical support.

Finally, the participating countries will support ongoing discussions regarding the environmental impacts and trade effects of relevant subsidies, as well as the WTO's role in addressing these issues.[81] Seventh, they will keep adjusting the TESSD work plan as necessary.[82] The consensus on these two points is highly commendable. The TESSD has arguably become the primary forum for advancing sustainability negotiations within the WTO. Although the plurilateral negotiations proposed by participating countries are still in their early stages, the mechanism ensures continued progress on existing negotiations and the exploration of new sustainability issues in the future.

7.4 Conclusion

The stagnation of WTO multilateral negotiations has not entirely halted the progress of WTO rule-making. Instead, WTO members are addressing outstanding trade-related issues through separate plurilateral negotiations. These discussions span a range of topics, including economic development, social rights, and environmental protection.

WTO members have delivered outcomes in negotiations on services domestic regulation, e-commerce, and investment facilitation, which have the potential to advance sustainable development. These plurilateral agreements could enhance the global economy and assist countries in reducing poverty and hunger. However, it is important to recognise that these negotiations do not directly target sustainable development issues and, therefore, contribute to sustainability only indirectly. Additionally, ensuring that the outcomes of these negotiations do not compromise social and environmental rights remains a significant challenge.

Regarding social rights, WTO members are currently engaged in negotiations solely focused on trade and gender.[83] The objective is to enhance women's participation in trade and promote female entrepreneurship by improving trade rules in alignment with SDG 5 on gender equality.[84] However, this plurilateral negotiation has notable limitations. First, it addresses gender equality, specifically between men and women, without extending to broader LGBTQ+ gender equality issues. Second, the negotiation primarily targets barriers to female entrepreneurship and does not tackle broader workplace gender disparities or labour rights. Consequently, this effort does not advance labour rights or address other critical aspects of gender

[81] Ibid., p. 3.

[82] Ibid., p. 3.

[83] See WTO (2017a); See also WTO (2021d).

[84] See WTO (2021f), para. 1.2; See also WTO (2021d), p. 2.

equality. As a result, WTO members are not pursuing substantive discussions on labour rights within this framework.

The remaining plurilateral negotiations focus on environmental sustainability, which is a pressing issue for sustainable development. Unfortunately, progress in this area has been notably slower compared to advancements in the economic sphere. While WTO members have initiated a formal plurilateral negotiation on environmental goods, this effort is currently on hold, and it is uncertain whether it will resume in the near future.[85] Additionally, discussions on addressing plastic pollution and reforming fossil fuel subsidies are still in the early stages. The TESSD represents an innovative approach to integrating sustainable development within the WTO framework. It functions as a prototype for a sustainable development club, with its members increasingly incorporating their practical experiences from outside the WTO into this organisation.[86] In this context, the TESSD's practices can be seen as constitutionalisation within the WTO.

In conclusion, WTO members have discovered a new approach to integrating sustainability into WTO law through plurilateral negotiations. While these ongoing negotiations do not yet incorporate the comprehensive TSDCs or analytical frameworks required for a robust sustainability test, they represent significant steps towards the WTO's constitutionalisation. If WTO members can reach a consensus on all necessary TSDCs and analytical frameworks and expand participation to include more members, WTO law could substantially contribute to achieving the SDGs.

References

Bibliography

Cottier T, Payosova T (2016) Common concern and the legitimacy of the WTO in dealing with climate change. In: Delimatsis P (ed) Research handbook on climate change and trade law. Edward Elgar, Cheltenham, pp 9–30

Jose R, Oeschger A (2022) The state of play for sustainable development in the joint statement initiative on investment facilitation for development. International Institute for Sustainable Development. April 2022. Available via https://www.iisd.org/system/files/2022-04/joint-statement-initiative-investment-facilitation-en.pdf. Accessed 21 July 2024

Wu M (2016) The WTO environmental goods agreements: from multilateralism to Plurilateralism. In: Delimatsis P (ed) Research handbook on climate change and trade law. Edward Elgar, Cheltenham, pp 279–301

[85] See WTO (2016) Progress made on Environmental Goods Agreement, setting stage for further talks. The WTO's 2016 News Items. 4 December 2016. https://www.wto.org/english/news_e/news16_e/ega_04dec16_e.htm. Accessed 21 July 2024.

[86] See Sect. 7.3.3.

Documents

WTO (2017a) Joint declaration on trade and women's economic empowerment on the occasion of the WTO Ministerial Conference in Buenos Aires in December 2017. December 2017
WTO (2017b) Joint ministerial statement on investment facilitation for development. WT/MIN (17)/59. 13 December 2017
WTO (2017c) Joint ministerial statement: declaration on the establishment of a WTO informal work programme for MSMES. WT/MIN (17)/58. 13 December 2017
WTO (2017d) Joint statement on electronic commerce. WT/MIN (17)/6013. December 2017
WTO (2019a) Joint ministerial statement on investment facilitation for development. WT/L/1072/Rev.1. 22 November 2019
WTO (2019b) Joint statement on electronic commerce communication from China. INF/ECOM/19. 24 April 2019
WTO (2019c) Joint statement on electronic commerce WT/L/1056. 25 January 2019
WTO (2019d) Joint statement on electronic commerce. INF/ECOM/21. 26 April 2019
WTO (2019e) Joint statement on electronic commerce. INF/ECOM/25. 30 April 2019
WTO (2019f) Joint statement on electronic commerce. INF/ECOM/27/Rev.1/Add.1. 7 October 2019
WTO (2019g) Joint statement on electronic commerce. INF/ECOM/27/Rev.1. 9 July 2019
WTO (2019h) Joint statement on electronic commerce. INF/ECOM/34. 11 June 2019
WTO (2019i) Joint statement on electronic commerce. INF/ECOM/36. 5 July 2019
WTO (2019j) Joint statement on electronic commerce: EU proposal for WTO disciplines and commitments relating to electronic commerce. INF/ECOM/22. 26 April 2019
WTO (2019k) Joint statement on electronic commerce: possible customs duties text. INF/ECOM/33. 11 June 2019
WTO (2020) Work programme on electronic commerce: reinvigorating the work under the 1998 Work Programme on Electronic Commerce. WT/GC/W/812. 4 December 2020
WTO (2021a) Final draft declaration on micro, small and medium-sized enterprises (MSMES). WT/MIN (21)/1. 27 October 2021
WTO (2021b) Informal dialogue on plastic pollution and environmentally sustainable plastic trade (IDP): ministerial statement on plastic pollution and environmentally sustainable plastic trade. WT/MIN (21)/8/Rev.2. 10 December 2021
WTO (2021c) Joint initiative on services domestic regulation, reference paper on services domestic regulation. INF/SDR/2. 26 November 2021
WTO (2021d) Joint ministerial declaration on the advancement of gender equality and women's economic empowerment within trade 12th WTO ministerial conference. WT/MIN (21)/4. 10 November 2021
WTO (2021e) Joint statement on electronic commerce: joint text proposal for the disciplines relating to telecommunications services. INF/ECOM/64. 14 April 2021
WTO (2021f) Proposal for a framework to deliver an ambitious ministerial outcome on trade and gender for MC12 and beyond. INF/TGE/W/3. 30 April 2021
WTO (2021g) Proposed fossil fuel subsidies ministerial statement. JOB/GC/264/Rev.3. 18 November 2021
WTO (2021h) Trade and Environmental Sustainability Discussions (TESSD): proposal on contributions to achieving global carbon neutrality at the WTO. INF/TE/SSD/W/10. 23 March 2021
WTO (2021i) Trade and Environmental Sustainability Structured Discussions (TESSD): ministerial statement on trade and environmental sustainability. WT/MIN (21)/6/Rev.2. 14 December 2021
WTO (2021j) Trade and Environmental Sustainability Structured Discussions (TESSD): communication by New Zealand. INF/TE/SSD/W/1. 15 February 2021
WTO (2021k) Trade and Environmental Sustainability Structured Discussions (TESSD) communication by Iceland. INF/TE/SSD/W/5. 15 February 2021
WTO (2021l) Trade and Environmental Sustainability Structured Discussions (TESSD) communication by the European Union. INF/TE/SSD/W/7. 16 February 2021

WTO (2021m) Trade and Environmental Sustainability Structured Discussions (TESSD) communication by Switzerland. INF/TE/SSD/W/4. 15 February 2021
WTO (2022a) Joint Statement: Agreement on Climate Change, Trade and Sustainability (ACCTS) at MC 12. 15 June 2022. Available via https://www.beehive.govt.nz/release/joint-statement-agreement-climate-change-trade-and-sustainability-accts-mc12. Accessed 10 July 2022
WTO (2022b) Work Programme on Electronic Commerce Ministerial Decision. WT/ MIN (22)/ 32 WT/ L/ 1143. 22 June 2022
WTO (2023) Investment Facilitation For Development (IDF) Agreement. INF/IFD/W/52. 27 November 2023. Available via https://tradebetablog.wordpress.com/wp-content/uploads/2024/01/2023-11-27_inf-ifd-w-52-final.pdf. Accessed 27 July 2024
WTO (2024a) Joint statement initiative on electronic commerce. INF/ECOM/87. 26 July 2024. Available via https://docs.wto.org/dol2fe/Pages/SS/directdoc.aspx?filename=q:/INF/ECOM/87.pdf&Open=True. Accessed 27 July 2024
WTO (2024b) Ministerial statement on plastic pollution and environmentally sustainable plastics trade. WT/MIN (24)/14. 23 February 2024. Available via https://docs.wto.org/dol2fe/Pages/SS/directdoc.aspx?filename=q:/WT/MIN24/14.pdf&Open=True. Accessed 27 July 2024
WTO (2024c) Fossil Fuel Subsidy Reform (FFSR) initiative background information and MC13 outcomes. Available via https://www.wto.org/english/tratop_e/envir_e/ffsr_background_note_mc13_e.pdf. Accessed 27 July 2024
WTO and OECD (2021) Services domestic regulation in the WTO: cutting red tape, slashing trade costs, and facilitating services trade. OECD-WTO Trade Policy Brief. 19 November 2021. Available via https://www.wto.org/english/news_e/news21_e/jssdr_26nov21_e.pdf. Accessed 21 July 2024
WTO Database on gender provisions in RTAs (2023). https://www.wto.org/english/tratop_e/womenandtrade_e/gender_responsive_trade_agreement_db_e.htm. Accessed 21 July 2024

Chapter 8
The Sustainable Development Club: A Practical Approach

Abstract The ongoing WTO plurilateral negotiations could break the stagnation of WTO multilateral negotiations and bring TSDCs into WTO law. However, these negotiations still leave much to be desired. This chapter argues that WTO members should establish a sustainable development club within the WTO to constitutionalise the organisation, thereby incorporating the TSDCs required for the sustainability test into WTO rules. Why is the sustainable development club necessary for constitutionalising the WTO? Would WTO members accept this approach? This chapter addresses these two questions. The chapter begins by introducing the concept of the sustainable development club and explaining its implications for the constitutionalisation of the WTO. It also recognises that the sustainable development club, as a plurilateral mechanism, could undermine the WTO's multilateral negotiation system. This chapter explains why WTO members must adopt this plurilateral strategy to facilitate negotiations on TSDCs. Additionally, it discusses how to minimise the side effects of this approach and anticipates the difficulties and opportunities that the sustainable development club will encounter.

8.1 A Plurilateral Strategy

A sustainable development club is a plurilateral strategy designed to break the stagnation of WTO multilateral negotiations on sustainable development. The idea is for like-minded members to establish a sustainable development club to recognise and enforce their own economic, social, and environmental rights and constitutional principles necessary for achieving sustainable development, such as the proportionality principle. In other words, WTO members within this exclusive group implement their consensus on sustainable development, regardless of other members' positions. The formation of this club can be seen as a step toward the constitutionalisation of the WTO. This club contributes to sustainable development by setting an example and exerting pressure on other members to change their trade and development policies.

X. Zhao, *Integrating the UN SDGs into WTO Law*, EYIEL Monographs - Studies in European and International Economic Law 41,
https://doi.org/10.1007/978-3-031-73876-0_8

To some extent, the concept of a sustainable development club is not entirely new and is rooted in the well-established theory of clubs.[1] In essence, a club is a voluntary group that derives mutual benefits from sharing the costs of producing an activity with public good characteristics.[2] Historically, countries have used this theory to establish military alliances,[3] with NATO being the most successful example.[4] However, following the end of the Cold War and the reduction of war threats, countries shifted focus from forming military clubs,[5] with the AUKUS pact being a rare exception,[6] to establishing trade partnerships to foster economic relationships.[7]

In addition to enhancing international trade prosperity, many economists believe that countries can use the club strategy to coordinate their trade policies in the fight against climate change, leading to the well-known climate club strategy.[8] Such a strategy could significantly reduce greenhouse gas emissions, addressing the lack of harmonised climate policies and free-riding that contributed to the failure of the Kyoto Protocol, the first major international climate treaty.[9] By eliminating free-riding,[10] the club theory ensures political solidarity in combating climate change and promotes broad participation.[11] Similarly, the club theory can contribute to achieving sustainable development. Sustainable development clubs use the same strategy as climate clubs but with a more complex political objective, as they must address a broader range of economic and social rights in addition to environmental goals.

---

[1] To know the concept of the theory of clubs, please see Buchanan (1965), pp. 1–14; See also Sandler and Tschirhart (1980), pp. 1481–1521.

[2] See Nordhaus (2015), p. 1340.

[3] Ibid., p. 1340.

[4] For a recent review of the NATO, please see Olsen (2020), pp. 60–72.

[5] Of course, people cannot ignore the role of these military alliance in deterring against the use of force and keeping peace, one of the essential UN SDGs. The current Russian invasion of the Ukraine demonstrates that countries shall devote to maintain the peace as they did in the past decades. Regarding the judgment on the Russian invasion, please see the ICJ Order, Allegations of Genocide under the Convention on the Prevention and Punishment of the Crime of Genocide (Ukraine v. Russian Federation). 16 March 2022. Available via https://www.icj-cij.org/public/files/case-related/182/182-20220316-ORD-01-00-EN.pdf. Accessed 21 July 2024.

[6] See The AUKUS Agreement. Available via https://researchbriefings.files.parliament.uk/documents/CBP-9335/CBP-9335.pdf. Accessed 21 July 2024.

[7] See Nordhaus (2015), p. 1340.

[8] Ibid., pp. 1341–1343; See also Tagliapietra and Wolf (2021b), pp. 526–528; See also Tagliapietra and Wolf (2021a), pp. 112527–112530; See also Paroussos et al. (2019), pp. 542–546; See also Falkner et al. (2021), pp. 236–247.

[9] See Nordhaus (2015), p. 1340.

[10] See Ibid., p. 1339. In the case of the international climate change policy, free-riding means that countries rely on the emissions reductions of others without taking proportionate domestic abatement.

[11] See Ibid., pp. 1340–1341; See also Hovi et al. (2019), p. 1073.

The success of a sustainable development club hinges on its ability to achieve sustainable development rights. Lessons from climate clubs indicate that their effectiveness is determined by two key criteria: participation and commitment.[12]

Participation is a crucial benchmark for assessing the viability of using the club strategy to achieve sustainable development.[13] A successful sustainable development club must not only contribute to sustainable development but also encourage non-member countries to align their trade policies with sustainable development principles. This encouragement can be reflected in the club's growing membership. An increasing number of participants helps disseminate the club's plurilateral outcomes at the multilateral level. If club members constitute two-thirds of WTO members, they can drive negotiations and adopt new rules through majority votes at the Ministerial Conference or General Council meetings.[14] Such developments would facilitate significant and desirable reforms in international trade policy.

Turning a concept into reality is never easy. Club theory offers an alternative to WTO multilateral negotiations, but many WTO members may not recognise the club's legitimacy. Consequently, club members must make significant efforts to persuade other countries to join. The foremost task is to take the moral high ground by committing to achieving the UN SDGs. This commitment would justify their plurilateral actions and encourage others to join.[15]

However, achieving such significant goals without the support of non-member countries is challenging. Club members must bear high costs to advance sustainable development across its economic, social, and environmental dimensions. Specifically, they need to drive global economic growth, uphold human rights for large populations,[16] and substantially reduce GHG emissions. The social costs associated with these efforts—in terms of labour, natural and financial resources, and technology—are undoubtedly immense.

Participation and commitment, the two criteria for the success of a sustainable development club, can sometimes be contradictory. Higher commitments to abatement often result in lower participation and vice versa. This dynamic can be observed in many international agreements.[17] Countries are generally less willing to reduce their GHG emissions when the social costs are extremely high.

Fortunately, this paradox is not insurmountable. The successful conclusion of the Paris Agreement among UN member states demonstrated that sufficient key benefits

[12] See Nordhaus (2015), pp. 1341–1343.

[13] Ibid., p. 1340.

[14] See the Marrakesh Agreement, Article IX.

[15] In the context of international law, normative values are one of the important sources of legitimacy. Given this, a club will have legitimacy if its members achieve the trade-related UN SDGs. See Langvatn and Squatrito (2017), pp. 55–57; See also Petersmann (2005), pp. 357–358.

[16] To know why WTO members shall fulfil human rights, please see Petersmann (2005), pp. 358–366.

[17] See Nordhaus (2015), p. 1339.

could ensure both participation and committed abatement.[18] In the context of the Paris Agreement, these key benefits included enhanced technological diffusion and the provision of low-cost climate finance, which reduced investment costs and enabled developing countries to benefit from technological advances fully.[19] These benefits significantly offset the social costs of reducing GHG emissions, encouraging countries to maintain a stable climate alliance.[20]

Therefore, the members of a sustainable development club must share sufficient benefits to foster cooperation aimed at incorporating TSDCs into international trade policies. For the club to attract WTO members, it must offer more economic advantages than the existing WTO Agreements.[21] This can be achieved by signing a cooperation treaty similar to mega free trade agreements. For instance, WTO members could create a Trade and Sustainable Development Agreement on Establishing the Sustainable Development Club (TSDA). This treaty would aim to incorporate sustainability into the contracting parties' trade policies while providing significant benefits to participants. In return for joining this agreement, contracting parties would enjoy lower tariff rates, increased technical and financial cooperation, greater access to economic sectors for investors from member countries, and improved supply chain efficiency.

8.2 Plurilateralism: A Necessary Evil in the WTO's Constitutionalisation

8.2.1 Plurilateral Strategies' Side Effects

There is a general sense of disillusionment with the existing WTO multilateral negotiations as members fail to reach a consensus on development issues.[22] Consequently, countries have shifted their focus to promoting free trade agreements (FTAs) negotiations.[23] Notable examples include NAFTA, EU FTAs, and the TPP (now CPTPP).[24] It is important to note that FTAs are pursued by both developed and developing countries. In recent years, developing countries have actively participated in negotiating large FTAs, such as the RCEP and CPTPP. These agreements

[18] See Paroussos et al. (2019), p. 543.

[19] See Ibid., p. 543.

[20] See Ibid., p. 543.

[21] See Nordhaus (2015), p. 1340.

[22] For example, see Froman (2013).

[23] Hoekman and Mavroidis (2015), p. 104.

[24] To know more information, please see WTO Regional Trade Agreement Database (2023). According to the database, as of 4 January 2019, 291 RTAs were in force. 200 out of 291 RTAs were signed off since 2001.

have established the world's largest free-trade zones. It can be said that WTO members have collectively driven the development of plurilateralism.

In addition to the emergence of FTAs, countries have employed plurilateral strategies to advance WTO negotiations. The WTO Agreements do not prohibit members from signing plurilateral agreements.[25] Furthermore, Annex 4 of the Marrakesh Agreement includes four plurilateral agreements as official WTO legal texts.[26] Consequently, plurilateral agreements have become a means for countries to promote the reform of WTO rules. As discussed in Chap. 7, WTO members are currently using plurilateral agreements to address various issues, including services domestic regulation, e-commerce, MSMEs, trade and gender, investment facilitation, plastic pollution, and fossil fuel subsidy reform.

Needless to say, plurilateral negotiations among countries, both inside and outside the WTO, have had a significant impact on the multilateral trading system. This impact has been largely positive, as plurilateral negotiations are gradually reshaping international trade by adding sustainability standards to trade rules. However, plurilateral strategies can also undermine the multilateral trading system in several ways. First, countries are increasingly using non-multilateral trade negotiations to create new trade rules. Second, they are turning to non-multilateral dispute settlement mechanisms, such as FTA panels, to resolve disputes arising from these new sustainability rules. As a result, plurilateral mechanisms have taken on much of the role previously played by the WTO. This disruptive nature is a side effect of the plurilateral strategy.

Many argue that plurilateral negotiations are inconsistent with the spirit of the WTO's Marrakesh Agreement. They present two main arguments. First, they contend that Article IX (1) of the Marrakesh Agreement prioritises multilateral negotiations.[27] Therefore, they argue that WTO members should resolve major trade issues through multilateral negotiations and reserve WTO plurilateral negotiations to address a small subset of trade issues related to goods and services.

Second, they argue that the WTO discourages the initiation of exclusive plurilateral negotiations. Some contend that a WTO member can only initiate an exclusive plurilateral negotiation if no other WTO member objects. This requirement significantly reduces the incentive for members to pursue plurilateral negotiations. The primary purpose of plurilateral negotiations is to prevent free-rider behaviour by ensuring that only participating members benefit from the outcomes. This exclusivity prevents non-participating members from enjoying the advantages of plurilateral negotiations without bearing the political costs involved.

[25]The Marrakesh Agreement, Article IX, para. 1 and 5; The GATT, Article XXIV; The GATS, Article V.

[26]The Annex 4 of the Marrakesh Agreement includes four plurilateral agreements: (1) Agreement on Trade in Civil Aircraft; (2) Agreement on Government Procurement; (3) International Dairy Agreement; (4) International Bovine Meat Agreement.

[27]See Hoekman et al. (2022), p. 8.

8.2.2 Stagnation of WTO Multilateral Negotiations: The Reason Why We Need Pragmatic Solutions

The need to pragmatically address sustainability issues within the WTO arises from the stagnation of multilateral negotiations, primarily caused by the lack of consensus among all WTO members.[28] The WTO, as a multilateral organisation,[29] relies on a decision-making process that reflects the principles of multilateralism.[30] This process involves two key features: consensus-based decision-making and the single undertaking requirement. In principle, WTO members should adopt decisions by consensus, adhering to a multilateral decision-making process.[31] Additionally, members should unanimously adopt or reject a package of negotiated decisions in one negotiation round.[32]

Given these multilateral features, reaching a consensus on incorporating entirely new and challenging elements, such as TSDCs, into WTO rules is a lengthy process.[33] The complexity of sustainable development issues further extends negotiation times. Delegates often navigate uncharted territory during these multilateral negotiations due to the lack of pre-existing political arrangements regarding sustainability. As a result, it is foreseeable that members may not refurbish WTO rules in a timely manner through multilateral negotiations to achieve UN SDGs.[34] Such delays, potentially spanning several decades, could significantly undermine global efforts to achieve sustainable development. In the worst-case scenario, this delay could prevent humanity from safeguarding sustainable development, hindering countries' ability to adapt to and mitigate the side effects of climate change on human development.[35]

---

[28] WTO members have never initiated new multilateral negotiations after the definite failure of the Doha Round Talks. See Delimatsis (2016), p. 3.

[29] See Ruggiero (2000), p. 1.

[30] The Marrakesh Agreement, Article X.

[31] Capling (2003), p. 53.

[32] See Wolfe (2009), p. 838; See also Mendoza and Wilke (2011), p. 486.

[33] I have discussed these rights and principles. Please see Chaps. 2, 4, and 5.

[34] WTO members only agreed to write sustainable development objectives into the preamble to the Marrakesh Agreement. After that, there are few advancements except for the reaffirmation of these objectives in the Doha agenda. First, members confirmed that they would negotiate the relationship between existing WTO rules and specific trade obligations set out in multilateral environmental agreements. Second, they confirmed that they would negotiate procedure for regular information exchange between multilateral environmental agreement secretariats and WTO committees and the criteria for the granting of observer status. Third, they confirmed that they would negotiate the reduction of tariff and non-tariff barriers to environmental goods and services. See McDonald (2003), pp. 158–159.

[35] WTO members, as the contracting parties of the Paris Agreement on Climate Change, have recognised that they must reduce their greenhouse gas emissions so that the temperature will not augment more than 2° C compared to pre-industrial levels. Otherwise, humanity will be no longer able to adapt and mitigate climate change. Sustainability rules in international trade are necessary for humanity to prevent this environmental disaster. See the Paris Agreement on Climate Change.

Although the stagnation of WTO multilateral negotiations reflects the lengthy debates inherent to the democratic nature of multilateralism, this stagnation must not be endless if these negotiations are to yield fruitful outcomes. Prolonged debates can impede the incorporation of sustainability into WTO rules, hindering the achievement of sustainable development. Such a consequence would undermine the substantive quality of the WTO's decision-making process and distort its fundamental objectives. The WTO's democratic decision-making process aims not only to ensure a fair procedure for negotiations[36] but also to contribute to the substantive realisation of shared democratic values that WTO members have committed to pursuing.[37] In other words, democratic means must lead to democratic ends.

Sustainability is unquestionably one of the objectives of the WTO's democratic decision-making process.[38] WTO members have explicitly committed to ensuring that their trade and economic relations allow for the optimal use of the world's resources in accordance with the objective of sustainable development. This includes protecting the environment and enhancing the means for doing so, consistent with their respective needs and concerns at different levels of economic development.[39] This commitment, embedded in the preamble to the Marrakesh Agreement, represents a multilateral obligation that every member must uphold.[40] It is a fundamental condition for qualifying as a WTO member.

There is no justification for delaying or obstructing sustainable development negotiations within the WTO. The democratic features of the WTO's decision-making process cannot be used as an excuse to hinder the integration of TSDCs necessary for achieving sustainable development into WTO Agreements. If these democratic features enable any member to impede sustainable development negotiations, the WTO's multilateral decision-making process fails to achieve its democratic ends. In such a case, members have an obligation to rectify this democratic distortion.

Furthermore, the WTO Agreements indicate that the WTO's democratic decision-making process should not be reduced to mere formalism. This process should ensure not only procedural rights but also the realisation of substantive democratic

---

[36]The WTO's constitutional institutional structure ensures this democratic procedure. The WTO is a constitutional institution whose operation relies on a balance between different power centres. Benefiting from its internal balance, the WTO can be prevented from the monopoly of a single power centre. This balance determines that each WTO member should have equal rights to be heard and to influence the WTO's decision-making. See Jackson (1999), p. 824.

[37]WTO members have agreed to realise a set of values and confirmed this consensus. See the preamble to the Marrakesh Agreement.

[38]For example, see WTO Appellate Body Report, US-Gasoline, WT/DS2/AB/R, adopted on 20 May 1996, p. 30; See also WTO Appellate Body Report, US-Shrimp, WT/DS58/AB/R, adopted on 6 November 1998, paras. 129–131, 152–153, and 155; See also WTO Appellate Body Reports, China-Raw Materials, WT/DS394/AB/R, adopted on 22 February 2012, para. 306.

[39]The Marrakesh Agreement, preamble.

[40]The WTO panel in EC-Tariff Preferences confirmed that members shall obey the policy objectives they write into the WTO Agreements. See WTO Panel Report, EC-Tariff Preference, WT/DS246/R, adopted on 20 April 2004, para. 7.52.

values through its democratic procedural features. Article III of the Marrakesh Agreement states:

> The WTO shall provide the forum for negotiations among its members concerning their multilateral trade relations in matters dealt with under the agreements in the Annexes to this Agreement. The WTO may also provide a forum for further negotiations among its members concerning their multilateral trade relations, and a framework for the implementation of the results of such negotiations, as may be decided by the Ministerial Conference.[41]

The phrase 'shall provide the forum for negotiations,' when read in conjunction with the preamble to the Marrakesh Agreement,[42] indicates that launching sustainable development negotiations within the WTO is an obligation, not a discretionary right for members. Even if WTO multilateral negotiations are at an impasse, members are still required to initiate negotiations on sustainable development to integrate these principles into WTO Agreements.

Members can propose negotiations on TSDCs at Ministerial Conference or General Council meetings,[43] where decisions can be made by majority vote.[44] However, political preparedness is essential for garnering support for such proposals. WTO members advocating for the adoption of sustainable development commitments must engage in extensive informal negotiations to align as many other members' positions as possible in advance.[45] Their collective stance should specify which TSDCs they agree to include as new commitments.

If members cannot agree on detailed rules for certain issues, particularly those related to uncertain or rapidly changing fields, they should reach a consensus on the extent of their domestic regulatory autonomy.[46] This shared position, addressing these crucial questions, will facilitate negotiations on TSDCs and their enforcement within the WTO. This enforcement should include dispute settlement rules that ensure the implementation of the sustainability test conceived in Chap. 4.[47] Such a common position can reinvigorate the WTO's engagement in sustainable development negotiations.

In addition, the stalled WTO multilateral negotiations pose a significant threat to the legitimacy of the WTO DSB, a crucial institution in international trade.[48] Given its essential role in addressing international trade disputes, all countries have a vested

---

[41] See the Marrakesh Agreement, Article III paragraph 2.

[42] Namely, the optimal use of the world's resources in accordance with the objective of sustainable development. See the preamble of the Marrakesh Agreement.

[43] See Marrakesh Agreement, Article IX paragraph 1.

[44] See Ibid., Article IX paragraph 1.

[45] The idea is similar to that of climate club. It aims to put pressure on countries so that they will commit to incorporate sustainability into their trade policies. To know more details about this strategy, please see Nordhaus (2015).

[46] WTO members need to incorporate the sustainability test into WTO dispute settlement rules to ensure their domestic regulatory autonomy. To know the concept of the sustainability test, please see Chap. 4.

[47] Ibid.

[48] See Jackson (2000), p. 67.

interest in safeguarding its legitimacy. A rule-based dispute settlement mechanism benefits every WTO member.[49]

Members' efforts to restart the AB and establish the Multi-Party Interim Appeal Arbitration Arrangement, a temporary alternative to the AB, demonstrate their profound concern for the proper functioning of the WTO DSB.[50] If the threat to its legitimacy is real, it should motivate members to overcome the stagnation in negotiations. The issue, however, is that many members have overlooked the adverse effects that stalled WTO multilateral negotiations have on the DSB's legitimacy.

People may ignore this threat due to a lack of attention to the impact of WTO negotiations on the DSB's legitimacy.[51] WTO negotiations significantly influence the rulings of WTO panels and the AB because delegates establish procedural and substantive dispute settlement rules through these negotiations.[52] One consequence of the stagnation of WTO multilateral negotiations is that WTO members cannot draft new rules to update WTO Agreements. Consequently, WTO panels and the AB lack the necessary rules to promote sustainable development through their decisions. The absence of such rules prevents the WTO DSB from effectively realising economic, social, and environmental rights as part of sustainable development, a commitment made by the members.[53] This situation arguably undermines the ability of the WTO DSB to ensure outcome justice.

However, existing international law establishes that outcome justice is essential for an international judicial institution, such as the WTO DSB, to maintain its legitimacy.[54] This means that, beyond adhering to procedural rules, international judicial institutions must also realise substantive rights through their decisions. This requirement for outcome justice can be understood by examining the development of the legitimacy concept in international criminal tribunals (ICTs).

International criminal law jurists have coined and refined this concept through several significant milestones, including the establishment of international military tribunals like the Nuremberg Tribunal, the International Criminal Tribunal for the Former Yugoslavia (ICTY), and the International Criminal Court (ICC).[55] Although

---

[49] See WTO General Council (2019). A group of 60 of developing and developed country members reaffirm that the dispute settlement system of the WTO is a central element in providing security and predictability to the multilateral trading system.

[50] See Starshinova (2021), pp. 792–794.

[51] See Malacrida and Marceau (2018), pp. 20–69. According to these authors, the framework used for assessing the legitimacy of the WTO dispute settlement body includes eight aspects: (1) the selection of the members of WTO adjudicating bodies and their independence; (2) relevant procedural rules; (3) fact-finding and standards of review; (4) the WTO adjudicating bodies' interpretative approach; (5) forum shopping; (6) implementation of adverse WTO rulings and interaction with national courts; and (7) WTO-specific legitimacy concerns.

[52] See the Marrakesh Agreement, Article IX and Article X.

[53] I have discussed these commitments. Please see Chap. 2.

[54] Bingham (2007), pp. 69–84.

[55] See May and Fyfe (2017).

there is no universally accepted definition of legitimacy,[56] jurists agree that a legitimate judgment, along with legitimate establishment and procedural fairness, is a crucial component of the legitimacy of international judicial institutions.[57] For instance, Silje Aambø Langvatn and Theresa Squatrito define the legitimacy of ICTs as follows:

> ICT legitimacy requires sufficient adherence to established rules and procedures and a sufficient degree of both procedural and outcome justice with regard to all three dimensions, or with regard to the ICT's pedigree and procedures as well as its results.[58]

This academic definition accurately reflects the practice of international judicial institutions. The legitimacy of ICTs, such as the Tribunal of Athena and the ICC, is fundamentally based on their capacity to settle disputes legitimately.[59] Take the ICC, for example. Before its establishment, countries extensively debated whether the tribunal should have universal jurisdiction over all citizens, regardless of whether their countries had ratified the Rome Statute.[60] Ultimately, countries agreed to the ICC's universal jurisdiction, affirming their belief in its legitimacy. They trusted that a legitimately established court could ensure both procedural and substantive legitimacy in its judgments.

Countries consent to the jurisdiction of any international judicial institution on the condition that it can guarantee outcome justice.[61] From the experience of ICTs, we can infer that the absence of rules and principles regarding sustainable development, resulting from the stagnation of WTO multilateral negotiations, may lead WTO members to view the WTO as illegitimate in its jurisdiction over their trade disputes. Today's trade disputes are no longer purely economic; future disputes will increasingly involve conflicts among economic, social, and environmental rights.[62] If WTO Agreements lack sustainable development rules, WTO panels and the AB will be unable to ensure outcome justice.

Moreover, the history of other international governmental organisations demonstrates that active negotiations are crucial for their continued existence. Inactivity in their negotiation functions often leads to their eventual dissolution.[63] Thus, the stagnation of WTO multilateral negotiations poses an existential threat to the WTO.

In conclusion, the stagnation of WTO multilateral negotiations necessitates pragmatic solutions to incorporate sustainability into WTO Agreements. There are

---

[56] Silje Aambø Langvatn and Theresa Squatrito categorize the accounts of the legitimacy of ICTs into three groups, namely abstract conception, monistic conception, and mixed conception, besides their political or multidimensional conception of legitimacy. See Langvatn and Squatrito (2017), pp. 49–52.

[57] See Ibid., pp. 55–57.

[58] See Ibid., p. 52.

[59] See May and Fyfe (2017), pp. 26–30.

[60] See Kaul (2003), pp. 23–25.

[61] See Langvatn and Squatrito (2017), p. 58.

[62] See McDonald (2003), pp. 146–151.

[63] See Eilstrup-Sangiovanni (2010), p. 348.

three primary reasons for this. First, the objective of sustainable development is explicitly stated in the preamble to the Marrakesh Agreement, obligating members to negotiate relevant rules and principles. Second, these rules and principles are essential for the DSB to maintain a rules-based multilateral trading system. Third, it is in the best interest of WTO members to overcome the stagnation of negotiations to ensure the continued functioning of the WTO, as this stagnation poses an existential threat to the institution.

8.3 Public Participation: A Remedy to the Side Effects of Plurilateralism

While WTO members may be unable to reform international trade rules further through multilateral negotiations, the WTO's multilateral trading system remains invaluable. Its value lies in upholding a democratic decision-making process for international trade rules, thus preventing individual countries from dominating the negotiation process. If members of the sustainable development club were to force other WTO members to accept their development and international trade policies by controlling the WTO's multilateral negotiations, this would devolve into power politics, undermining the WTO's legitimacy under international law.

International institutions do not inherently possess legitimacy simply by being established[64]; their legitimacy must be recognised by states.[65] Legitimacy based solely on coercion, self-interest, or specific outcomes is insufficient to secure the recognition of states.[66] In this context, legitimacy encompasses both normative and sociological aspects.[67] It means that an institution not only has the right to rule but is also widely perceived as having that right.

Therefore, the plurilateral actions of the sustainable development club must avoid undermining the WTO's legitimacy. To achieve this, the sustainable development club must adopt a democratic decision-making mechanism that ensures its decisions are representative of all international trade participants' interests.[68] Only with such

---

[64] See Chapman (2009), p. 739; See also Van den Brink (2019), pp. 293–318.

[65] Keller (2008), p. 257; Wolfrum (2008), pp. 10–19; Simmons (1999), pp. 739–771. Simmons explains the definition of legitimacy in the light of social contract approach, which also builds on the consent or authorization of States to the right to rule of particular institutions; See also Wincott (2002).

[66] Bodansky (2008), pp. 310–315.

[67] Buchanan and Keohane (2008), p. 25; See also D'Amato (2008), p. 83; See also Treves (2008), p. 169; See also Erman (2018), p. 138.

[68] Keller (2008), pp. 259–298; Von Bogdandy (2008), pp. 306–307; Bodansky (2008), p. 310. Bodansky believes that legitimacy represents a third basis of compliance. In his opinion, an individual or State might comply with a directive, not because of the fear of sanctions or because it is rationally persuaded that the decision is correct, but rather because it accepts the decision-making process as legitimate.

representativeness can the decisions of the sustainable development club evolve into constitutional norms applicable to international trade. However, plurilateral mechanisms often fail to meet these requirements, as they cannot include all countries as participants. Consequently, under traditional state-centred international law, states cannot develop international constitutions applicable to all through plurilateral mechanisms.

However, SDGs differ from other matters that countries negotiate internationally. Their unique nature lies in their encompassing economic, social, and environmental values that transcend traditional national interests.[69] Instead, these values reflect the interests of every individual,[70] expressing instinctive human needs that do not vary by nationality.[71] Achieving sustainable development, therefore, is a bottom-up action involving all human beings.[72] The success of this action hinges on realising each individual's economic, social, and environmental rights.

These particularities enable sustainable development club members to enhance the representativeness of their decisions by increasing individual participation in the decision-making process. Ideally, all individuals, regardless of nationality, gender, race, or colour, should participate in the sustainable development club's decision-making process through civil society fora.[73] This participation should also extend to monitoring the club's actions and dispute settlement mechanisms. Civil society fora should encompass all forms of citizen participation in public discourse, such as parliaments, citizen groups, unions, and NGOs.[74] If some countries prohibit their citizens from participating in events held by the sustainable development club, a bottom-up, individual-driven democratic decision-making mechanism can still mitigate the interference of national interests in law-making.

Public participation not only compensates for the lack of representation in plurilateral negotiations but also ensures the legitimacy of the sustainable development club's decision-making process. By involving the public, the club can ensure that its policies and regulations align with the goals of achieving the UN SDGs. Policymakers in the sustainable development club must understand the real demands of individuals regarding their economic, social, and environmental rights to develop appropriate policies. Public participation maximises the accuracy and timeliness of information exchange between individuals and policymakers, reflecting real social changes as Sciulli argues.[75]

Furthermore, public participation plays a crucial monitoring role, ensuring that the sustainable development club genuinely promotes and implements sustainable development policies and laws. This involvement is essential for maintaining the

[69] See Browne and Weiss (2014), p. 22.

[70] See Fonseca et al. (2020), pp. 3361–3362.

[71] See Reinert (2020), pp. 128–129.

[72] See Fox and Stoett (2016), pp. 560–561.

[73] Ibid., p. 568.

[74] Ibid., p. 568.

[75] Sciulli (1991), p. 187.

legitimacy of the club's decision-making process. Modern legal legitimacy encompasses more than procedural justice, as described by Habermas,[76] and extends to the morality of law, as defined by Fuller.[77] Laws must be designed, enacted, and enforced to satisfy procedural requirements such as democracy and common social obligations. Only then does the law have the legitimacy to demand compliance from citizens.[78] In this context, the WTO's multilateral decision-making process is often incompatible with the need to formulate effective sustainable development and trade policies. Its state-driven model rarely reflects individuals' aspirations for sustainable development. If the sustainable development club can fully involve citizens in its decision-making, policy implementation, and dispute settlement processes, its plurilateral negotiation mechanism will significantly address the legitimacy gaps in multilateral decision-making processes like those of the WTO.

8.4 Prospect of the Sustainable Development Club

8.4.1 Difficulties of Establishing the Sustainable Development Club

The decade-long stagnation of WTO multilateral negotiations epitomises the challenges of establishing a large sustainable development club with two-thirds of WTO members. This difficulty primarily stems from a well-known split between developing and developed countries.[79] Since the Doha Round negotiations, these two groups have been divided on sustainable development issues, forming distinct camps with differing views, especially on trade-related intellectual property rights and the elimination of agricultural subsidies.[80] Developing countries have emphasised the adverse effects of intellectual property rights and agricultural subsidies on their citizens' health and food security, while developed countries have been unwilling to compromise.[81]

The requirement for consensus in adopting new rules, along with the single undertaking principle, which mandates that all negotiated issues must be

[76] Habermas (1973), pp. 157–189; Habermas (1981); Sciulli (1991), pp. 386–388; See also Ophir (1989), pp. 218–221.

[77] Fuller (1969), pp. 46–84; See also Sciulli (1991), pp. 390–392.

[78] Sciulli (1991), p. 395.

[79] See Alessandrini (2010), pp. 202–203.

[80] In agricultural negotiations, developing countries launched their own proposals to counter the EU-US proposal. G20's proposal was one of them, which raised the level of ambition with respect to the reductions in domestic support and the elimination of all forms of export subsidies and lowered the level of ambition with respect to improvements in access to the markets of developing countries. See Harbinson (2005), p. 123. Concerning negotiations on intellectual property rights, please see Abbott (2005), pp. 325–333.

[81] See Tangermann (2005), pp. 99–100; See also Vanduzer (2005), pp. 167–169.

unanimously adopted or declined, has led to a stalemate in these crucial negotiations.[82] This impasse ultimately thwarted the reform of outdated WTO rules. Furthermore, the consensus approach prevents the inclusion of advanced rules developed outside the WTO framework into WTO rules, hindering the organisation's ability to adapt to contemporary global challenges.[83] Scholars criticise this negotiation model as inadequate for addressing the evolving needs of international trade and sustainable development.[84]

Of course, WTO members had no choice but to apply rigid, multilateral decision-making rules when launching the Doha Round negotiations. Notably, developing members were extremely disappointed after the bitter failures of the Singapore and Seattle Ministerial Conferences.[85] The exclusion from key trade negotiations, particularly during the Uruguay Round,[86] made it impossible for developing countries to accept the old negotiation model, whether bilateral negotiations or a flexible multilateral approach. In the latter case, the US and the EU often conducted green room negotiations to form a common position, which they then requested other members to accept.[87] Given this, developing countries insisted on applying these famous multilateral decision-making rules to ensure their participation.[88] They hoped that these rules would prevent the US and the EU from dominating the Doha Round negotiations.[89]

In this sense, the consensus approach was initially successful. Developing countries effectively countered the proposals of the US and the EU as they had hoped. However, this approach proved to be a double-edged sword. Delegates from both developing and developed countries ended up mired in prolonged negotiations. Today, bridging the gap between the positions of developing and developed countries remains crucial for increasing consensus on issues related to sustainable development.

Second, beyond the split between developing and developed countries, the divergence of views among nations that have already implemented trade and sustainable development policies also complicates the creation of a sustainable development club within the WTO. Addressing this issue is even more critical because these countries are likely to be the founders of such a club. If irreconcilable conflicts arise among them, their actions will become uncoordinated. This lack of cooperation will significantly undermine the collective efficiency of the club members.

---

[82] See Ehlermann and Ehring (2005), pp. 510–512.

[83] See Ibid., p. 512.

[84] See Odell (2005), p. 490.

[85] See Blackhurst and Hartridge (2005), p. 456.

[86] See Dunoff (2003), pp. 60–63.

[87] To know more details about green room negotiations, please see Blackhurst and Hartridge (2005), p. 457.

[88] For an overview about the use of the existing decision-making rules, please see Ehlermann and Ehring (2005), pp. 510–512.

[89] See Odell (2005), p. 469; See also Ehlermann and Ehring (2005), pp. 515–519.

Therefore, certain bilateral relationships are crucial to the stability of the sustainable development alliance, with the most significant being the relationship between the EU and the US. Although this relationship could form the foundation for establishing a sustainable development club, it remains fragile. Numerous trade disputes highlight the contradictions between these two major WTO members. The biotech dispute over the approval and marketing of genetically modified food is one of the most notable examples.[90] Additionally, there have been several other conflicts, such as the dispute over the use of hormones in livestock growth[91] and the anti-dumping duties on steel products.[92] These cleavages, particularly during the Trump administration, weakened the trans-Atlantic bond, making it difficult to use this relationship as a basis for a sustainable development club. Furthermore, trade frictions between other members cannot be ignored, as they can significantly impact the solidarity of potential club members and the efficacy of their joint actions.[93]

In summary, trade disputes among potential members of the sustainable development club are not always directly related to sustainable development issues. Nevertheless, the outcomes of these trade disputes can significantly affect the willingness of these like-minded countries to cooperate. Consequently, establishing a sustainable development club within even a small group of countries that share sustainable development values is not simple for WTO members.

8.4.2 New Impetus for Establishing the Sustainable Development Club

Our world is undergoing a series of dramatic changes that have reinvigorated negotiations on sustainable development. Since the end of 2019, the COVID-19 pandemic has erupted worldwide, significantly altering people's lives and heavily impacting international trade and development policies. The pandemic's ripple effects are numerous, but perhaps the most striking is the impact on the 2020 US election. President Trump's bid for re-election was unsuccessful, largely due to his handling of the pandemic. His successor, President Joe Biden, holds markedly different views on foreign policy, particularly his support for combating climate

[90] See the WTO dispute EC-Biotech. For an overview of this dispute, please see Conrad (2007), pp. 233–248.

[91] See WTO dispute EC-Hormones. For an overview of this dispute, please see Pauwelyn (1999), pp. 655–657.

[92] To know more details about these disputes, please see Hoekman and Wauters (2011), pp. 5–43.

[93] There are also disputes between other countries that would participate in a sustainable development club. For example, the dispute between Japan and Canada over regulatory measures affecting the renewable energy generation sector. See WTO dispute Canada-Renewable Energy. For an overview of this dispute, please see Espa and Durán (2018), pp. 621–653.

change. Early in his term, Biden signed the decision for the US to rejoin the Paris Agreement,[94] providing a crucial opportunity to mend trans-Atlantic relations.[95]

Moreover, while the pandemic has adversely affected the achievement of the UN SDGs, it also presents an opportunity to reform international trade and development policies. In 2020, international trade came to a significant halt, prompting international trade rule-makers to discuss how to revive trade in the post-pandemic era. At relevant WTO meetings, sustainable development emerged as a central focus,[96] a rarity in previous discussions. WTO officials and member representatives committed to sustainable development aimed to leverage this opportunity to restart international trade in a way that reshapes and enhances countries' trade activities.[97]

The pandemic's impact on international trade underscores the urgency of adopting more stringent environmental policies. In 2020, global carbon emissions fell sharply due to lockdowns in major economies. Surprisingly, countries still fell short of projected carbon reduction standards even with these reductions.[98] This harsh reality highlights the need for strict carbon reduction regulations. Under current rules, countries are unlikely to achieve their committed GHG emissions reductions. In 2021, the EU enacted the Carbon Border Adjustment Mechanism (CBAM), a new emission reduction measure. Although the pandemic did not directly cause this regulation, it arguably made it easier for countries to understand the EU's commitment to implementing such regimes. The pandemic has revealed the inadequacy of current emission reduction measures. Additionally, the election of Joe Biden, influenced by the pandemic's effects, has bolstered confidence in implementing the Paris Agreement. It is foreseeable that other countries may adopt similar measures to the EU's in the future, increasing acceptance of the CBAM.

These changes in domestic policy are injecting new energy into the stalled multilateral climate negotiations under the United Nations Framework Convention on Climate Change. Countries are currently advancing agreements on biodiversity protection, plastic waste management, and other critical issues.[99] While these efforts alone may not substantially achieve sustainable development, they could gradually build a consensus on integrating sustainability measures into trade negotiations.

Of course, these positive changes come with a bittersweet undertone. Countries have paid a heavy price in lives and economic costs to gain this new impetus for advancing sustainable development negotiations. Hopefully, they will seize the opportunity presented by the relaunch of international trade to reform their trade

---

[94] See Blinken (2021).

[95] See Latici (2021).

[96] See Lim et al. (2022).

[97] See Ibid.

[98] See Liu et al. (2020), pp. 5172–5183.

[99] See United Nations, UN Environment Assembly Concludes with 14 Resolutions to Curb Pollution, Protect and Restore Nature Worldwide. UN Environment Programme. https://www.unep.org/news-and-stories/press-release/un-environment-assembly-concludes-14-resolutions-curb-pollution. Accessed 21 July 2024.

and development policies, contributing to sustainable development in the post-pandemic era.

8.5 Conclusion

Establishing a sustainable development club is a plurilateral strategy aimed at breaking the stagnation of WTO multilateral negotiations on sustainable development.[100] The concept involves like-minded members forming a club to recognise and enforce their own TSDCs to achieve sustainable development.[101] As a promising way to eliminate free-riding, the club theory ensures political solidarity in trade and sustainable development policies and encourages broad participation.[102] The experience gained from operating climate clubs demonstrates that the success of such a club hinges on active participation and commitment.[103]

However, many WTO members may not recognise the legitimacy of a sustainable development club, viewing it as undermining the WTO's multilateral negotiation system. Therefore, club members must make significant commitments to achieving UN SDGs to justify their plurilateral actions and encourage others to join.[104] Additionally, the sustainable development club must offer greater economic benefits than those provided by existing WTO Agreements to attract participants.[105] This objective can be attained by signing a TSDA similar to mega free trade agreements.[106]

Despite concerns that the sustainable development club may erode the results of multilateral negotiations, WTO members must accelerate its establishment to reform international trade policy and achieve the UN SDGs.[107] Plurilateral strategies, while not ideal, are a necessary step in the WTO's evolution.[108] From a realist perspective, integrating the SDGs into international trade rules requires countries to accept the adverse effects of plurilateral strategies on the multilateral trading system. Furthermore, public participation can help address the lack of justice in plurilateral

[100] See Sect. 8.1.

[101] Ibid.

[102] Ibid.

[103] See Nordhaus (2015), pp. 1341–1343.

[104] In the context of international law, normative values are one of the important sources of legitimacy. Given this, a club will have legitimacy if its members achieve the trade-related UN SDGs. See Langvatn and Squatrito (2017), pp. 55–57; See also Petersmann (2005), pp. 357–358.

[105] See Nordhaus (2015), p. 1340.

[106] See Sect. 8.1.

[107] See Sect. 8.2.1.

[108] See Sect. 8.2.

negotiations, ensuring that these negotiations are governed by the rule of law rather than political power.[109]

The establishment of a sustainable development club within the WTO is undoubtedly challenging, and club members will face many difficulties in forming and developing their alliance.[110] Nonetheless, recent dramatic global changes have provided new impetus for creating such a club.[111] As a result, there is a promising outlook for integrating sustainability into WTO rules by establishing a sustainable development club.

Jurisprudence

ICJ Order, Allegations of Genocide under the Convention on the Prevention and Punishment of the Crime of Genocide (Ukraine v. Russian Federation). 16 March 2022. Available via https://www.icj-cij.org/public/files/case-related/182/182-20220316-ORD-01-00-EN.pdf. Accessed 21 July 2024

WTO Appellate Body Report, US-Gasoline, WT/DS2/AB/R, adopted on 20 May 1996

WTO Appellate Body Report, US-Shrimp, WT/DS58/AB/R, adopted on 6 November 1998

WTO Appellate Body Reports, China-Raw Materials, WT/DS394/AB/R, adopted on 22 February 2012

References

Bibliography

Abbott FM (2005) Are the competition rules in the WTO agreement on trade-related aspects of intellectual property rights adequate? In: Petersmann E-U (ed) Reforming the world trading system: legitimacy, efficiency, and democratic governance. Oxford University Press, New York, pp 317–334

Alessandrini D (2010) Developing countries and the multilateral trade regime: the failure and promise of the WTO's development Mission. Hart, Oxford

Bingham T (2007) The rule of law. Cambridge Law J 66(1):67–85

Blackhurst R, Hartridge D (2005) Improving the capacity of WTO institutions to Fulfil their mandate. In: Petersmann E-U (ed) Reforming the world trading system: legitimacy, efficiency, and democratic governance. Oxford University Press, New York, pp 455–468

[109] See Sect. 8.3.

[110] See Sect. 8.4.1.

[111] See Sect. 8.4.2.

Bodansky D (2008) The concept of legitimacy in international law. In: Wolfrum R, Röben V (eds) Legitimacy in international law. Springer, Berlin, pp 309–317
Browne S, Weiss TG (eds) (2014) Post-2015 UN Development: Making Change Happen? Routledge, New York
Buchanan JM (1965) An economic theory of clubs. Economica 32(125):1–14
Buchanan A, Keohane RO (2008) The legitimacy of global governance institutions. In: Wolfrum R, Röben V (eds) Legitimacy in international law. Springer, Berlin, pp 25–62
Capling A (2003) The multilateral trading system at risk? Three challenges to the World Trade Organization. In: Buckley RP (ed) The WTO and the Doha round: the changing face of world trade. Kluwer Law International, The Hague, pp 37–58
Chapman TL (2009) Audience beliefs and international organization legitimacy. Int Organ 63(4): 733–764
Conrad CR (2007) The EC-biotech dispute and applicability of the SPS agreement: are the Panel's findings built on shaky ground? World Trade Rev 6(2):233–248
D'Amato A (2008) On the legitimacy of international institution. In: Wolfrum R, Röben V (eds) Legitimacy in international law. Springer, Berlin, pp 83–92
Delimatsis P (2016) Introduction: climate change and trade law challenges for governance and coordination. In: Delimatsis P (ed) Research handbook on climate change and trade law. Edward Elgar, Cheltenham, pp 1–8
Dunoff JL (2003) The post-Doha trade agenda: questions about constituents, competence, and coherence. In: Buckley RP (ed) The WTO and the Doha round: the changing face of world trade. Kluwer Law International, The Hague, pp 59–88
Ehlermann C-D, Ehring L (2005) Are WTO decision-making procedures adequate for making revising, and implementing worldwide and 'Plurilateral' rules? In: Petersmann E-U (ed) Reforming the world trading system: legitimacy, efficiency, and democratic governance. Oxford University Press, New York, pp 497–522
Eilstrup-Sangiovanni M (2010) Death of international organizations. The organizational ecology of intergovernmental organizations, 1815–2015. Rev Int Organ 15(2):339–370
Erman E (2018) The political legitimacy of global governance and the proper role of civil society actors. Res Publica 24:133–155
Espa I, Durán GM (2018) Renewable energy subsidies and WTO law: time to rethink the case for reform beyond Canada-renewable energy/fit program. J Int Econ Law 21(3):621–653
Falkner R et al (2021) Climate clubs: politically feasible and desirable? Clim Pol 22(2):236–247
Fonseca LM et al (2020) Mapping the sustainable development goals relationships (2020). Sustain For 12(8):3359–3373
Fox O, Stoett P (2016) Citizen participation in the UN sustainable development goals consultation process: toward global democratic governance? Glob Gov 22(4):555–573
Fuller LL (1969) The morality of law, revised edition. Yale University Press, New Haven
Habermas J (1973) A postscript to knowledge and human interests. Philos Soc Sci 3:157–189
Habermas J (1981) The theory of communicative action: reason and the rationalization of society. Polity Press, Cambridge
Harbinson S (2005) The agriculture negotiations: the road from Doha and how to keep the negotiations on a positive track. In: Petersmann E-U (ed) Reforming the world trading system: legitimacy, efficiency, and democratic governance. Oxford University Press, New York, pp 121–126
Hoekman BM, Mavroidis PC (2015) Embracing diversity: Plurilateral agreements and the trading system. World Trade Rev 14(1):101–116
Hoekman B, Wauters J (2011) US-compliance with WTO rulings on zeroing in anti-dumping. World Trade Rev 10(1):5–43
Hovi J et al (2019) The Club approach: a gateway to effective climate co-operation? Br J Polit Sci 49(3):1071–1096
Jackson JH (1999) Fragmentation or unification among international institutions: the World Trade Organization. Int Law Politics 31:823–831

Jackson JH (2000) Dispute settlement and the WTO: emerging problems. In: Secretariat TWTO (ed) From GATT to the WTO: the multilateral trading system in the new millennium. Kluwer Law International, The Hague, pp 67–82
Kaul H-P (2003) The international criminal court-current perspective. In: Zimmermann A (ed) International criminal law and the current development of public international law. Duncker and Humblot, Berlin, pp 15–26
Keller H (2008) Codes of conduct and their implementation: the question of legitimacy. In: Wolfrum R, Röben V (eds) Legitimacy in international law. Springer, Berlin, pp 219–298
Langvatn SA, Squatrito T (2017) Conceptualising and measuring the legitimacy of international criminal tribunals. In: Hayashi N, Bailliet CM (eds) The legitimacy of international criminal tribunal. Cambridge University Press, Cambridge, pp 41–65
Liu Z et al (2020) Near-real-time monitoring of global CO2 emissions reveals the effects of the COVID-19 pandemic. Nat Commun 11:5172–5183
Malacrida R, Marceau G (2018) The WTO adjudicating bodies. In: Howse R et al (eds) The legitimacy of international trade courts and tribunals. Cambridge University Press, New York, pp 20–69
May L, Fyfe S (2017) The legitimacy of international criminal tribunals. In: Hayashi N et al (eds) The legitimacy of international criminal tribunals. Cambridge University Press, Cambridge, pp 25–40
McDonald J (2003) It's not easy being green: trade and environment linkages beyond Doha. In: Buckley RP (ed) The WTO and the Doha round: the changing face of world trade. Kluwer Law International, The Hague, pp 145–168
Mendoza MR, Wilke M (2011) Revisiting the single undertaking: towards a more balanced approach to WTO negotiations. In: Birkbeck CD (ed) Making global trade governance work for development. Cambridge University Press, New York, pp 486–506
Nordhaus W (2015) Climate clubs: overcoming free-riding in international climate policy. Am Econ Rev 105(4):1339–1370
Odell JS (2005) Chairing a WTO negotiation. In: Petersmann E-U (ed) Reforming the world trading system: legitimacy, efficiency, and democratic governance. Oxford University Press, New York, pp 469–496
Olsen JA (2020) Understanding NATO. RUSI J 165(3):60–72
Ophir A (1989) The ideal speech situation: neo-Kantian ethics in Habermas and Apel. In: Yovel Y (ed) Kant's practical philosophy reconsidered. Springer, Dordrecht, pp 213–234
Paroussos L et al (2019) Climate clubs and the macro-economic benefits of international cooperation on climate policy. Nat Clim Chang 9:542–546
Pauwelyn J (1999) The WTO agreement on sanitary and Phytosanitary (SPS) measures as applied in the first three SPS disputes. EC-hormones, Australia-Salmon and Japan-varietals. J Int Econ Law 2(4):641–664
Petersmann E-U (2005) The 'human rights approach' advocated by the UN high commissioner for human rights and by the international labour organizations: is it relevant for WTO law and policy? In: Petersmann E-U, Harrison J (eds) Reforming the world trading system: legitimacy, efficiency, and democratic governance. Oxford University Press, New York, pp 357–380
Reinert KA (2020) From sustainable development goals to basic development goals. Ethics Int Aff 34(2):125–137
Ruggiero R (2000) Opening remarks at the 50th anniversary symposium. In: The WTO Secretariat (ed) From GATT to the WTO: the multilateral trading system in the new millennium. Kluwer Law International, The Hague, pp 1–4
Sandler T, Tschirhart JT (1980) The economic theory of clubs: an evaluative survey. J Econ Lit 18(4):1481–1521
Sciulli D (1991) Theory of societal constitutionalism: foundations of a non-Marxist critical theory. Cambridge University Press, New York
Simmons AJ (1999) Justification and legitimacy. Ethics 109(4):739–771

Starshinova O (2021) Is the MPIA a solution to the WTO appellate body crisis? J World Trade 55(5):787–804
Tagliapietra S, Wolf GB (2021a) Conditions are ideal for a new climate Club. Energy Policy 158: 112527–112530
Tagliapietra S, Wolf GB (2021b) Form a climate Club: United States, European Union and China. Nature 591(7851):526–528
Tangermann S (2005) How to forge a compromise in the agriculture negotiations. In: Petersmann E-U (ed) Reforming the world trading system: legitimacy, efficiency, and democratic governance. Oxford University Press, New York, pp 99–120
Treves T (2008) Aspects of legitimacy of decisions of international courts and tribunals. In: Wolfrum R, Röben V (eds) Legitimacy in international law. Springer, Berlin, pp 189–201
Van den Brink M (2019) Justice, legitimacy and the Authority of Legislation within the European Union. Modern Law Rev 82(2):293–318
Vanduzer JA (2005) Navigating between the poles: unpacking the debate on the implications for development of GATS obligations relating to health and educations services. In: Petersmann E-U (ed) Reforming the world trading system: legitimacy, efficiency, and democratic governance. Oxford University Press, New York, pp 167–204
Von Bogdandy A (2008) Codes of conduct and the legitimacy of international law. In: Wolfrum R, Röben V (eds) Legitimacy in international law. Springer, Berlin, pp 299–307
Wincott D (2002) National States, European Union and changing dynamics in the quest for legitimacy. In: Arnull A, Wincott D (eds) Accountability and legitimacy in the European Union. Oxford University Press, New York, pp 487–488
Wolfe R (2009) The WTO single undertaking as negotiating technique and constitutive metaphor. J Int Econ Law 12(4):835–858
Wolfrum R (2008) Legitimacy of international law from a legal perspective: some introductory considerations. In: Wolfrum R, Röben V (eds) Legitimacy in international law. Springer, Berlin, pp 1–24

Documents

Blinken AJ (2021) The United States officially rejoins the Paris Agreement. Press Statement. 19 February 2021. Available via https://www.state.gov/the-united-states-officially-rejoins-the-paris-agreement/. Accessed 21 July 2024
Froman M (2013) Keynote Remarks by U.S. Trade Representative Michael Froman at the World Trade Organization Public Forum on Innovation and the Global Trading System. WTO Public Forum, Geneva. Available via https://ustr.gov/about-us/policy-offices/press-office/speeches/transcripts/2013/october/froman-wto-innovation-global-trade. Accessed 21 July 2024
Hoekman B et al (2022) Revitalizing multilateral governance at the World Trade Organization: report of the high-level board of experts on the future of global trade governance. High-level board of experts on the future of global trade governance. 6 May 2022. Available via https://www.wto.org/english/news_e/news18_e/bertelsmann_rpt_e.pdf. Accessed 21 July 2024
Latici T (2021) Harnessing the new momentum in transatlantic relations: potential areas for common action during the biden presidency. European Parliamentary Research Service. June 2021. Available via https://www.europarl.europa.eu/RegData/etudes/IDAN/2021/690660/EPRS_IDA(2021)690660_EN.pdf. Accessed 21 July 2024

Lim AH et al (2022) Where do WTO trade and environmental sustainability initiatives stand today?. IISD Policy Analysis. 2 March 2022. Available via https://www.iisd.org/articles/policy-analysis/where-do-wto-trade-and-environmental-sustainability-initiatives-stand. Accessed 21 July 2024

WTO General Council (2019) Statement in support of the rules-based multilateral trading system. WT/GC/203. 10 December 2019

WTO Regional Trade Agreement Database (2023) WTO. http://rtais.wto.org/UI/PublicMaintainRTAHome.aspx. Accessed 21 July 2024

Chapter 9
Potential Trade Rules and Policies of the Sustainable Development Club

Abstract WTO members can draft the Trade and Sustainable Development Agreement (TSDA) based on international and domestic trade rules and sustainable development policies. These members can build on existing frameworks to reach a consensus on establishing a sustainable development club, thereby progressively driving the reform of international trade rules and development policies. This process will eventually lead to the revision of WTO rules, effectively constitutionalising the organisation. EU trade and development policies and mega Free Trade Agreements (FTAs) offer valuable lessons for reforming international trade rules. As the most active advocate for sustainable development, the EU has incorporated numerous sustainability provisions into its trade policies. The quantity and quality of these provisions are outstanding compared to those of other national policies. Furthermore, mega FTAs represent significant efforts by the EU and other countries to integrate sustainable development policies into international trade rules. This chapter details the contents of the TSDA (i.e., the potential trade rules and policies of the sustainable development club) by highlighting the sustainability measures embedded in EU trade and development policies and mega FTAs.

9.1 EU Trade and Development Policy: As an Example of Domestic Trade and Development Policy

9.1.1 Introduction

Throughout the long history of humanity, trade has been nearly synonymous with human prosperity and closely tied to human development.[1] From the earliest acts of bartering to sophisticated commercial transactions, everyone has needed to trade to obtain life's necessities. Those who made trade their profession became merchants, amassing wealth through their commercial endeavours. The rise of the merchant

[1] For the relationship between trade and human prosperity, please see European Commission (2010b); See also Panagariya (2019).

X. Zhao, *Integrating the UN SDGs into WTO Law*, EYIEL Monographs - Studies in European and International Economic Law 41,
https://doi.org/10.1007/978-3-031-73876-0_9

class fuelled urban development, leading to the formation of large cities where trade thrived and populations grew. In these cities, people had easy access to food, water, textiles, and a wide range of other commodities. The flourishing of trade spurred the creation of additional industries. Beyond marketplaces, restaurants, bars, cafés, hotels, cinemas, and other entertainment venues emerged, attracting more residents and wealth-seeking businessmen, thereby expanding the city further. This evolution epitomises the history of nearly every present-day metropolis. It can be said that trade has significantly shaped the pattern of human development. Consequently, trade regulations have a profound impact on the development of human society.

Over the past few decades, more trade regulators have come to believe that human development must be sustainable, using trade policy to address the negative externalities of trade, such as environmental pollution and the violation of labour rights.[2] However, most countries' trade policies still lack norms related to sustainable development.[3] Trade policymakers in these countries have traditionally not considered environmental protection, human rights, and social rights as integral to trade policies. Is this also the case for EU trade policy? Quite the contrary. The EU has a long history of formulating trade policies grounded in sustainable development.[4] The values of sustainable development, embedded across various trade rules, reflect the EU's commitment to sustainability in its trade policy. The EU aims to use these rules to address non-trade concerns, which some scholars, such as John H. Jackson, refer to as international trade's 'link issues.'[5] These encompass human rights, economic sanctions, labour, agriculture, fisheries, and trade in goods and services.[6]

To address non-trade concerns, the EU enacted the European Development Policy in 2017. This policy contains guiding principles for sustainable development,[7] formulated based on normative values and citizens' interests. These two core values confer legitimacy on the European Development Policy and ensure its alignment with the needs of countries worldwide. This universality is essential for WTO members to use the European Development Policy as a foundation for establishing a sustainable development club.

According to the principles of the European Development Policy (PEDP), the EU and its member states must achieve the 17 UN SDGs at both domestic and international levels.[8] Specifically, the EU must ensure human development and dignity, protect the environment, manage natural resources, and tackle climate change.[9] It

[2] See Bellmann and Tipping (2015).

[3] For example, the RCEP, one of the largest FTAs does not have a chapter on sustainable development. Please see Hsieh (2022), p. 146.

[4] See Ziegler (1996), pp. 220–242; Reid (2016), pp. 353–361; See also Kilian and Elgström (2010), pp. 260–269.

[5] See Jackson (1996), p. 41.

[6] See Ibid., p. 41.

[7] See European Commission (2017), pp. 7–8.

[8] See Ibid., p. 6.

[9] See Ibid., p. 7.

must also ensure inclusive and sustainable growth and jobs, peaceful and inclusive societies, democracy, effective and accountable institutions, and the rule of law and human rights for all.[10] To this end, the EU is committed to integrating the PEDP into its key policies in foreign affairs, trade, and other areas.[11] In essence, the development of EU sustainable development policies and related legislative work in all aspects are guided by the PEDP.

In line with the PEDP, the EU must promote sustainable development both within and beyond its borders in accordance with the UN SDGs. Domestically, the EU needs to enhance its legislation on sustainable development.[12] These laws will increase corporate responsibility for the environment and society, promote gender equality, uphold all human rights, and strengthen the participation of civil society organisations (CSOs) in promoting the rule of law, social justice, and human rights.[13] Currently, these sustainable development rules are scattered across EU regional laws, policy documents, directives, regulations, decisions, and EU member states' national laws and regulations.[14]

Externally, the European Development Policy mandates that the EU integrate the PEDP into its foreign policy, including international trade policy.[15] Accordingly, the EU must promote democracy, the rule of law, the principles of equality and solidarity, and the universality and indivisibility of human rights and fundamental freedoms.[16] It must also respect human dignity and the principles of the United Nations Charter and international law.[17] The EU regards these universal values as the core of the 2030 Agenda.[18] It is evident that the EU is reshaping its international trade policy based on EU laws, including environmental law, human rights law, and other regulations related to sustainable development. The Green Deal and the Carbon Border Adjustment Mechanism are the most representative examples.

While the UN 2030 Agenda for Sustainable Development outlines the contents and contours of the SDGs that countries need to achieve, it does not specify the means to accomplish them. Therefore, national policymakers must design their strategies to attain these SDGs. From the existing European Development Policy, it is evident that the EU has adopted three strategies to achieve these goals: adopting higher international standards, taking a comprehensive approach to sustainable development, and employing a principled pragmatism approach. These strategies guide EU policymakers in refining the EU's international trade policy in specific areas in line with the PEDP. The Advisory Opinion 2/15 of the CJEU also confirms

---

[10] See Ibid., p. 7.

[11] See Ibid., p. 6.

[12] See Ibid., p. 33.

[13] See Ibid., pp. 7–8.

[14] See De Sadeleer (2014), pp. 175–176.

[15] See European Commission (2017), p. 51.

[16] See Ibid., p. 7.

[17] See Ibid., p. 7.

[18] See Ibid., p. 9.

that the objective of sustainable development is an integral part of the EU's common commercial policy.[19] WTO members can refer to the EU's strategies when formulating the common policy for the sustainable development club.

The EU has integrated various sustainability provisions into its Free Trade Agreements (FTAs) to implement its development policies. As the EU reforms its international trade policy, it has progressively included more sustainability provisions in FTAs through negotiations with its trade partners. Consequently, the sustainability provisions in EU FTAs negotiated at different times vary in quantity and quality, depending on the specific circumstances of each negotiation. For instance, when negotiating FTAs with countries seeking EU membership, the EU may require these countries to fully align their national laws with the EU's environmental, human rights, and trade laws.[20] Conversely, the EU may lower sustainability standards when negotiating with other countries. The EU's negotiation power in a trade relationship arguably influences how it incorporates sustainability provisions into its FTAs.

Finally, it is crucial to recognise the gaps in EU trade policy. This section highlights the need for EU trade officials to substantially enhance the European Development Policy to ensure the sustainability of the EU's economic development model. It suggests that WTO members should address these deficiencies when drafting the TSDA.

9.1.2 EU Development Policy

9.1.2.1 The Core Values of the European Development Policy

9.1.2.1.1 Normative Values

The EU's normative values encompass democracy, the rule of law, human rights,[21] equality, solidarity, the principles of the United Nations Charter, and international law.[22] These values are not solely derived from EU law but are also embedded in the UN Charter and various UN conventions and protocols.[23] Members of the international community, particularly UN member states, generally recognise these values[24] and continue to pursue them through the UN 2030 Agenda for Sustainable

[19] CJEU's Advisory Opinion 2/15, ECLI:EU:C:2017:376, 16 May 2017, para. 147.

[20] For example, see EU-Ukraine Association Agreement, Article 290.

[21] The ECtHR has now recognised the right to environmental protection as a fundamental human right. See Fleurke (2016), p. 393.

[22] See European Commission (2017), p. 7.

[23] See United Nations, The UN's Core International Human Rights Instruments. https://www.ohchr.org/en/core-international-human-rights-instruments-and-their-monitoring-bodies. Accessed 21 July 2024; See also Ramcharan (2015), pp. 13–54.

[24] See Ibid., p. 18.

Development.[25] However, it is important to acknowledge that some individual countries have not fully embraced and implemented these norms.

The EU has a strong record of implementing these norms and often sets higher standards than those established by the UN in its domestic laws and regulations, striving to achieve a greater realisation of these universal values within its territory.[26] Before adopting the 2030 Agenda, the EU had already integrated similar normative values into the Treaty on European Union (TEU). Article 21 of the TEU mandates the EU to adopt external policies that promote good global governance. This article requires the EU to consider human rights, sustainability, food security, economic growth, environmental protection, the rule of law, and peace when shaping its global strategy.[27]

EU law effectively embodies the economic, social, and environmental dimensions of the UN SDGs. Consequently, the constraints imposed by these EU laws ensure that both the EU and its member states must promote sustainability, protect human rights, and uphold the international rule of law in their policy design and application. Therefore, the EU's normative values guarantee that its commitment to achieving the UN SDGs is not merely a vague political aspiration but a clear and enforceable legal obligation for the EU and its member states.

The European Development Policy emphasises three core normative values: sustainability, human rights, and a rules-based global legal order.[28] To achieve the 17 UN SDGs, the EU must uphold these values rigorously. Sustainability involves meeting the goals outlined in the UN 2030 Agenda for Sustainable Development and adhering to existing multilateral conventions related to sustainable development, such as environmental agreements. The fulfilment of human rights requires the EU to enforce robust human rights laws domestically[29] and advance human rights globally.[30] Through trade and technical assistance, particularly via the Generalised Scheme of Preferences (GSP),[31] the EU helps emerging democracies to reform their political and legal systems.[32] This, in turn, helps their citizens fully enjoy human and social rights. These countries can reduce poverty and create jobs while benefiting from duty-free access to the EU market.[33] A rules-based global legal order mandates the EU to promote the international rule of law in diplomatic

[25] For the relationship between human rights and the SDGs, please see Rattray (2019); See Feiring et al. (2016), pp. 11–15; See also UNGA (2015). Goal 16, 'Peace, Justice and Strong Institutions', aims to ensure the international rule of law and justice. Goal 10, 'Reduce Inequality', aims to reduce inequality within and among countries. Goal 17, 'Partnerships for the Goals', aims to revitalise the global partnership and maintain solidarity.

[26] See European Commission (2007).

[27] See Ibid., pp. 7–14; See also Durán and Morgera (2012).

[28] See European Commission (2017), pp. 8–37.

[29] See Ibid., pp. 9–19.

[30] See Zamfir and Dobreva (2019), p. 1.

[31] See Zamfir and Dobreva (2019), p. 7; See also Ziegler (2022), p. 16.

[32] See Zamfir and Dobreva (2019), p. 7.

[33] Take Sri Lanka as an example. See European Commission (2018c).

negotiations.[34] The EU views the enhancement of the international rule of law as a critical condition for achieving the 17 UN SDGs. Beyond being a specific SDG, the international rule of law is crucial for ensuring peace and implementing numerous international conventions.[35]

9.1.2.1.2 Citizens' Interests

The second core normative value of EU development policy is the protection of citizens' interests. This obligation mandates that the EU ensure the well-being of its citizens both within and beyond its borders.[36] The legitimacy of the EU's existence hinges on fulfilling this duty. Consequently, in the face of threats such as climate change, pandemics, and conflict, the EU must take political responsibility to safeguard its citizens from these dangers and, at a minimum, work to mitigate their impacts. To achieve this, the EU must enhance its sustainable development policy and prepare for both visible and latent risks. As Mogherini has observed, given the current international political landscape, the EU's foreign and security policy aims to build a stronger Union capable of addressing its citizens' needs and serving as a global security provider.[37] Therefore, the EU must assume responsibility for protecting its citizens' interests and responding effectively to both internal and external crises, including those related to sustainable development, such as climate change.

Despite the complexities of the challenges it faces, the EU's sustainable development policy is proactive rather than conservative. Recognising the interconnectedness of internal and external security, the EU's development policy seeks not merely to mitigate these external risks but to positively engage with the current turbulent era.[38] While the effectiveness of this approach will need to be assessed on a case-by-case basis, it is evident that the EU views the achievement of the 17 UN SDGs as an opportunity to address security concerns.

By attaining these SDGs, the EU can enhance the well-being of its citizens through improved domestic laws and also foster a peaceful and stable international environment by addressing development crises in other countries. Therefore, the EU must realise the core normative values of its development policy both domestically and internationally. This approach is crucial for protecting its citizens from threats such as climate change, environmental pollution, and conflict. Thus, the EU must ensure the realisation of its core normative values to effectively achieve the second core value of its development policy.

---

[34] See European Commission (2017), p. 7.

[35] See Keith (2015), pp. 407–415.

[36] See European Commission (2017), pp. 49 and 55.

[37] European External Action Service (2017), p. 3.

[38] See Ibid., p. 17.

According to the EU Global Strategy (EUGS), EU civic interests can be divided into two categories. The first category is the preservation of democratic values and governance within European countries.[39] To safeguard this interest, the EU must act both internally and through its external policies. The EU's security plan mandates that it not only uphold its existing normative values but also contribute as much as possible to global sustainable development, thereby fostering sustainable security.[40] This approach aims to address the root causes of conflict and poverty in the international community, reducing threats to the well-being of EU citizens.

The second category of EU civic interests is ensuring overall peace and security for its citizens. This involves mitigating various risks, such as economic instability, cyber-attacks, energy crises and conflicts both within and beyond the EU's borders.[41] To achieve this, the EU must promote fair and open markets,[42] support humanitarian development,[43] ensure a free and secure internet,[44] protect media freedom from disinformation,[45] diversify energy sources,[46] implement effective migration policies,[47] strengthen counter-terrorism efforts,[48] maintain a robust military,[49] uphold a rules-based global order, respect the law of the sea,[50] and advance global disarmament.[51]

As noted earlier, the EU has embedded the interests of its citizens into its development policy, yielding two significant benefits. First, prioritising citizens' interests ensures that the EU's sustainable development policies are genuinely focused on realising the economic, social, and environmental rights of European citizens. This is crucial because if policymakers deviate from this focus, they may not fully understand or address the real needs of citizens. A misalignment in recognising these needs can lead to policies that fail to balance economic, social, and environmental rights effectively. When policies prioritise one right over another, they can undermine the benefits intended by the UN SDGs and may even create new problems for citizens.

Second, a development policy is legitimate when it centres on citizens' interests. Since sustainable development is a universal value, the EU's legislation aligned with these values will generally be compatible with other countries' sustainable

[39] Ibid., p. 8.

[40] Ibid., pp. 23–24.

[41] Ibid., p. 9.

[42] Ibid., p. 15.

[43] Ibid., p. 31.

[44] Ibid., p. 15.

[45] Ibid., p. 23.

[46] Ibid., p. 22.

[47] Ibid., p. 27.

[48] Ibid., p. 21.

[49] Ibid., pp. 20–21.

[50] Ibid., pp. 33 and 38.

[51] Ibid., pp. 33 and 39–40.

development policies. This compatibility is essential for achieving the UN SDGs globally, as countries need to collaborate on harmonious and sustainable development measures. Conversely, if a country's development policy is centred more on perceived national interests rather than the genuine needs of its citizens, it may diverge from the universal principles of sustainable development. Such a policy might overly focus on specific aspects of sustainable development, potentially undermining global efforts to achieve the UN SDGs. In extreme cases, a government could use selective achievement of SDGs as a political tool to consolidate authoritarian control, thereby violating fundamental human rights and damaging the credibility of the UN SDGs. This, in turn, could diminish citizens' support for these goals and hinder global progress.

9.1.2.1.3 Conclusion

In conclusion, the two core values—normative values and citizens' interests—provide legitimacy and rationality to the EU's development policy. By aligning development policies with citizens' interests, policymakers can comprehensively and equitably realize economic, social, and environmental rights, thus achieving sustainable development. Additionally, basing development policies on universal values helps to eliminate conflicts between national and international norms and interests. This value compatibility is crucial for advancing global sustainable development policies, as countries must share common normative values and civic interests to form a sustainable development coalition and drive the reform of international trade and development policies.

The universality of the EU's core values indicates that it is feasible for countries to reach a consensus on fundamental values and interests in development policy. Consequently, the EU's development policy can serve as a model for promoting a unified development approach among the member states of a sustainable development club. While there are some drawbacks to the EU's development policy,[52] these issues can be addressed and do not significantly impede the achievement of the UN SDGs.

9.1.2.2 Strategies of the EU's Development Policy

9.1.2.2.1 Higher Sustainable Development Standards

The EU's first strategy involves setting sustainable development standards that exceed international benchmarks. As countries work towards achieving the SDGs, they must adhere to established international standards such as sanitary and

[52] See Gouritin (2016); See also Bronckers and Gruni (2021), pp. 26–30.

phytosanitary regulations,[53] technical standards,[54] and guidelines set by organisations like the WHO[55]and FAO.[56] These international standards are crucial for the successful implementation of the UN SDGs. They assist policymakers in developing and enforcing laws and regulations necessary to meet these goals.

International standards play a key role in minimising disputes over the interpretation of the SDGs. By aligning with these standards, countries can achieve consensus on the fundamental principles of the SDGs, reducing regulatory differences to technical matters.[57] Disputes over technical issues, such as those related to quarantine or food safety, can often be resolved through expert consultations, which facilitates flexible and scientific coordination among countries.[58] WTO practices demonstrate that many technical disputes related to sustainable development measures can be resolved through such consultations.[59] A robust and adaptable consultation mechanism enhances the legitimacy and binding nature of international standards. National policies must align with the overarching principles of international conventions and adhere to expert interpretations on specific technical matters. This science-driven approach mitigates political interference in policymaking, thus avoiding irreconcilable conflicts between states.

Furthermore, international standards are instrumental in monitoring and evaluating progress towards the UN SDGs. They provide clear metrics for assessing advancements in the 17 SDGs, allowing countries to track their implementation of sustainability standards. This transparency fosters peer pressure among nations, motivating them to enhance their efforts to achieve the SDGs and improve their international standing. Such peer pressure also encourages countries to contribute to refining scientific standards. For instance, the EU has established a Joint Research Centre to offer independent scientific advice to EU policymakers, illustrating the commitment to evidence-based policymaking.[60]

However, it is important to acknowledge that existing international sustainability standards have significant room for improvement, and strict adherence to these

[53] See Commission on Phytosanitary Measures, Adopted International Standards for Phytosanitary Measures (ISPMs). https://www.ippc.int/en/core-activities/standards-setting/ispms/. Accessed 21 July 2024.

[54] For example, see ITU, The International Telecommunication Union's Technical Standards. https://www.itu.int/en/ITU-T/publications/Pages/recs.aspx. Accessed 21 July 2024.

[55] See WHO, WHO Guidelines. https://www.who.int/publications/who-guidelines. Accessed 21 July 2023.

[56] See Codex Alimentarius Commission, Codex Standards. https://www.fao.org/fao-who-codexalimentarius/codex-texts/en/. Accessed 21 July 2024.

[57] See Peel (2010), p. 241.

[58] See Ibid., p. 263.

[59] When countries have disputes on their different SPS measures or TBT measures, they first need to resolve their disputes through consultations within the SPS Committee or TBT Committee to find a mutually satisfactory solution.

[60] To know the work of the Joint Research Centre, please see European Commission (2020b).

standards alone is insufficient to achieve the UN SDGs. There are two main reasons for this limitation.

First, the level of scientific development is often inadequate. The world is rapidly evolving, and countries need to continually advance their science and technology to enhance their sustainability standards. This progress is crucial for policymakers to design effective policies that meet the demands of sustainable development. For example, in *EC-Hormones I and II*, the European Commission expanded its scientific research to better align its SDG implementation measures with international standards.[61]

Secondly, ideological differences between countries contribute to discrepancies in human rights standards, such as social rights. These differences often lead to conflicts, causing the UN SDGs to adopt more generalised standards in these areas.[62] To effectively promote sustainable development, the EU must develop more detailed and advanced standards that build on the principles of the UN 2030 Agenda. This approach will enable the EU to advance sustainable development both within its borders and globally. Currently, the EU's European Development Policy (PEDP) incorporates not only the principles of the 2030 Agenda but also numerous localised principles crafted by EU lawmakers.

These EU principles are aligned with various SDGs. For Goal 3, 'Good Health and Well-being,' the EU adheres to the principle of aid effectiveness.[63] This principle emphasises strengthening health systems, enhancing coordination, monitoring, and capacity building.[64] For Goal 2, 'Zero Hunger,' the EU translates its principles into actionable measures within the critical domain of food security. The EU acknowledges that investments in agriculture should respect human rights, livelihoods, and resources to increase food availability.[65] This principle mandates that EU member states adhere to the guidelines set out in the Land Policy Guidelines. Additionally, the EU and its member states are encouraged to support research and innovation for smallholder farmers, provided that biodiversity is preserved. The EU also advocates for adopting new technologies and developing incentive schemes that promote synergy between climate change adaptation and mitigation.

Regarding food access, the EU aims to establish a political and legal framework based on the 'right-to-food' principle, as outlined in the Voluntary Guidelines. This approach supports the progressive realisation of national food security. Other key principles include enhancing the nutritional adequacy of food intake, integrating

---

[61] See WTO Panel Report, United States-Continued Suspension of Obligations in the EC-Hormones Dispute, WT/DS320/R, adopted on 14 November 2008, para. 2.3.

[62] See Knox (2015), p. 536.

[63] See European Commission (2017), p. 8.

[64] See European Commission (2010a), pp. 9–10

[65] See European Commission (2010a), pp. 5–6.

relief, rehabilitation, and development efforts, and maximising the effectiveness of food security investments.[66]

9.1.2.2.2 Comprehensive Approach

The EU's second strategy is to adopt a comprehensive approach to sustainable development.[67] This strategy emphasises the need for policymakers to consider the interconnectedness of policies across various sectors to effectively and holistically achieve the SDGs. It reflects the EU's understanding of the intricate linkages between different public goods essential for sustainable development.

In practice, the EU addresses these interdependencies by treating them as cross-cutting elements within its sustainable development policies. For instance, water is identified as a cross-cutting element in the EU's development strategy. The European Commission's reform agenda highlights the mutually reinforcing relationship between water and other public goods, such as energy, security, markets, infrastructure, and transboundary cooperation.[68] Similarly, the European Commission acknowledges the significant impact of agriculture and energy on the economy, environmental protection, and climate change mitigation and adaptation.[69] Another key example of this comprehensive approach is the EU's recognition of the critical role that product and energy supply chains play in achieving sustainable development. By integrating these cross-cutting elements into its policy framework, the EU aims to address the complex and interrelated challenges of sustainable development effectively.[70]

Moreover, the European Commission has incorporated the private sector into its overarching policies.[71] This inclusion aims to enhance companies' social responsibility regarding environmental protection and human rights. By integrating businesses into its strategy, the EU leverages the private sector to advance sustainable development comprehensively. This approach ensures that sustainability efforts are embedded across all levels of social life, both within and beyond the EU's borders.

To effectively implement this multifaceted strategy, the EU must establish a broad international partnership. As highlighted by the European Commission:

---

[66] Ibid., pp. 7–8. These principles include the principle of support policy development capacity and inter-sectoral coordination mechanism, the principles of the Paris Declaration on Aid Effectiveness, and the principle of close collaboration with the Rome-based agencies. These diverse principles create a complicated normative network relying on the norms of international law, providing the legitimacy of the EU's outward development policy.

[67] See European Commission (2017), pp. 5 and 8.

[68] See European Commission (2011), p. 8.

[69] Ibid., p. 7.

[70] See European Commission (2022).

[71] See, for example, European Commission (2018b).

> When violent conflicts erupt, our shared vital interests are threatened. The EU will engage in a practical and principled way in peacebuilding and foster human security through an integrated approach. Implementing the comprehensive approach to conflicts and crises through a coherent use of all policies at the EU's disposal is essential. But the meaning and scope of the 'comprehensive approach' will be expanded. . .Sustainable peace can only be achieved through comprehensive agreements rooted in broad, deep and durable regional and international partnerships, which the EU will foster and support.[72]

To establish such a partnership, the EU emphasises the importance of mutually beneficial cooperation. According to Section 3.4 of the EUGS, titled 'Cooperative Regional Orders,' the European Commission asserts that:

> Voluntary forms of regional governance offer states and peoples the opportunity to better manage security concerns, reap the economic gains of globalisation, express more fully cultures and identities, and project influence in world affairs. This is a fundamental rationale for the EU's own peace and development in the 21st century. This is why we will promote and support cooperative regional orders worldwide, including in the most divided areas. Regional orders do not take a single form. Where possible and when in line with our interests, the EU will support regional organisations. We will not strive to export our model, but rather seek reciprocal inspiration from different regional experiences.[73]

According to this guiding principle, the EU is committed to fostering regional cooperation with other countries in the economic, social, and human rights fields.[74] The EU's development policy encompasses numerous cooperation projects, including initiatives such as the EU Fisheries and Aquaculture Projects,[75] the Global Europe Project,[76] and Official Development Assistance projects.[77]

A key aspect of these EU cooperation projects is the provision of financial and technical assistance to developing and least-developed countries. For example, in alignment with Goal 1, 'No Poverty,' the EU plans to utilise its geographical and thematic funds to support the eradication of poverty in the world's poorest nations. Through these initiatives, the EU aims to assist developing countries and vulnerable regions, such as small island states grappling with severe climate change impacts, in advancing their economies and ensuring their citizens' social, cultural, and political rights.

Moreover, this support serves a dual purpose: it not only aids these countries in their development efforts but also aligns with the EU's own objectives. By enhancing global well-being, the EU strives to meet its specific needs and contribute effectively to achieving the SDGs.

---

[72] European External Action Service (2017), pp. 9–10.

[73] Ibid., p. 32.

[74] Ibid., p. 18.

[75] To know EU Fisheries and Aquaculture Projects, please see European Commission, Oceans and Fisheries. https://oceans-and-fisheries.ec.europa.eu/index_en. Accessed 21 July 2024.

[76] See European Commission (2018a).

[77] See EU Commission, Publication of Preliminary Figures on 2020 Official Development Assistance Annex: Tables and Graphs. https://ec.europa.eu/commission/presscorner/detail/ro/qanda_21_1704. Accessed 21 July 2024.

Undoubtedly, this reciprocal relationship has the potential to maximise global public goods, thereby advancing the achievement of the UN SDGs on a global scale. In my view, the comprehensive approach offers flexibility for sustainable development policies, requiring policymakers to address and implement the SDGs in a holistic and balanced manner.[78] This approach enables countries to collaborate effectively, achieving SDGs that they might not reach independently. For instance, a country can assist others in meeting their SDGs with the expectation of reciprocal support for its own objectives. Such a cooperative dynamic not only fosters mutual progress but also serves as a compelling incentive for countries to join the sustainable development coalition.

9.1.2.2.3 Principled Pragmatism Approach

The EU's third strategy is to employ a principled pragmatism approach. According to the European Commission:

> The EU will promote a rules-based global order. We have an interest in promoting agreed rules to provide global public goods and contribute to a peaceful and sustainable world. The EU will promote a rules-based global order with multilateralism as its key principle and the United Nations at its core. We will be guided by clear principles. These stem as much from a realistic assessment of the current strategic environment as from an idealistic aspiration to advance a better world. Principled pragmatism will guide our external action in the years ahead.[79]

The EU's strategic aim is to position itself midway between the extremes of isolationism and reckless interventionism.[80] Specifically, the EU strives to balance engaging with countries that have conflicting interests to achieve the SDGs while firmly maintaining its stance on crucial issues such as global security. This balanced approach enables the EU to advance the SDGs effectively. It is important to distinguish principled pragmatism from realism. This strategy does not allow the EU to sacrifice the pursuit of SDGs in other countries for its own objectives, such as energy security and economic development. A notable example is the EU's diplomatic stance toward Russia. To uphold global peace and the international rule of law, the EU cannot endorse Russia's aggressive actions against Ukraine, including the annexation of Crimea in 2014, the secession of the Donbas and Luhansk regions, and the full-scale invasion of Ukraine in 2022.[81] Although the EU might achieve energy security and meet its clean energy targets by importing Russian gas, it cannot engage in such cooperation if it undermines international security and the rule of law, which are central to the UN SDGs.

---

[78] See European Commission (2017), para. 1. The EU recognises the balanced understanding of sustainable development.

[79] See European External Action Service (2017), p. 8.

[80] See Ibid., p. 16.

[81] Ibid., p. 33.

It can be argued that this strategy offers the EU flexibility in its sustainable development policy while preserving core sustainable development principles. On the one hand, constructive negotiations remain possible even amidst significant geographical or ideological conflicts between the EU and other countries. In other words, the EU is committed to opening avenues for negotiation and cooperation despite severe conflicts of interest. This approach is crucial for advancing the UN SDGs, as it prevents individual disputes from hindering progress on global issues like climate change, which threatens the well-being of humanity as a whole. Adopting such a pragmatic approach allows for continued collaboration and increases the likelihood of achieving the SDGs.

On the other hand, this strategy ensures that the EU does not overlook international injustices and atrocities while pursuing specific SDGs. The EU must account for the impact of international politics on sustainable development when seeking cooperation. In situations where severe human rights violations or military aggression render the fulfilment of economic, social, and environmental rights unattainable, the EU must lead efforts to restore these rights for affected populations. This approach reflects the principle of proportionality in international constitutional law,[82] meaning that the pursuit of one right should not come at the expense of another.[83]

9.1.2.3 Green Deal

To address the most pressing environmental issues, such as global warming, the EU enacted the Green Deal in 2021.[84] Building on the PEDP, this policy document outlines the EU's strategy for environmental protection and combating climate change.[85] The policy aims to transform the EU into a fair and prosperous society with a modern, resource-efficient, and competitive economy, achieving net-zero GHG emissions by 2050 and decoupling economic growth from resource use.[86] Additionally, it seeks to protect, conserve, and enhance the EU's natural capital while safeguarding citizens' health and well-being from environmental risks and impacts.[87]

This policy requires the EU to revise its legislation across various sectors, including industrial production, energy, mobility patterns, chemical products, ecosystems and biodiversity, and agriculture.[88] By developing specific policies and laws

[82] See citations in Sect. 4.2.2.

[83] See Giovanella (2017), pp. 10–11.

[84] See UNFCCC, Copenhagen Accord, COP Dec.2/CP.15, UNFCCCOR, UN Doc. FCCC/CP/2009/11/Add.1, para. 1.

[85] See European Commission (2019b).

[86] See Ibid., p. 2.

[87] See Ibid., p. 2.

[88] See Ibid., p. 3.

in these areas, the EU aims to meet its emission reduction targets as committed in the Paris Agreement and to mitigate the damage to biodiversity caused by trade.[89] While these policies are not directly related to EU trade policy, achieving these objectives will transform the EU's economy for a sustainable future, fundamentally impacting its trade policy.[90] This impact will primarily be seen in the EU's rules on market access for imported goods.

Existing policies indicate that the EU will mandate producers within its territory to reduce carbon emissions from production processes by using clean energy or emission reduction equipment.[91] This requirement will alter the demand for imported products in the EU market, leading to an increase in imports of green products and a gradual reduction and elimination of fossil energy imports. Similarly, policies on clean mobility, chemical products, and agriculture will also influence the EU's market access policies. According to the Green Deal, the EU will legislate to ban the production and use of industrial and agricultural products with high carbon emissions and toxic substances.[92] For instance, high-emission cars,[93] foodstuffs with a significant carbon footprint, and environmentally harmful chemical products will be progressively excluded from the EU market.

To prevent unfair competition from imports while achieving its policy objectives, the EU will likely impose trade barriers on environmentally unfriendly products through various regulatory measures. These measures may include increasing import tariffs, introducing border adjustment taxes, raising quarantine standards for products, and implementing labelling requirements. Additionally, the EU will introduce preferential import policies for green products and clean energy sources to meet consumer demands in its single market. The EU will selectively apply different trade measures to different imports, taking into account the pace of its own policy changes and specific trade interests. It will also adjust its policies as circumstances evolve.

9.1.2.4 Domestic Emission Reduction Mechanisms

On July 14, 2021, the European Commission officially launched a proposal for the Carbon Border Adjustment Mechanism (CBAM), which entered its transitional period on October 1, 2023.[94] This initiative is part of the EU's efforts to fulfil its international environmental commitments, particularly under the Paris Agreement, with the goal of achieving a climate-neutral EU by 2050 and ensuring net-zero

[89] See Ibid., pp. 20–21.

[90] See Ibid., p. 4.

[91] See Ibid., pp. 6–7. See also European Commission (2019a).

[92] See European Commission (2019b), pp. 14–15.

[93] The EU approved effective ban on new fossil fuel cars from 2035. See European Parliament (2022).

[94] European Commission (2024).

carbon dioxide emissions within the territory of EU member states.[95] The establishment of the CBAM also demonstrates the EU's commitment to greening international trade rules.

One reason the CBAM as attracted significant attention is its relationship with WTO rules. As a border adjustment measure, the CBAM will inevitably impact goods imported into the EU internal market.[96] While the mechanism aims to reduce GHG emissions and combat climate change, the EU must ensure that the effects of the CBAM on the flow of imported goods do not violate its WTO free trade commitments. Therefore, the EU needs to implement its emission reduction measures in a way that complies with WTO rules. In this context, the CBAM can be seen as a legal tool for the EU to align its domestic emission reduction measures with international trade regulations.

In addition, the CBAM serves as a measure for the EU to enhance its existing emission reduction mechanism. In 2005, the EU established the EU Emissions Trading System (ETS) to create a new governance model for air pollution within the territories of its member states.[97] Instead of dictating specific emission reduction targets for each entity, this system uses price signals to encourage producers to decarbonise their production processes.[98] National competent authorities in EU member states review the qualifications of GHG emitters in terms of their ability to monitor and report emissions.[99] Qualified operators of industrial installations must purchase allowances corresponding to their GHG emissions through auctions organised by the member states.[100] This internalises the cost of decarbonisation into the cost of products, incentivising producers to adopt cleaner production methods to avoid paying carbon taxes.

However, the ETS has a significant drawback: it cannot achieve zero carbon emissions. Member states allocate free allowances to specific industry sectors to level the playing field within the EU single market. This policy allows certain producers to emit GHGs without cost,[101] which, while preventing unfair competition from imported goods, contradicts the objective of achieving climate neutrality.

The CBAM addresses a key gap in the ETS by extending the carbon tax to imported goods. By internalising the cost of decarbonisation in the prices of imports, the CBAM ensures that the carbon tax applies not only to domestic producers but also to those outside the EU. This provision creates a level playing field within the EU single market by imposing a carbon tax on imports, thereby reducing the need for free carbon allowances for domestic producers. As a result, the use of free

[95] Ibid., Explanatory Memorandum.

[96] Weko et al. (2020).

[97] See European Commission, EU Emissions Trading System (EU ETS). https://climate.ec.europa.eu/eu-action/eu-emissions-trading-system-eu-ets_en. Accessed 21 July 2024.

[98] Anke et al. (2020), pp. 111647–111657.

[99] The European Parliament and the Council (2003), Article 6.

[100] Ibid., Article 10.

[101] Ibid., p. 3.

allowances will no longer be necessary for maintaining market fairness. Over time, this could allow EU member states to phase out all ETS allowances, ultimately helping the EU achieve its goal of zero GHG emissions.

Despite its potential, the CBAM still has some limitations. First, it does not cover all product-related carbon emissions. For example, the EU ETS currently excludes emissions from mining activities and non-air transport.[102] To avoid discriminating against foreign producers and thereby breaching WTO rules, the EU must ensure consistency between the ETS and CBAM in terms of scope. Consequently, the CBAM currently only addresses direct emissions from the production of basic materials and their products up to the point of import, as well as related indirect emissions when they are significant. This limitation may reduce the effectiveness of the CBAM in driving emission reductions. To achieve its goal of zero GHG emissions, the EU will need to implement further legislation to ensure that domestic industries also achieve zero emissions.[103] This step would be a necessary precursor for expanding the scope of the CBAM and enhancing its impact on global emission reductions.

Second, the administrative procedures for the CBAM are quite complex. According to Article 36 of the CBAM regulation, EU importers must begin fulfilling their CBAM obligations by January 1, 2026.[104] To import foreign goods, they will need to register with their national registry,[105] open an account for CBAM certificate transactions,[106] and purchase these certificates.[107] The European Commission will determine the price of CBAM certificates based on the average closing prices of EU ETS allowances on the common auction platform for each calendar week.[108] Additionally, as authorised declarants, EU importers must ensure that the number of CBAM certificates in their national registry account at the end of each quarter represents at least 80 per cent of the GHGs embedded in their imported goods (measured in CO2e).[109] By May 31 each year, importers must submit an annual CBAM declaration and surrender CBAM certificates corresponding to their total annual CO2e.[110] The total CO2e must be verified by an accredited verifier from the national accreditation body of the importing country.[111]

---

[102] See European Commission (2021b), p. 5; See also Euromines (2020).

[103] There is already a proposal for reforming the ETS. See European Commission (2021a).

[104] Ibid., Article 36.

[105] Ibid., Article 14.

[106] Ibid., Article 16.

[107] Ibid., Article 20.

[108] Ibid., Article 21.

[109] Ibid., Article 22 (2).

[110] Ibid., Article 22 (1).

[111] Ibid., Article 18 (2).

These requirements can result in significant costs for EU trade partners.[112] The high administrative burden and carbon tax could potentially lead some trade partners to exit the EU market. Furthermore, the substantial cost of compliance might deter countries with smaller markets from adopting similar emission reduction mechanisms, as their markets are not as attractive as the EU's. This challenge could hinder the EU's efforts to promote the CBAM globally, highlighting a potential downside of the mechanism.

9.1.3 Inclusion of Sustainability in EU FTAs

9.1.3.1 Labour Rights

Given that workers are integral to goods production, labour rights are inherently connected to trade regulations.[113] Today, labour rights encompass both the right to full employment and the right to be treated according to international standards.[114] These standards are grounded in the ILO's fundamental principles and rights at work,[115] which include the freedom of association, the right to collective bargaining, the elimination of all forms of forced or compulsory labour, the effective abolition of child labour, and the eradication of discrimination in employment and occupation.[116] The EU upholds these rights through a dedicated labour clause in its free trade agreements (FTAs).[117]

Current trade-related labour disputes often centre on human rights discrepancies resulting from varying labour practices across different countries.[118] These disparities can manifest in differences in minimum wage rates, maximum working hours, and access to job security measures.[119] Companies may exploit these lower standards to reduce production costs, gaining an unfair competitive advantage over their trade partners. The EU's free trade agreements (FTAs) aim to address this issue by obligating contracting parties to enhance their labour practices in line with ILO standards, harmonise domestic standards, and prevent unfair competition progressively.[120]

---

[112] See Lydgate (2021), p. 3.

[113] See Cottier and Caplazi (2011).

[114] See OECD (1996), pp. 11–15.

[115] See Kaufmann (2007), p. 71.

[116] See Ibid., pp. 53–67.

[117] See Article 13.4 of the EU-Korea FTA, Article 23.3 of the CETA, Article 13.2 of the EU-Singapore FTA, Article 16.3 of the EU-Japan FTA, and Article 13.4 of the EU-Vietnam FTA.

[118] For example, US-Guatemala Labour Dispute. See the Report on Guatemala-Issues Relating to the Obligations Under Article 16.2.1 (a) of the CAFTA-DR; See also De Schutter (2015), p. 7.

[119] See Murry (2001); See also Panel of Experts Proceeding Constituted Under Article 13.15 of the EU-Korea Free Trade Agreement, Report of the Panel of Experts, 20 January 2021, para. 81.

[120] See, for example, Panel of Experts Proceeding Constituted Under Article 13.15 of the EU-Korea Free Trade Agreement, Report of the Panel of Experts, 20 January 2021, paras. 80-2.

In the EU-Korea labour dispute, the EU filed a complaint against Korea for failing to improve worker treatment as required.[121] Following unsuccessful consultations, the EU requested a panel of experts to investigate the issue.[122] The panel ultimately concluded that the labour clause in the EU-Korea FTA was legally binding[123] and that Korea must fulfil its commitment to uphold the freedom of association and ratify essential ILO conventions.[124]

The panel report on the EU-Korea labour dispute is currently the only decision addressing labour provisions in EU free trade agreements (FTAs). This decision has significant implications. First, it establishes that the labour provisions in EU FTAs are legally binding, which adds substantial value to the decision.[125] To date, the EU has only clarified the legally binding nature of the labour provisions in the Comprehensive Economic and Trade Agreement (CETA).[126] Consequently, there has been room for EU FTA parties to challenge the enforceability of these provisions, potentially viewing them as merely exhortatory or encouraging. However, the panel report eliminates this ambiguity by clearly stating that contracting parties must ratify fundamental ILO conventions and enhance the treatment of domestic workers.[127]

However, this decision might pose challenges for the EU in future trade negotiations. It could deter potential trade partners who may be unwilling to commit to improving labour standards and aligning their national laws with EU regulations.[128] Such reluctance could lead these partners to resist including labour provisions and similar clauses, such as those related to environmental protection, in their FTAs. The full impact of this decision on future EU trade policy remains to be seen.

9.1.3.2 Environmental Protection

The EU places a high priority on environmental protection and requires its trade partners to recognise and adhere to multilateral environmental agreements as part of trade policy.[129] Moreover, the EU is committed to enhancing the role of trade in environmental protection by aligning its policies with international standards.[130] The

[121] See Ibid., para. 55.

[122] See Ibid., paras. 4–5.

[123] See Ibid., para. 277.

[124] See Ibid., para. 293.

[125] See Ibid., para. 277.

[126] The CETA, Article 23.11 (3).

[127] See Panel of Experts Proceeding Constituted Under Article 13.15 of the EU-Korea Free Trade Agreement, Report of the Panel of Experts, 20 January 2021, paras. 277 and 293.

[128] See Han (2021), p. 543.

[129] See Article 13.5 of the EU-Korea FTA, Article 12.6 of the EU-Singapore FTA, Article 16.4 of the EU-Japan FTA, Article 13.5 of the EU-Vietnam FTA, and Article 24.4 of the CETA.

[130] See European Commission (2019b), p. 22.

EU's trade policy, recognising both the positive and negative impacts of trade on the environment, seeks to advance existing international trade rules in these areas. Specifically, the EU promotes and facilitates trade in environmental products and services through its free trade agreements (FTAs),[131] including clauses favouring environmental protection and sustainable development.[132]

Secondly, the EU mandates environmental impact assessments for trade to mitigate or prevent negative environmental effects, such as pollution, over-harvesting, or excessive capture of plant and animal resources.[133] Some EU free trade agreements (FTAs) provide more detailed requirements for these assessments. For instance, the EU-Singapore FTA, CETA, EU-Japan FTA, and EU-Vietnam FTA include provisions that require parties to prevent the overexploitation of forestry and water resources and to combat overfishing, thereby promoting the conservation of these resources.[134] Additionally, the EU-Japan FTA and the EU-Vietnam FTA contain specific provisions on biological diversity, committing parties to conserve and sustainably use biodiversity in line with the Convention on Biological Diversity and its protocols, as well as the Convention on International Trade in Endangered Species of Wild Fauna and Flora.[135] These provisions comprehensively address the protection needs of various plant and animal species. If fully implemented, these agreements will help ensure that international trade practices are conducted in a manner that promotes the sustainable use of the planet's resources.

9.1.3.3 Climate Change

Addressing climate change is central to the EU's green trade policy.[136] Under the Paris Agreement, the EU and its member states have committed to reducing emissions by at least 55% by 2030,[137] achieving carbon neutrality by 2050,[138] and ultimately phasing out greenhouse gases to reach negative GHG emissions.[139]

---

[131] See Ibid., p. 7.

[132] See Article 13.6 of the EU-Korea FTA, Article 12.11 of the EU-Singapore FTA, Article 16.5 of the EU-Japan FTA, Article 13.10 of the EU-Vietnam FTA, and Article 24.9 of the CETA.

[133] See Article 13.10 of the EU-Korea FTA, Article 12.14 of the EU-Singapore FTA, Article 16.11 of the EU-Japan FTA, Article 13.13 of the EU-Vietnam FTA.

[134] See Articles 12.7 and 12.8 of the EU-Singapore FTA, Articles 1.9 (2), 24.10 and 24.11 of the CETA, Articles 16.7 and 16.8 of the EU-Japan FTA, and Article 13.9 and 13.9 of the EU-Vietnam FTA.

[135] See Article 16.6 of the EU-Japan FTA and Article 13.7 of the EU-Vietnam FTA.

[136] See European Commission (2019b), p. 2.

[137] See Ibid., p. 4.

[138] Regulation (EU) 2021/1119 of the European Parliament and of the Council of 30 June 2021 establishing the framework for achieving climate neutrality and amending Regulations (EC) No 401/2009 and (EU) 2018/1999 ('European Climate Law'), Article 2.

[139] Ibid., Article 2.

However, the EU's climate ambitions exceed those of many of its trade partners. Among its existing free trade agreements (FTAs), the EU has included only a specific climate change clause in the EU-Vietnam FTA.[140] Article 13.6 of the EU-Vietnam FTA mandates that the parties pursue the ultimate objective of the United Nations Framework Convention on Climate Change (UNFCCC) and adhere to both the Kyoto Protocol and the Paris Agreement. This provision also requires establishing domestic emission reduction mechanisms, such as carbon pricing systems.[141] Clearly, the EU aims to use this clause to extend its emission reduction framework to Vietnam.

For the EU's trade partners, the absence of climate change provisions in their agreements with the EU does not exempt their exports from high carbon taxes. From its original ETS to the latest CBAM, the EU has consistently implemented legislation to influence the carbon costs of products entering its domestic market. The EU aims to use price signals to encourage both domestic and foreign manufacturers to reduce the carbon emissions associated with their products, including those generated during production.[142] These carbon emissions reduction mechanisms may encounter significant challenges. A primary concern is the possibility that major trade partners might challenge these mechanisms through the WTO's dispute settlement process.[143] This is particularly concerning because many countries have not explicitly agreed to support the EU's climate policies, including the CBAM, in their bilateral agreements.

9.1.4 The Gaps in EU Trade Policy

9.1.4.1 The Gaps in Domestic Legislation

It is important to recognise that advancements in the EU's environmental and human rights laws significantly influence how EU trade regulators shape trade policies. EU trade policy is not an isolated domain but is intricately connected to the EU's broader environmental and human rights agendas. Historically, shortcomings in the EU's human rights and environmental policies, including inadequate legislation and poor enforcement,[144] have hindered the ability of trade regulators to integrate human rights and environmental standards into trade policies effectively.

The CJEU's Advisory Opinion 2/15 establishes that the EU holds exclusive competence to incorporate sustainable development provisions—such as

[140] See the EU-Vietnam FTA, Article 13.6.

[141] Ibid., Article 13.6.

[142] See European Commission, Emissions Trading-Putting a Price on Carbon https://ec.europa.eu/commission/presscorner/detail/en/qanda_21_3542. Accessed 21 July 2024.

[143] See Bacchus (2021).

[144] See Gouritin (2016), pp. 341–382; See also Bronckers and Gruni (2021), pp. 26–30.

commitments to protect labour rights and the environment—into its FTAs, provided these provisions do not undermine the regulatory autonomy of EU member states. Member states retain the authority to advance their environmental and social protection standards and to adopt or amend relevant laws and policies in alignment with their international commitments in these areas.[145] Consequently, national legislation remains crucial for enhancing sustainability standards within EU member states.

Issues related to human rights and trade primarily stem from challenges in implementation. As a core constitutional value, human rights must be upheld by all EU member states,[146] which means that human rights concerns in other countries consistently influence EU trade policy. However, EU leaders have often been cautious about imposing trade sanctions on nations that severely violate human rights, aiming to minimise the economic impact of such sanctions. This issue has been partially addressed by enhancing the European Parliament's oversight over the European Commission,[147] which has strengthened the connection between EU trade policy and human rights, ensuring that trade policies increasingly reflect human rights values.

Historically, the EU has struggled to develop a cohesive environmental policy due to the division of legislative power between the EU and its member states.[148] This fragmentation frequently required political negotiations or dependence on rulings and advisory opinions from the European Court of Justice to enable the European Commission to engage in international environmental negotiations, such as climate talks.[149] Efforts to establish robust European environmental laws and emission reduction regimes were often limited, resulting in compromised targets in order to secure support from member states. These constraints hindered the progress of EU environmental legislation and the integration of environmental provisions into the EU's FTAs.

However, recent years have seen significant progress.[150] The EU adopted the ambitious Green Deal in 2019,[151] leading to new European environmental laws and the establishment of the CBAM. These regulations set specific targets for emission reductions, promote the use of clean energy, support industrial strategies for a clean and circular economy, advocate for sustainable and smart mobility, reform the Common Agricultural Policy (through the Farm to Fork Strategy), and focus on preserving biodiversity.[152] These advancements provide a solid legal foundation for EU trade regulators to embed environmental rules into the EU's Common

[145] CJEU's Advisory Opinion 2/15, ECLI:EU:C:2017:376, 16 May 2017, paras. 164–165 and 167.

[146] About the binding nature of human rights in EU law, please see Douglas-Scott (2011), pp. 646–647.

[147] See Vara (2019), p. 63; See also Andrade (2019), p. 115.

[148] See TFEU, Article 4.

[149] See Macleod et al. (1996), p. 48.

[150] See Fermeglia (2021), pp. 313–323.

[151] European Commission (2019b).

[152] Ibid., p. 3.

Commercial Policy. Consequently, we can anticipate a gradual increase in climate and environmental provisions within the EU's FTAs, transforming initially vague sustainable development goals into explicit legal norms.

9.1.4.2 Enforcement of Sustainability Provisions in EU FTAs

The effectiveness of sustainability provisions hinges on the adherence of parties and the presence of a binding dispute resolution mechanism.[153] EU FTAs include clauses mandating compliance with sustainability standards,[154] with parties expected to uphold their commitments under the sustainable development chapter in good faith. Within the existing EU FTAs, these clauses generally take one of two forms. The first type is non-derogation clauses, which prohibit parties from undermining established sustainability standards. For instance, Article 13.7 of the EU-Korea FTA stipulates that:

1. A party shall not fail to effectively enforce its environmental and labour laws, through a sustained or recurring course of action or inaction, in a manner affecting trade or investment between the parties;
2. A party shall not weaken or reduce the environmental or labour protections afforded in its laws to encourage trade or investment, by waiving or otherwise derogating from, or offering to waive or otherwise derogate from, its laws, regulations or standards, in a manner affecting trade or investment between the parties.

The second type is non-regression clauses, which not only prohibit parties from undermining existing sustainability standards but also mandate that they actively enhance these standards.[155] For instance, Articles 23.2 and 24.2 of the CETA state:

> Each party shall seek to ensure those laws and policies provide for and encourage high levels of labour protection (and environmental protection) and shall strive to continue to improve such laws and policies with the goal of providing high levels of labour protection (or shall strive to continue to improve such laws and policies and their underlying levels of protection).

The second type of clause imposes a more stringent obligation on the parties. These clauses not only seek to maintain existing sustainability standards but also require active efforts to elevate them. While such clauses can effectively drive improvements in sustainability standards, incorporating them into EU FTAs is often challenging. This difficulty reflects the tendency of the EU to compromise on sustainability provisions during trade negotiations in exchange for economic gains.

Furthermore, compromises in sustainability provisions are evident not only in specific clauses but also in the overall structure of treaties and their dispute settlement mechanisms. The latter is crucial as it determines the enforceability of these provisions. Most existing EU FTAs include a dedicated sustainable development

[153] See Bronckers and Gruni (2021), pp. 36–37.

[154] See Lydgate (2019), pp. 37–39.

[155] See Ibid., p. 36.

chapter to address sustainability issues, such as the EU-Korea FTA, the EU-Singapore FTA, the EU-Japan FTA, and the EU-Vietnam FTA.[156] The EU-Korea FTA, ratified in 2015, was the first to feature such a chapter.[157] This chapter, known as Chapter Thirteen, stipulates that disputes arising from sustainability provisions should not be resolved under the FTA's general dispute settlement procedures.[158] Instead, parties are required to attempt a resolution through government consultations first.[159] If these consultations do not yield a satisfactory resolution, a party may then request a panel of experts to review the issues in dispute.[160] Notably, while the panel can offer recommendations, these are not binding,[161] representing a significant departure from the binding nature of the general dispute settlement rules.[162]

Thus, the sustainability provisions in the EU-Korea FTA are relatively weak in terms of enforceability.[163] The effectiveness of these provisions relies on the parties' willingness to implement the panel's recommendations in good faith. This limitation significantly undermines the impact of sustainability provisions in EU FTAs. In contrast, the CETA, which provisionally entered into force in 2017, represents a step forward in strengthening the enforcement of sustainability provisions. In this agreement, the EU has separated labour and environmental provisions into distinct chapters, each with its own dispute settlement rules.[164] This separation allows the EU to enhance the binding effect of these provisions by including additional enforcement measures. For instance, in CETA, the parties agreed that the panel's recommendations on labour disputes are binding.[165] Article 23.11(3) explicitly states that 'the parties understand that the obligations included under this chapter are binding and enforceable through the procedures for the resolution of disputes provided in Article 23.10.' However, this binding clause was not included in the environmental chapter.[166] Despite this, CETA's approach to dispute settlement marks a significant improvement, particularly in reinforcing the enforceability of labour provisions.

However, it is unfortunate that this enhanced legal framework has not been adopted in the EU-Japan FTA (which entered into force in 2019), the EU-Singapore FTA (also effective from 2019), and the EU-Vietnam FTA (which began in 2020). In these agreements, the EU has reverted to the dispute settlement

156 See Titievskaia (2021), p. 2.
157 Ibid., p. 2.
158 The EU-Korea FTA, Article 13.16.
159 Ibid., Article 13.14.
160 Ibid., Article 13.15 (1).
161 Ibid., Article 13.15 (2).
162 See Ibid., Chapter fourteen.
163 See Bronckers and Gruni (2021), p. 29; See also Han (2021), p. 542.
164 See the CETA, Chapters 23 and 24.
165 See Ibid., Article 23.11 (3).
166 See Ibid., Chapter 24.

mechanism used in the EU-Korea FTA. Under these agreements, only the Committee on Trade and Sustainable Development is responsible for overseeing the implementation of a panel's recommendations.[167] This reliance on a less robust dispute settlement process significantly undermines the enforceability of sustainability provisions, representing a notable step backwards. It is particularly disappointing that the EU has failed to reach a consensus on dispute settlement rules in its Comprehensive Agreement on Investment (CAI) negotiations with China.

Examining the EU's trade negotiations over the past decade reveals a concerning trend: the EU has been undermining the enforceability of its sustainability rules while promoting its global values on sustainable development. This approach is like drinking poison to quench thirst—unenforceable provisions ultimately undermine the very sustainability commitments they are meant to uphold. The enforcement of these provisions thus relies entirely on dispute settlement mechanisms within the FTA framework that may not provide timely solutions.

9.2 Mega Free Trade Agreements

9.2.1 Introduction

WTO members are increasingly using FTAs to integrate sustainability into international trade rules.[168] By incorporating as many TSDCs as possible, countries can enhance both the integration and enforcement of these provisions.[169] WTO members could sign a TSDA similar to an FTA to harmonise their trade policies and establish a sustainable development club. This club could then use its common rules to reform the WTO Agreements into a more sustainable framework. I will elaborate on this point in Chap. 10. Before that, this section will outline the sustainability provisions and dispute settlement rules of existing FTAs, which can serve as a basis for drafting the TSDA.[170] Additionally, the findings in this section will highlight necessary improvements to these rules for the future TSDA.

Existing FTAs have made sustainability an essential element of international trade rules, often including provisions related to the environment, labour rights, ecosystem biodiversity, resource conservation, and climate change. These sustainable development rules comprehensively cover the economic, social, and environmental dimensions of sustainability. The primary driver behind the creation of these rules is the collective will of the FTA signatories. Unlike WTO Agreements, FTAs

[167] See the EU-Japan FTA, Article 16.19. See the EU-Singapore FTA, Article 12.15. See also EU-Vietnam FTA, Article 13.17(9).

[168] See Moïse and Rubínová (2021).

[169] See Bronckers and Gruni (2021), pp. 25–26.

[170] For the sake of clarity, I hereby state that I will not discuss any FTA investment rules and procedures in the following sections.

are concluded between a small number of like-minded trading partners, reflecting their shared values—particularly those of the leading parties in the negotiations. Consequently, WTO members that prioritise social and environmental rights can effectively promote the inclusion of sustainability provisions in FTAs they lead.

It is worth noting that the quantity and quality of sustainability provisions included in FTAs vary widely. Differences between the sustainability rules in these agreements lead to a lack of coherence in countries' governance models. This variation is understandable given the different national contexts and specific concerns of trade negotiators about sustainable trade issues, which shape distinct policies in each country.

Climate change measures illustrate these differences well. The EU promotes the Paris Agreement through internalising carbon costs by imposing carbon taxes on products.[171] In contrast, the Biden administration in the US aims to meet the Paris Agreement's emissions reduction targets by phasing out polluting energy sources, such as eliminating the Keystone pipeline and the Nord Stream pipeline.[172] Canada prefers developing and using green production facilities to reduce carbon emissions.[173] While these countries share a commitment to sustainable trade policies, their approaches reflect different aspirations.

Fortunately, their governance models do not differ significantly. To optimise sustainable trade, WTO members must enhance the compatibility of their regulatory frameworks. Adopting uniform legislative standards for sustainability rules would help harmonise national development and trade policies in terms of substantive norms and procedural requirements. This approach can enable FTA expert panels to gradually eliminate unreasonable elements of national sustainable trade policies through dispute settlement proceedings, ultimately forming a cohesive body of sustainable international trade rules.

Section 9.2.2 summarises three categories of sustainability provisions present in contemporary FTAs: sustainability standards provisions, adjustment provisions, and support provisions. The sustainability standards provisions enhance international binding rules for sustainable trade by clarifying and concretising the WTO's commitment to sustainability. Adjustment provisions further strengthen these rules by requiring FTA members to align their domestic laws and regulations with international standards. Support provisions encourage FTA members to adopt internationally binding sustainable trade rules by preventing disguised trade barriers. These provisions reflect the commonality in national legislation regarding sustainable trade rules. It is crucial for WTO members to incorporate these provisions into the TSDA to ensure a cohesive and effective approach to sustainable trade.

---

[171] Recent researches on the EU's carbon tax policy include, among others, the following publications: Zachariadis (2020); Hájek et al. (2019), pp. 110955–110965; Brink et al. (2016), pp. 603–617.

[172] Shear and Davenport (2021); Hunnicutt and Holland (2021).

[173] See Global Affairs Canada (2020), pp. 7–32.

In addition to sustainability provisions, the dispute settlement mechanisms of FTAs are crucial for drafting the TSDA. Existing international trade rules still lack specific implementing measures for the UN SDGs,[174] prompting many major WTO members to use FTA dispute settlement mechanisms to enforce TSDCs. FTAs can effectively ensure the implementation of sustainability rules by incorporating provisions that differ from WTO rules and establishing separate dispute settlement mechanisms. This ensures that disputes arising from an FTA's sustainability provisions are resolved within the framework of that FTA.

Under this framework, parties must determine compliance with sustainability provisions based on the agreed-upon standards within the FTA. As a result, a party cannot seek a favourable decision through the WTO dispute settlement mechanism. An agreement on FTA jurisdiction would thus eliminate the impact of WTO rules on the implementation of FTA sustainability provisions. To ensure the enforcement of sustainability rules, a contracting party should explicitly incorporate these rules into its FTA and include strict dispute settlement provisions to ensure their enforceability. These dispute settlement provisions should guarantee two key aspects: first, that the FTA's dispute settlement mechanism has exclusive jurisdiction over the relevant disputes, and second, that the decisions of this mechanism are legally binding. This approach would solidify the enforceability of sustainability rules within FTAs.

However, many FTAs' dispute settlement mechanisms do not meet the necessary requirements for effective enforcement of sustainability provisions. There are three basic models of FTA jurisdictional clauses, which illustrate this variability. First, some FTAs mandate compulsory jurisdiction, requiring parties to refer their disputes to an institution established by the treaty. Second, certain FTAs allow for forum shopping, enabling the settlement of disputes either in the FTA forum or in the WTO forum at the discretion of the complaining party. Finally, other FTAs contain exclusive forum clauses, stipulating that once a matter is brought before either forum, the chosen procedure must be followed, excluding any other forum.[175]

Additionally, FTAs often apply different dispute settlement rules to different chapters,[176] further complicating their jurisdictional frameworks. This section will describe the dispute settlement rules in four key FTAs: the RCEP, the CETA (as an example of EU FTAs), the USMCA, and the CPTPP, highlighting their approaches to sustainability provisions and the complexities of their jurisdictional rules.

On the other hand, many FTAs' sustainability provisions are non-mandatory.[177] In particular, labour provisions in most FTAs merely encourage parties to respect fundamental labour rights without imposing binding obligations for their implementation. This lack of enforceability complicates efforts to ensure adherence to sustainability standards. Furthermore, even when FTAs include specific dispute

[174] See Bronckers and Gruni (2021), pp. 25–26; See also van't Wout (2022), pp. 81–98.

[175] Kwak and Marceau (2006), p. 476.

[176] See Bronckers and Gruni (2021), p. 36.

[177] See Ibid., pp. 37–38.

settlement mechanisms for sustainability issues, these mechanisms often dilute the binding effect of decisions or recommendations made by FTA panels. As a result, the enforcement of sustainability provisions in FTAs is significantly weakened.

Several factors may influence a country's decision to adopt lenient dispute settlement rules during FTA negotiations. These factors can include concerns about the effectiveness of FTA dispute resolution, the significance of the issues at stake, and the value placed on multilateral mechanisms. This section will delve into the dispute settlement rules related to sustainable development in major FTAs, examining how these rules affect the implementation of sustainability provisions and identifying the challenges they present.

9.2.2 *Sustainability Provisions in FTAs*

9.2.2.1 Sustainability Standards Provisions

The sustainability standards provisions in FTAs can be broadly categorised into two main types. The first category includes exception clauses. These clauses permit contracting parties to deviate from trade obligations in order to implement measures that address social rights and environmental protection. Exception clauses represent one of the most traditional forms of sustainability standards in FTAs, with their origins in GATT Article XX.

The second category of sustainability standards provisions in FTAs encompasses social and environmental clauses, which are prevalent in most agreements. These provisions outline a range of sustainability criteria, addressing issues such as environmental protection, labour rights, and public health. Concerning public health, contemporary FTAs typically include provisions related to SPS measures and intellectual property rights pertaining to medicines.

SPS measures are designed to detect and manage pests, diseases, viruses, and invasive species in imported goods, preventing them from entering the country.[178] On the other hand, intellectual property provisions related to medicines enable a contracting party to permit domestic pharmaceutical companies to produce patented drugs during a public health crisis.[179] These provisions, which allow for the production of generic medicines, play a crucial role in saving lives, particularly in developing countries.[180]

Social and environmental provisions are essential components of sustainability standards clauses in FTAs. They ensure that a country's trade policy aligns with its environmental and human rights policies, thereby supporting the social and

[178] See Peel (2010), p. 245; See also SPS Agreement, Annex A (1), paras. (a)–(d).

[179] See Pusceddu (2018), p. 1072.

[180] The access to medicines in developing countries depends mostly on the ability of these countries to produce, export and import generic medicines. Please see Jurua (2017), p. 102.

environmental dimensions of sustainable development. These provisions reflect the sustainability standards to which contracting parties are committed. However, not all FTAs incorporate these crucial provisions. For instance, the RCEP, which is the world's largest FTA, covering 30% of global GDP, 27% of global trade in goods, over 18% of trade in services, and 19% of FDI outflows,[181] notably lacks chapters on labour rights and environmental protection.[182] The absence of such provisions in this significant agreement could impede progress towards sustainable trade. While current FTAs include social and environmental provisions to varying extents, their presence in these agreements sets the stage for future negotiations to enhance sustainability standards. The following sections will examine the current state of these provisions in FTA texts.

9.2.2.1.1 Exception Clauses

Exception clauses are fundamental to sustainability standards in FTAs. Given that WTO rules already offer well-developed exception clauses and a robust body of jurisprudence, countries often incorporate these clauses directly into their FTA texts. Typically, FTAs include a general exception clause modelled after GATT Article XX.[183] Additionally, FTAs embed these exceptions indirectly by acknowledging WTO rules within their chapters on SPS measures,[184] technical regulations,[185] and intellectual property rights.[186] For instance, Article 5.4 of the RCEP affirms each party's rights and obligations under the SPS Agreement, while Article 1.5 of the CETA confirms the parties' adherence to the WTO Agreements and other relevant treaties. Similarly, Article 5.2(c) of the CETA states that the chapter aims to further the implementation of the SPS Agreement. These provisions collectively demonstrate that contracting parties recognise their right to implement regulatory measures that protect social rights and the environment in line with existing WTO rules. Therefore, FTAs indirectly incorporate the exception clauses of the WTO Agreements.

---

[181] Cali (2020).

[182] See Hsieh (2022), p. 146.

[183] For example, see Article 17.12 of the RCEP, Article 2.15 of the EU-Korea FTA, Article 28.3 of the CETA, Article 32.1 of the USMCA, and Article 29.1 of the CPTPP.

[184] Examples are Chapter 5 of the CETA, Chapter 9 of the USMCA, Chapter 7 of the CPTPP, Chapter 5 of the RCEP, and Chapter 5 of the EU-Korea FTA.

[185] Examples are Chapter 4 of the CETA, Chapter 11 of the USMCA, Chapter 8 of the CPTPP, Chapter 6 of the RCEP, and Chapter 4 of the EU-Korea FTA. See also the following recent research on the integration of TBT provisions in FTAs: Romanchyshyna (2020); Piermartini and Buedetta (2009).

[186] Examples are Chapter 20 of the CETA, Chapter 20 of the USMCA, Chapter 18 of the CPTPP, Chapter 11 of the RCEP, and Chapter 10 of the EU-Korea FTA; See also Johnson (2021); See also Antons and Hilty (2015).

9.2.2.1.2 Environmental Provisions

Environmental provisions in FTAs encompass a range of measures aimed at promoting green trade. One key aspect is the facilitation of trade in environmentally friendly products.[187] These provisions require contracting parties to ease the import and export of green goods, enabling countries to acquire the necessary equipment and technology for energy conservation and GHG emissions reduction. For example, Article 24.9(2) of the CETA mandates that the contracting parties work to eliminate barriers to trade and investment in goods and services essential for climate change mitigation and adaptation.[188] This provision aligns with the perspectives of WTO panels in cases such as *Canada-Renewable Energy* and *India-Solar Cells*, which have determined that domestic content requirements can act as trade barriers and violate WTO rules.[189]

Secondly, FTAs often include provisions focused on the conservation of natural resources, addressing areas such as forestry, water resources, and biodiversity.[190] Some agreements specifically require compliance with international conventions such as the Convention on Biological Diversity and its protocols, as well as the Convention on International Trade in Endangered Species of Wild Fauna and Flora, to safeguard ecosystem diversity.[191]

In addition to the EU FTAs mentioned earlier,[192] the USMCA and the CPTPP also contain provisions aimed at conserving natural resources.[193] Notably, the USMCA includes detailed measures for fishery resource conservation. Article 24.40 of the USMCA requires member countries to control, reduce, and ultimately

[187] See EU FTA's provisions cited in section 7.2.3 (b). See also Article 24.24 'Environmental Goods and Services' of the USMCA and Article 20.18 'Environmental Goods and Services' of the CPTPP; See also Grübler (2021), p. 10.

[188] Article 24.9 (2) of the CETA.

[189] WTO Panel Report, Canada-Certain Measures Affecting the Renewable Energy Generation Sector (Canada-Renewable Energy), WT/DS412/R (WT/DS426/R), 19 December 2012; WTO Appellate Body Report, Canada-Certain Measures Affecting the Renewable Energy Generation Section (Canada-Renewable Energy), WT/DS412/AB/R, adopted on 24 May 2013; WTO Panel Report, India-Certain Measures Relating to Solar Cells and Solar Modules, WT/DS456/R, 24 February 2016; WTO Appellate Body Report, India-Solar Cells, WT/DS456/AB/R, adopted on 14 October 2016.

[190] For example, Articles 24.10 and 24.11 of the CETA; See also Hoggard (1993), pp. 635–681.

[191] See, for example, Article 24.10 of the CETA.

[192] See Sect. 9.2.3.2.

[193] See Article 24.15 'Trade and Biodiversity' of the USMCA, Article 24.17 'Marine Wild Capture Fisheries' of the USMCA, Article 24.18 'Sustainable Fisheries Management' of the USMCA, Article 24.19 'Conservation of Marine Species' of the USMCA, Article 24.20 'Fisheries Subsidies' of the USMCA, Article 24.21 'Illegal, Unreported, and Unregulated (IUU) Fishing' of the USMCA, Article 24.22 Conservation and Trade of the USMCA, and Article 24.23 'Sustainable Forest Management and Trade' of the USMCA; See also Article 20.13 'Trade and Biodiversity' of the CPTPP, Article 20.16 'Marine Capture Fisheries' of the CPTPP, and Article 20.17 'Conservation and Trade' of the CPTPP.

eliminate all fisheries subsidies to prevent overfishing and support the recovery of depleted fish stocks. Furthermore, Article 24.21 of the USMCA mandates that member states prevent illegal, unreported, and unregulated fishing, thereby establishing sustainability standards for the production processes of fishery products. This provision supports the restriction of products derived from illegally caught fish.

The USMCA's approach reflects a consensus among the US, Canada, and Mexico on fisheries and related trade policies. This is particularly significant given the historical disputes the US has had with other countries over fisheries, which were unsuccessful in the WTO dispute settlement mechanism.[194] Article 24.21 of the USMCA is designed to mitigate such conflicts among the US, Canada, and Mexico, promoting more effective and cooperative fisheries management.

Thirdly, many FTAs include provisions for environmental impact assessments. For instance, Article 24.7 of the USMCA seeks to enhance public scrutiny of the sustainability of trade and investment projects by increasing their transparency. The primary objective is to prevent or mitigate any adverse environmental impacts resulting from these projects.

Fourth, some FTAs include specific climate-related provisions. These provisions primarily aim to ensure that parties fulfil their emission reduction commitments under international agreements like the Paris Agreement. For instance, the EU-Vietnam FTA incorporates environmental provisions to support the parties' obligations under the Paris Agreement.[195]

US-led FTAs, such as the USMCA and the CPTPP, also contain an extensive range of environmental provisions. Although the US withdrew from the CPTPP negotiations, it significantly influenced the drafting of its environmental chapter.[196] The USMCA, for example, includes provisions not only for meeting the usual environmental standards but also for protecting the ozone layer,[197] safeguarding the marine environment from ship pollution,[198] addressing air pollution,[199] reducing and preventing marine litter,[200] ensuring corporate social responsibility,[201] and preventing invasive alien species.[202] Many of these provisions, such as those related to ozone layer protection,[203] marine pollution control,[204] corporate social

[194] See Trachtman (2017), pp. 273–275.

[195] Article 13.6 of the EU-Vietnam FTA.

[196] To know more details about the background and negotiations, please see Foreign Trade Information System: background and negotiations http://www.sice.oas.org/tpd/tpp/tpp_e.asp. Accessed 21 July 2024.

[197] Article 24.9 of the USMCA; See also Malhotra and Babu (2020), p. 285.

[198] Article 24.10 of the USMCA; See also Malhotra and Babu (2020), p. 284.

[199] Article 24.11 of the USMCA; See also Malhotra and Babu (2020), pp. 283–284.

[200] Article 24.12 of the USMCA; See also Malhotra and Babu (2020), p. 284.

[201] Article 24.13 of the USMCA; See also Malhotra and Babu (2020), p. 284.

[202] Article 24.16 of the USMCA.

[203] Article 20.5 of the CPTPP; See also O'Toole (2021), p. 640.

[204] Article 20.6 of the CPTPP; See also O'Toole (2021), p. 640.

responsibility,[205] and invasive species management,[206] are also present in the CPTPP. While these clauses might seem to replicate obligations found in multilateral environmental agreements, they play a crucial role in clarifying the regulatory autonomy of the contracting parties. The specific language in these provisions allows domestic regulators to implement various measures, such as enacting domestic laws and regulations to mitigate the environmental impacts of trade and enhance public information disclosure.

9.2.2.1.3 Labour Provisions

FTAs' labour provisions primarily reinforce the obligation of parties to adhere to ILO conventions.[207] However, some agreements go beyond these basic commitments to highlight workers' rights in specific areas. A few notable examples include: First, Article 23.3(3) of the CETA introduces the precautionary principle to safeguard workers' health.[208] This provision mandates that policymakers must consider all potential risks when developing laws and regulations, and take proactive measures to prevent worker exposure to these risks. It emphasises that the absence of complete scientific certainty should not be used as an excuse to delay cost-effective protective measures. This principle is particularly characteristic of the EU, exemplified by the *EC-Asbestos* case,[209] where the WTO DSB upheld the EU's policy based on this principle.[210]

Second, the USMCA incorporates several distinctive labour provisions. Article 23.2 of the USMCA stipulates that trade in goods is only permissible if the production of those goods complies with the labour standards outlined in the USMCA's labour chapter. This provision is notably rare in contemporary international trade agreements, as it explicitly ties labour rights compliance to trade eligibility. Additionally, Article 23.7 of the USMCA introduces a relatively uncommon labour rights provision aimed at protecting workers from violence. This provision is designed to combat forced labour by addressing both overt and covert forms of coercion. In the workplace, coercion does not always manifest as physical violence; threats and intimidation also play a significant role. These more subtle forms of coercion can be particularly insidious. By explicitly prohibiting such practices, the USMCA ensures a higher standard of protection for workers' freedom

[205] Article 20.10 of the CPTPP.

[206] Article 20.14 of the CPTPP.

[207] See Harrison et al. (2019), p. 638.

[208] Article 23.3 (3) of the CETA.

[209] In EC-Asbestos, the panel and AB supported the EC's claim. See WTO Panel Report, EC-Asbestos, WT/DS135/R, 18 September 2000; WTO Appellate Body Report, EC-Asbestos, WT/DS135/AB/R, adopted on 5 April 2001.

[210] See Horn and Weiler (2003), p. 32.

and fundamental rights. In this regard, the USMCA stands out as more progressive compared to many other FTAs in its approach to labour rights protection.

9.2.2.1.4 Sanitary and Phytosanitary Measures

Regarding sanitary and phytosanitary measures, most FTAs adhere to the principles outlined in the SPS Agreement, emphasising the obligation for countries to align with international standards.[211] However, existing international standards still have significant gaps. Countries must implement more scientifically grounded quarantine rules to mitigate the adverse effects of quarantine measures on legitimate trade.

One of the fundamental concepts designed to improve these quarantine measures is the zoning rule.[212] According to WTO regulations, this concept is explained as follows:

> A pest-free or disease-free area may surround, be surrounded by or be adjacent to an area-either in part of a country or in a geographical area comprising part or all of several countries-where a specific pest or disease is known to occur, provided that area control measures are taken, such as establishing protection, surveillance and buffer zones to limit or eliminate the pest or disease in question.[213]

Zoning rules assist trade regulators in importing countries by distinguishing between infected and non-infected areas, thereby minimising the impact of quarantine measures on agricultural trade. According to WTO jurisprudence, stringent trade restrictions, such as import bans, can only be imposed on products from infected areas.[214] Modern Free Trade Agreements (FTAs) offer more comprehensive zoning provisions. For instance, Annex 5-B of the CETA outlines detailed rules for determining infected zones, which further refine the WTO's zoning regulations and reduce disputes. These enhanced zoning provisions in FTAs are designed to help contracting parties implement quarantine measures more effectively while mitigating their adverse effects on trade.

---

[211] See, for example, Rudloff and Simons (2004), p. 2; See also Fukunaga (2021), p. 253.

[212] Examples are Article 5.5 (1) (a) of the CETA and Understanding on Regulation to Zoning, Urban Planning and Environmental Protection of the EU-Korea FTA, Article 7.7(1) of the CPTPP, Article 9.8 of the USMCA, and Article 5.6 of the RCEP. While Article 5.6 of the RCEP does not use the conception of zoning, it recognises a similar conception, that is, the concept of regional conditions.

[213] See The SPS Agreement, the note to Article 6 of Annex A.

[214] See WTO Panel Report, India-Agricultural Products, WT/DS430/R, adopted on 19 June 2015, para. 7.689; See also WTO Panel Report, United States-Measures Affecting the Importation of Animals, Meat and other Animal Products from Argentina, WT/DS447/R, 24 July 2015, para. 7.657.

9.2.2.1.5 Intellectual Property Provisions on Medicinal Products

Currently, intellectual property provisions related to medicinal products vary across FTAs. For instance, the RCEP agreement primarily reaffirms the Doha Declaration on the TRIPS Agreement and Public Health without incorporating additional provisions such as compulsory licensing directly into the FTA text.[215] However, the Doha Declaration's paragraph 5 acknowledges WTO members' rights to grant compulsory licenses and provides the flexibility to determine the conditions under which these licenses are issued.

The EU places significant emphasis on safeguarding intellectual property rights in the pharmaceutical sector. In the EU-Japan FTA, EU-Singapore FTA, and EU-Vietnam FTA,[216] the EU generally acknowledges the parties' rights under the Declaration on the TRIPS Agreement and Public Health. Furthermore, the EU includes specific provisions in the EU-Korea FTA and the CETA to ensure access to generic medicines.[217]

Article 1 of Annex 2-D of the EU-Korea FTA stipulates that contracting parties should facilitate access to high-quality patented generic pharmaceutical products and medical devices.[218] In the CETA, the EU has introduced a specific protection clause, Article 20.27 (9), designed to counteract the undue reduction of patent terms.[219] This provision mirrors the EU's domestic supplementary protection certificate (SPC) system,[220] which aims to compensate for the loss of effective patent protection caused by the time required for testing, clinical trials, and marketing authorisation. The SPC system provides pharmaceutical industries with incentives to innovate by extending patent protection.[221]

Additionally, certain exemptions to SPC rights are included, such as the Bolar exemption and the SPC manufacturing waiver.[222] The Bolar exemption permits the manufacture of generics during the patent and SPC term for clinical trial purposes,[223] while the SPC manufacturing waiver allows for the manufacture of generics under specific conditions during the SPC term for export or storage.[224] These provisions in the EU-Korea FTA and CETA ensure that parties can maintain access to generic medicines from the EU market during public health crises.

---

[215] Articles 11.4 &11.8 of the RCEP and Article 20.3 of the CETA.

[216] See Article 14.34 of the EU Japan FTA, Article 10.3 of the EU-Singapore FTA, and Article 12.39 of the EU-Vietnam FTA.

[217] See Articles 10.34 and Article 1 of the Annex 2-D of the EU-Korea FTA.

[218] Article 1 of the Annex 2-D of the EU-Korea FTA.

[219] Article 20.27 (9) of the CETA.

[220] See Vidal-Quadras (2019), pp. 971–1005.

[221] See European Commission (2020a), p. 7.

[222] Ibid., p. 8.

[223] Ibid., p. 8.

[224] Ibid., p. 8.

The USMCA and CPTPP offer more extensive exemptions for pharmaceutical intellectual property rights compared to the EU FTAs. Both agreements explicitly recognise the Bolar exception, which includes the regulatory review exception.[225] Additionally, the CPTPP incorporates a compulsory licensing clause for medicinal products, outlining the specific conditions under which it can be invoked.[226] Consequently, CPTPP member countries can more readily grant licenses to domestic pharmaceutical manufacturers for producing generic medicines and authorise their sale within their borders during public health crises.

9.2.2.2 Adjustment Provisions

Countries can use adjustment provisions to integrate sustainability standards and related procedural rules into FTAs. Based on my observations, there are three main types of adjustment provisions commonly found in contemporary FTAs: compatibility provisions, audit provisions, and localisation provisions.

9.2.2.2.1 Compatibility Provisions

Compatibility provisions require FTA members to harmonise their regulatory measures. Specifically, a contracting party must align its domestic regulations with those of another party that sets higher sustainability standards.[227]

The scope of application for compatibility provisions is quite limited. Many countries, if not all, either entirely or partially exclude these provisions from their FTAs. In the USMCA, compatibility requirements apply to SPS measures, TBT measures, and air quality monitoring methodologies.[228] The CPTPP, on the other hand, restricts compatibility requirements to domestic regulatory measures for pharmaceutical products and medical devices.[229] In the RCEP, such requirements are found exclusively in the chapter on technical regulations, specifically relating to TBT measures.[230]

Early EU FTAs also lack compatibility provisions. For instance, the EU-Korea FTA explicitly states that each contracting party is not required to harmonise its labour and environmental standards.[231] This means that if one party adopts higher standards in these areas, the other party is not obligated to comply with them. This

[225] Articles 20.39 and 20.47 of the USMCA and Articles 18.40 and 18.49 of the CPTPP.

[226] Article 18.6 (1) of the CPTPP.

[227] See, for example, Article 21.5 of the CETA.

[228] Article 9.7 (2) of the USMCA, Articles 11.5 (4) and Article 12.A.4 (4) and 12.A.4 (5), Article 24.11 (4) of the USMCA.

[229] Articles 7 and 16 of Annex 8-C of the CPTPP, Article 7 of Annex 8-E of the CPTPP.

[230] Article 6.5 of the RCEP.

[231] Article 13.1 (3) of the EU-Korea FTA.

principle was highlighted in a recent panel report on the EU-Korea labour dispute, where the panel acknowledged that Korea was not required to align its minimum wage rates, maximum working hours, or job security procedures with EU standards, as outlined in Article 13.1 (3) of the EU-Korea FTA.[232]

Only the CETA adopts a comprehensive approach to compatibility provisions. It contains a dedicated chapter on regulatory cooperation that systematically requires signatories to harmonise their domestic regulatory measures.[233] Consequently, compatibility requirements under the CETA extend to all types of domestic regulations, including SPS measures, TBT measures, trade in services, pharmaceutical patent regulations, labour rights, and environmental standards.[234]

9.2.2.2.2 Audit Provisions

Audit provisions enable FTA members to review other members' domestic regulatory measures. An auditing party has the right to conduct on-site investigations in a mutually agreed upon manner.[235] After the investigation, the auditing party provides a report, and the audited party has the opportunity to respond to the findings.[236] This arrangement is valuable for verifying the equivalence of regulatory measures and for providing evidence that can be used in dispute resolution.

Most FTAs establish standards and procedural rules for auditing within the SPS measures chapter.[237] The USMCA and the CPTPP are notable exceptions, extending these auditing provisions to domestic regulatory measures beyond SPS measures.[238] However, these audit provisions are typically not mandatory; a contracting party can only audit another party's domestic regulations if the audited party agrees to it voluntarily,[239] and the auditing methods must also be approved by the audited party.[240] For example, auditing of environmental measures is conducted on a voluntary basis.[241] This approach respects the autonomy of each party's domestic regulations, though it may somewhat limit the effectiveness of the audit provisions.

[232] Panel of Experts Proceeding Constituted Under Article 13.15 of the EU-Korea Free Trade Agreement, Report of the Panel of Experts, 20 January 2021, para. 81.

[233] According to Article 29.15 of the CETA, compatibility of domestic regulatory measures is subject to dispute settlement.

[234] Article 21.1 of the CETA.

[235] See, for example, Article 9.10 of the USMCA.

[236] Article 5.7 of the CETA, Article 5.8 of the RCEP, Article 7.10 of the CPTPP, Article 6.11 of the EU-Korea FTA.

[237] Article 9.10 of the USMCA, Article 7.10 of the CPTPP, Articles 5.7(3) (b) and 5.8 of the CETA, and Article 5.8 of the RCEP.

[238] See Article 24.14 of the USMCA and Article 20.11 of the CPTPP.

[239] Article 24.14 of the USMCA and Article 20.11 of the CPTPP.

[240] See, for example, Article 9.10 (4) of the USMCA.

[241] See Article 24.14 of the USMCA and Article 20.11 of the CPTPP.

Nevertheless, this balance between respecting regulatory autonomy and ensuring audit effectiveness remains one of the most effective approaches available.

9.2.2.2.3 Localisation Provisions

Localisation provisions are similar to the provisions of WTO accession protocols. They oblige a country to incorporate certain sustainability standards into its domestic laws and regulations before joining an FTA.[242] Currently, these clauses serve to harmonise the existing domestic laws and regulations of FTA members. In this regard, countries do not do their best to ensure synergies among their domestic rules.

Most FTAs do not include them. One notable example of an agreement that includes such provisions is the USMCA. Annex 23-A of the USMCA includes localisation rules that mandate Mexico to ensure workers' representation in collective bargaining. Under this clause, Mexico must amend its labour laws, establish trade unions and independent labour tribunals, and enact procedural laws for collective bargaining before it can fully participate in the FTA.

9.2.2.3 Support Provisions

Existing support provisions in FTAs include protection level provisions, notification provisions,[243] communication provisions,[244] and trade remedies provisions.[245] Although these provisions do not establish sustainability standards directly, they play a crucial role in facilitating the effective implementation of sustainable trade rules.[246] As such, support provisions are vital for integrating sustainability into FTAs.

9.2.2.3.1 Protection Level Provisions

Protection level provisions require parties to uphold their sustainability standards in both legislation and enforcement while also promoting sustainability legislation to some extent. This type of provision is found in EU FTAs, the USMCA, and the CPTPP. Based on their purpose, these provisions can be categorised into non-derogation and non-regression provisions.[247] Non-derogation provisions are more common, requiring parties not to violate the sustainability standards they

[242] See, for example, Annex 23-A of the USMCA.

[243] Articles 9.13 (5) and 9.13 (13) of the USMCA and Articles 7.11 (8)-(10) of the CPTPP.

[244] Article 9.13 (5) of the USMCA and Articles 7.13 (6)-(9) and 7.14 of the CPTPP.

[245] Article 20.46 of the USMCA and Article 18.48 of the CPTPP.

[246] Segger (2006), pp. 337–338.

[247] See Lydgate (2019), pp. 37–39.

have committed to in FTAs,[248] thereby enhancing the enforceability of these provisions. A notable example, in addition to Article 13.7 of the EU-Korea FTA, is Article 23.4 of the USMCA, which states:

> It is inappropriate to encourage trade or investment by weakening or reducing the protections afforded in each party's labour laws. Accordingly, no party shall waive or otherwise derogate from, or offer to waive or otherwise derogate from, its statutes or regulations. . . .

Similarly, Article 19.4 of the CPTPP labour chapter contains the same non-derogation rules as Article 23.4 of the USMCA. This wording is also found in Article 13.7(2) of the EU-Korea FTA. However, in the EU-Korea FTA, the non-derogation clause applies to both labour and environmental provisions,[249] whereas in the USMCA and CPTPP, it applies only to labour provisions.[250] For environmental provisions, these two FTAs adopt a non-regression provision similar to those in Articles 23.2 and 24.2 of the CETA.[251] These non-regression clauses require each party to strive to ensure that its environmental and labour laws provide for and encourage high levels of protection and continue to improve these standards.[252]

In addition to complying with existing sustainability standards, non-regression provisions impose additional obligations on parties to legislate for higher standards. Therefore, agreeing to non-regression provisions implies a commitment to further advancements. Conversely, neither the USMCA nor the CPTPP mandates that parties further upgrade existing labour standards, thus limiting their obligations in this area.

9.2.2.3.2 Notification and Communication Provisions

Notification and communication provisions require parties to implement sustainability rules according to established procedures. These provisions mandate that one party promptly inform the other of any sustainability measures it has taken and maintain ongoing communication. For example, Article 5.11 (1) of the CETA requires a party to notify the other without undue delay of significant changes to pest or disease status, such as those listed in Annex 5-B. Similarly, Article 9.5 of the USMCA outlines requirements for contact points, ensuring continuous communication to maintain the transparency of domestic laws and regulations. These provisions obligate contracting parties to establish competent authorities and contact points, facilitating the prompt exchange of essential information. This includes details about significant sanitary or phytosanitary risks related to the export of goods, substantial

[248] See Ibid.

[249] See Article 13.7 of the EU-Korea FTA.

[250] See Article 23.4 of the USMCA and Article 19.4 of the CPTPP.

[251] See Articles 24.3 of the USMCA and Article 20.3 (3) of the CPTPP.

[252] See Article 24.3 of the USMCA and Article 20.3(3) of the CPTPP.

changes in the status of regional pests or diseases, and new scientific findings that affect regulatory responses concerning food safety, pests, or diseases.[253]

Notification and communication provisions are valuable because they enhance transparency in the implementation of sustainability rules by contracting parties and prevent the misuse of these measures as trade barriers. Many trade policymakers are concerned that other countries' regulators might exploit sustainability provisions to restrict their exports.[254] This concern has been a major factor in opposition to including sustainability standards in international trade rules.[255]

Moreover, notification and communication provisions can assist contracting parties in defending their trade interests through dispute settlement mechanisms. These rules make it easier for parties to identify and challenge disguised trade barriers. If a domestic regulatory measure is adopted and implemented in a manner clearly contrary to the procedural rules outlined in the notification and communication provisions, the importing party's sustainability measure will be in violation of the FTA.

Furthermore, notification and communication provisions can help exporting countries minimise economic losses caused by disguised trade barriers. If an importing country misuses sustainability rules to restrict imports but still adheres to the procedures set out in the notification and communication provisions, it must complete the notification process and respond to the exporting party's comments.[256] This requirement grants the exporting country a transition period, providing the opportunity to seek new overseas markets.

9.2.2.3.3 Trade Remedies Provisions

Trade remedy provisions are essential for implementing sustainability measures within FTAs. For example, if a party to an FTA sells domestically produced green products, such as solar panels and crystalline silicon, to another party at unfairly low prices (dumping), this can create unfair competition for the domestic environmental industries in that country.[257] In such cases, trade remedies are a vital tool for supporting clean energy industries. By applying trade remedies, an FTA party can limit the influx of dumped green products, thereby preventing severe commercial losses or even bankruptcy for its environmental industries.

Furthermore, to promote the development and innovation of green industries, it is crucial to prevent a monopoly on global green products by a few producers.

---

[253] See, for example, Article 9.13 (12) of the USMCA.

[254] To know environmental requirements' impact on market access, please see OECD (2005).

[255] Kim (2012).

[256] See, for example, Article 9.11 (8)-(10) of the USMCA.

[257] For example, see the China-US PV products dispute. To know more details, please see WTO Panel Report, United States-Safeguard Measure on Imports of Crystalline Silicon Photovoltaic Products, WT/DS562/R, circulated 2 September 2021.

Therefore, FTA parties adopt trade remedy measures not only to protect their environmental industries but also to foster the healthy development of these industries on a global scale.

From a legal perspective, it is well documented that parties are entitled to use remedies to protect their environmental industries from dumped products. In the 2021 WTO case, *US-Safeguard Measure on PV Products*, the WTO panel upheld the US safeguard measure on Chinese photovoltaic products dumped into its domestic market.[258] This decision by the WTO panel clearly indicated that trade remedies can play an essential role in promoting the healthy development of environmental industries. In light of this, trade remedy provisions in FTAs will also play a significant role in promoting sustainable development.

To maximise the effectiveness of trade remedy provisions, FTA parties should enhance the existing WTO trade remedy provisions within their agreements. Currently, most FTAs simply adopt the trade remedy provisions of the WTO Agreements, such as Article 3.4 of the CETA. However, some FTAs include specific provisions that go beyond the WTO framework. The USMCA, for instance, contains notable provisions for preventing duty evasion.[259]

Section C of Chapter Ten, 'Trade Remedies,' of the USMCA includes provisions aimed at enhancing cooperation on trade remedy laws to prevent duty evasion. Article 10.6 explicitly requires the contracting parties to strengthen and expand their customs and trade enforcement efforts to address duty evasion. Article 10.7 (2) further mandates that the parties share information about imports, exports, and transit transactions to combat duty evasion effectively. These provisions help prevent an exporting country from evading trade remedy measures by routing exports through third countries. Consequently, they enable the parties to use trade remedy measures more effectively, ensuring the healthy development of their environmental industries.

Another notable type of provision found in the RCEP and CPTPP is the transitional safeguard measure.[260] This provision specifies that during the transitional period outlined in the FTA, parties are prohibited from using safeguard measures such as tariff-rate quotas and quantitative restrictions.[261] Instead, a contracting party may suspend future reductions in the customs duty rate for the originating good or increase the rate of customs duty on the originating good, but only up to a level that does not exceed the lesser of the most-favoured-nation applied rate of customs duty.[262]

These provisions significantly mitigate the impact of remedy measures on the exporting party, potentially favouring trade within the free trade area. From an international trade perspective, they may support the development of industries in

[258] See Ibid., para. 8.1.

[259] See Marcoux and Bjorklund (2022), pp. 217–218.

[260] See Article 7.2 of the RCEP and Article 6.3 of the CPTPP.

[261] See Article 7.2 (2) of the RCEP and Article 6.3 (2) of the CPTPP.

[262] See Articles 7.2 (1) (a) and 7.2 (1) (b) and Article 6.3 (2) (a) and 6.3 (2) (b).

member countries, particularly those in developing nations. However, they also limit the ability of these countries to use trade remedies to protect their environmental industries. Thus, these provisions can be seen as a double-edged sword, and their effects on the environmental industries of developing country members in the RCEP and CPTPP will need to be evaluated over time.

9.2.3 Dispute Settlement Mechanisms for Sustainable Development Chapters in FTAs

9.2.3.1 The RCEP

The RCEP introduces the first FTA dispute settlement mechanism model applicable to sustainable development-related disputes. Compared to existing FTAs, the RCEP stands out due to its distinctive dispute settlement mechanism. Typically, FTA parties design dispute settlement mechanisms to ensure the implementation of specific commitments, often incorporating comprehensive elements such as consultation mechanisms, expert panels, and arbitration processes.[263] These mechanisms are usually given jurisdiction over specific provisions, including those related to sustainability.[264] However, the RCEP deviates from this approach. RCEP parties have opted not to include such detailed dispute settlement rules within the agreement itself. Instead, they rely on the WTO dispute settlement mechanism, which is the most notable feature of the RCEP's approach to dispute resolution. In the following section, I will examine RCEP's dispute settlement rules specifically concerning disputes arising from sustainability provisions.

First, it is important to note that the RCEP does not include dedicated sustainable development chapters, such as those addressing labour or the environment.[265] Consequently, the RCEP's trade-related sustainability provisions are found only within some traditional FTA chapters, including Chapter 4 'Customs Procedures and Trade Facilitation,[266]' Chapter 5 'Sanitary and Phytosanitary Measures,' Chapter 6 'Standards, Technical Regulations, and Conformity Assessment Procedures,' Chapter 7 'Trade Remedies,' and Chapter 11 'Intellectual Property.'

Interestingly, the RCEP dispute settlement mechanism currently has jurisdiction only over Chapters 4 and 11. This limitation is because the parties to the RCEP have stipulated that the dispute settlement mechanism does not apply to the other mentioned chapters. Article 19.3 (1) of the RCEP states that the general dispute

[263] See Bercero (2006), pp. 389–399.

[264] See Bronckers and Gruni (2021), p. 36.

[265] See Hsieh (2022), p. 146.

[266] This chapter does not contain sustainability provisions explicitly. However, the contracting parties can use this chapter's provisions, particularly those regarding trade facilitation measures, to ensure the provision of the necessities of life, food, medicines, and green products. See Article 4.13 of the RCEP.

settlement rules apply to disputes regarding the interpretation and application of the agreement unless otherwise specified. However, the parties have explicitly excluded the jurisdiction of the RCEP dispute settlement mechanism over disputes arising from the provisions of Chapters 5, 6, and 7.[267]

Therefore, if parties have a dispute concerning the sustainability provisions in these chapters, they can only seek resolution through consultations as outlined in the respective chapters. Should these consultations fail to resolve the disputes, the WTO dispute settlement mechanism remains the sole international judicial institution available to them.

It is important to note that the provision for the non-application of dispute settlement rules in Chapters 5 and 6 is an interim arrangement. The parties are required to review these rules two years after the RCEP enters into force.[268] Specifically, the rules regarding dispute settlement in these chapters differ. Regarding the dispute settlement rules applicable to SPS issues, Article 5.17 (2) states that:

> Such a review shall be completed within three years from the date of entry into force of this agreement. After which those parties that are ready shall proceed to apply chapter 19 (dispute settlement) to this chapter as between one another. A party that is not ready will consult other parties and may apply chapter 19 (dispute settlement) to this chapter when it becomes party to any future free trade agreement or economic agreement in which it takes on a similar obligation.

This provision is considered a soft rule, as it does not mandate all parties to apply it. Instead, it encourages parties to begin using the RCEP dispute settlement rules to resolve disputes arising from the SPS provisions within three years of the agreement's entry into force. Countries that are not ready to do so may opt to use the WTO dispute settlement mechanism for resolving disputes, in accordance with Article 19.5 (Choice of Forum) of the RCEP. Regarding the dispute settlement rules applicable to TBT (Technical Barriers to Trade) issues, Article 6.14 of the RCEP specifies only that the review of the non-application of dispute settlement rules must be completed within three years from the date the agreement enters into force. However, it does not clarify whether the RCEP's dispute settlement rules should apply to Chapter 6 following this review.

In contrast to the interim nature of the dispute settlement provisions in Chapters 5 and 6, Article 7.16 of the RCEP regarding Chapter 7 appears to be a permanent rule. The contracting parties have decided to review this provision only five years after the RCEP enters into force.[269] This review is mandated by the RCEP's general review clause, Article 20.8, which requires a comprehensive review of the agreement every five years. As a result, Article 7.16 does not impose a specific obligation on parties to use the RCEP's dispute settlement rules for trade remedies disputes. From a practical standpoint, there seems to be a lack of strong intent among the parties to apply the RCEP's dispute settlement rules to trade remedies issues. It is likely that

[267] See Articles 5.17, 6.14, and 7.16 of the RCEP.

[268] See Articles 5.17(2) and 6.14 of the RCEP.

[269] See Articles 7.16 and 20.8 of the RCEP.

RCEP countries will prefer to address trade remedies disputes through the WTO's DSB.

Additionally, while RCEP parties have agreed to use the RCEP dispute settlement rules for disputes related to trade in goods and services, including those under Chapters 4 and 11, they still have the option to refer such disputes to the WTO DSB. Article 19.5 of the RCEP allows the complaining party to choose its forum for dispute resolution. Therefore, disputes arising from these chapters can be addressed either through the RCEP dispute settlement mechanism or other fora, such as the WTO DSB. However, once the complaining party selects a forum, it cannot subsequently bring the dispute before a different forum.[270]

It is evident that the RCEP's dispute settlement rules have diminished the enforcement of sustainability provisions, contrary to the goal of using FTAs to enhance the enforcement of sustainability standards. Additionally, the RCEP model significantly underuse the potential of FTA dispute settlement mechanisms. Today's FTA dispute settlement procedures are generally more efficient and less time-consuming compared to the WTO dispute settlement process.[271] Although not all scholars agree that a shorter timeframe for dispute resolution is preferable,[272] RCEP members could indeed benefit from leveraging the RCEP's dispute settlement mechanism to expedite the resolution of disputes.

From a pragmatic perspective, there is some justification for the RCEP parties adopting the current dispute settlement rules. Compared to the WTO dispute settlement mechanism, FTA dispute settlement mechanisms often incorporate more bilateral elements.[273] Countries with smaller economies are frequently hesitant to accept such bilateral mechanisms, preferring instead the multilateral framework provided by the DSB. The WTO's rules—such as the involvement of third parties, the negative consensus rule,[274] and the appeal stage[275]—are designed to ensure fairness in the dispute settlement process and effective enforcement of panel reports.

Given that most RCEP members are smaller economies, this membership structure may inhibit the RCEP from supporting a robust dispute settlement mechanism. It can be argued that these smaller economies are not yet fully prepared to implement a more comprehensive FTA dispute settlement system.

[270] See Article 19.5 (1) of the RCEP.

[271] Kennedy (2011), pp. 221–253.

[272] There are different points of view on the timeframe issues. Scholars (like Joost Pauwelyn) suggest shortening the timeframe of dispute settlement. Please see Pauwelyn (2020). Ernst-Ulrich Petersmann and others hold different ideas. See Petersmann (2019), p. 509; Roessler (2017), pp. 110–111; Furculita (2020), p. 97.

[273] Furculita notes that there are no co-complainants, third parties, and appeal stage in the CETA and EUJEPA proceedings. See Furculita (2020), p. 96.

[274] A ruling by a WTO panel or the Appellate Body will stand unless there is a consensus among WTO members against the decision. See Barfield (2001), p. 407.

[275] Although the AB is blocked, the WTO still has a multi-party interim appeal arrangement (MPIA) as appeal institution.

9.2.3.2 The CETA: As an Example of EU FTAs

The EU has utilised its FTA dispute settlement mechanisms to enhance the enforcement of sustainability rules by minimising WTO jurisdiction over sustainability provisions in its agreements. This approach represents the second model of FTA dispute settlement mechanisms for addressing sustainable development-related disputes. While this model is similar to the first in that it allows the complaining party to choose the WTO DSB for dispute resolution through the FTA's universal jurisdiction clause,[276] it differs in that it explicitly excludes sustainability provisions from this universal jurisdiction clause.[277] As a result, disputes arising from the FTA's sustainability provisions can only be resolved through the FTA's own dispute settlement mechanism, creating a closed system independent of the WTO dispute settlement system.[278] However, the EU has not fully implemented this model. This will be further clarified by examining the rules formerly applied under NAFTA. Before that, I will use the EU's CETA as an example to illustrate this approach in detail.

CETA includes numerous provisions related to trade and sustainable development. These provisions are integrated into traditional FTA chapters, such as Chapter 3 'Trade Remedies,' Chapter 4 'Technical Barriers to Trade,' Chapter 5 'Sanitary and Phytosanitary Measures,' Chapter 6 'Customs and Trade Facilitation,' Chapter 7 'Subsidies,' and Chapter 20 'Intellectual Property.' Additionally, CETA includes three dedicated chapters on sustainable development: Chapter 22, 'Trade and Sustainable Development,' and two sub-chapters, Chapter 23, 'Trade and Labour,' and Chapter 24, 'Trade and Environment.'[279]

The CETA outlines three distinct jurisdictional models for handling sustainable development-related disputes. First, both the trade remedies and subsidies chapters explicitly exclude the CETA dispute settlement mechanism from jurisdiction over disputes in these areas.[280] As a result, disputes arising from these chapters can be referred to the WTO DSB rather than being resolved through the CETA dispute settlement mechanism.

Secondly, the CETA's universal jurisdiction clause applies to Chapters 4, 5, 6, and 20.[281] This clause permits the complaining party to choose between the CETA dispute settlement mechanism and the WTO DSB for resolving disputes.[282] Consequently, disputes arising from these chapters can be addressed through either

[276] See Article 29.3 of the CETA.

[277] See Bronckers and Gruni (2021), p. 36.

[278] See Articles 23.11(1) and 24.16 (1) of the CETA.

[279] In terms of legislative structure, Chapters twenty-three and twenty-four do not subordinate to Chapter twenty-two. However, these two chapters refine the content of Chapter twenty-two. In this sense, I consider that these two chapters are Chapter twenty-two's sub-chapters.

[280] See Articles 3.7 and 7.9 of the CETA.

[281] See Article 29.2 of the CETA.

[282] See Article 29.3 of the CETA.

the CETA's dispute settlement system or the WTO DSB, depending on the preference of the complaining party. If the party prefers the rules and procedures of CETA, it can opt to use the CETA dispute settlement mechanism instead.

Finally, the CETA dispute settlement mechanism holds exclusive jurisdiction over disputes arising from the provisions in the labour and environment chapters.[283] This means that CETA parties cannot refer these disputes to the WTO DSB.[284] By doing so, the EU and Canada, as parties to CETA, can enhance the enforcement of labour rights and environmental standards within the FTA framework.

However, these jurisdictional rules do not fully exploit the potential of FTAs to enforce sustainability provisions. Under these rules, the CETA dispute settlement mechanism has exclusive jurisdiction only over labour and environmental provisions. CETA parties could have extended this exclusivity to a broader range of sustainability provisions. For comparison, the former NAFTA framework granted its FTA dispute settlement mechanism exclusive jurisdiction over labour and environmental provisions as well as SPS measures and TBT measures.[285]

It is important to note that NAFTA's jurisdictional rules allowed the responding party to veto the WTO's jurisdiction over NAFTA sustainability provisions.[286] In contrast, other FTAs typically grant only the complaining party the authority to select the forum for resolving sustainable development-related disputes. Articles 2005 (1) and (2) of NAFTA stipulate that if the parties cannot agree on a single forum, disputes arising from sustainability provisions must be referred to the NAFTA dispute settlement mechanism.[287] This arrangement effectively ensures that NAFTA's dispute settlement mechanism holds exclusive jurisdiction over sustainability issues. Had the EU and Canada adopted similar rules from NAFTA into CETA, the CETA dispute settlement mechanism could have played a more significant role in strengthening the enforcement of sustainability provisions.

Furthermore, although the CETA dispute settlement mechanism has exclusive jurisdiction over disputes related to both labour rights and environmental provisions, there is a notable difference in the binding effect of the decisions made by CETA panels in these areas. When comparing the dispute settlement rules for environmental and labour issues, it becomes evident that CETA panels' decisions on environmental disputes are more binding than those on labour disputes. Article 24.15 (11) of CETA specifies that:

> If the final report of the panel of experts determines that a party has not conformed with its obligations under this chapter, the parties shall engage in discussions and shall endeavour,

---

[283] See Articles 23.11 and 24.16 of the CETA.

[284] Other EU FTAs have the same arrangement. For example, the Chapter thirteen 'Trade and Sustainable Development' of the EU-Korea FTA, which includes labour rights and environmental standards is solely subject to the dispute settlement rules provided by this chapter itself. See Articles 13.14 and 13.15 of the EU-Korea FTA.

[285] See Articles 2005 (3) and 2005 (4) of the NAFTA.

[286] See Article 2005 (2) of the NAFTA.

[287] Articles 2005 (1) and 2005 (2) of the NAFTA.

within three months of the delivery of the final report, to identify an appropriate measure or, if appropriate, to decide upon a mutually satisfactory action plan.

Although a similar provision exists in the labour chapter, Article 23.11 (4) notably reduces the binding effect of the decisions made by CETA panels in labour disputes. This provision permits consultations on the dispute resolution procedures outlined in the labour chapter, with the potential for amending these rules if a party is dissatisfied with the current procedures. The responding party could also leverage this review clause to pressure the complaining party, potentially deterring it from referring a dispute regarding labour provisions to the CETA dispute settlement mechanism.

Once again, the bilateral nature of the CETA dispute settlement mechanism may be responsible for the issues mentioned above. It is reasonable to assume that Canada's reluctance to adopt more binding CETA dispute settlement rules stems from its greater confidence in multilateral mechanisms, such as the WTO DSB.

9.2.3.3 The USMCA

The USMCA, signed by the United States, Canada, and Mexico, introduces a third model for FTA dispute settlement mechanisms applicable to sustainable development-related disputes.[288] This approach emphasises cooperative mechanisms, such as consultations, to prevent or resolve disputes. While parties have the option to establish a panel under the rules and procedures of Chapter 31, 'Dispute Settlement,[289]' these general dispute settlement rules apply to the USMCA's sustainable development-related chapters, except Chapter 10, 'Trade Remedies.'[290] The relevant chapters include Chapter 7, 'Customs Administration and Trade Facilitation,' Chapter 9, 'Sanitary and Phytosanitary Measures,' Chapter 11, 'Technical Barriers to Trade,' Chapter 20, 'Intellectual Property,' Chapter 23, 'Labour,' and Chapter 24, 'Environment.' According to Article 31.3 of the USMCA, the complaining party may choose to refer disputes arising from these chapters either to the USMCA dispute settlement mechanism or to the WTO DSB. The dispute settlement provisions for the trade remedies chapter will be discussed separately.

It is important to note that the USMCA's dispute settlement rules place specific conditions on the application of its universal dispute settlement procedures to the labour and environment chapters. Specifically, a party must first complete a consultation process before invoking the USMCA's universal dispute settlement rules.[291] This consultation process involves multiple high-level consultative sessions.[292]

[288] See Hart (2021), p. 21.

[289] See Ibid., p. 2.

[290] See Article 10.12 of the USMCA.

[291] See Hart (2021), p. 2; See Article 23.17 of the USMCA and Articles 24.29–24.31 of the USMCA.

[292] Ibid.

Consequently, parties must navigate a lengthy and intensive consultation process before they can refer disputes to the USMCA's dispute settlement mechanism.

The labour consultation process under Article 23.17 requires that parties first attempt to resolve issues arising from the labour chapter through cooperation and dialogue. During this phase, the parties must make every effort to reach a mutually satisfactory resolution.[293] If the initial consultations fail to resolve the issue, a party may then request that the relevant ministers or their designees convene to address the matter.[294] These ministers are also tasked with resolving the issue and may utilise procedures,[295] such as good offices, conciliation, or mediation.[296] If the issue remains unresolved following ministerial consultations, the parties can then establish a panel of experts and refer the dispute to the USMCA's dispute settlement mechanism under Chapter 31, 'Dispute Settlement'.[297]

The environmental consultation process closely mirrors the labour consultation process but is more extensive, involving three levels of consultation. Initially, the parties must attempt to resolve issues arising from the environment chapter through a general consultation process.[298] If this initial consultation does not resolve the matter, the parties must then move to a senior representative consultation,[299] where representatives from the environmental committees work to address the issue.[300] Should the issue remain unresolved after this second level, a ministerial consultation may be convened as the final step in the environmental consultation process.[301] If the matter is still not resolved, the parties may then establish a panel of experts and refer the dispute to the USMCA's dispute settlement mechanism under Chapter 31, 'Dispute Settlement'.[302]

USMCA parties may still refer disputes to the WTO's DSB after completing the consultation procedures.[303] However, the consultation provisions in the labour and environmental chapters require parties to first resolve issues arising from these chapters through amicable and cooperative means within the USMCA framework. This rule ensures that parties cannot bypass the USMCA's labour and environmental provisions by directly referring disputes to the WTO DSB. In this regard, the consultation rules significantly strengthen the implementation of the USMCA's human rights and environmental standards.

---

[293] Article 23.17 (1) of the USMCA.

[294] Article 23.17 (6) of the USMCA.

[295] Ibid.

[296] Ibid.

[297] Article 23.17 (8) of the USMCA.

[298] See Article 24.29 (1) of the USMCA.

[299] See Article 24.30 (1) of the USMCA.

[300] See Article 24.30 (2) of the USMCA.

[301] See Article 24.31 of the USMCA.

[302] See Article 24.32 (1) of the USMCA.

[303] See Hart (2021), p. 21.

Finally, I will briefly outline the rules applicable to Chapter 10 'Trade Remedies.' The USMCA excludes the trade remedies chapter from its general dispute settlement procedures. According to Article 10.5 (3) of the USMCA, no party may resort to dispute settlement under the agreement for issues arising under this section or Annex 10-A, which pertains to antidumping and countervailing duty practices. Consequently, disputes related to trade remedies must be addressed exclusively through the WTO's DSB.

9.2.3.4 The CPTPP

The CPTPP's dispute settlement mechanism for sustainable development-related disputes closely mirrors that of the USMCA. There are several key similarities between the two agreements. First, disputes related to trade remedies under the CPTPP can only be referred to the WTO's DSB.[304] Second, the CPTPP's dispute settlement mechanism covers sustainable development chapters except for the trade remedies chapter.[305] Disputes arising from these sustainable development chapters can be addressed through either the CPTPP's dispute settlement mechanism or the WTO DSB.[306]

Furthermore, CPTPP parties must complete a consultation process before referring labour rights and environmental disputes to the CPTPP's dispute settlement mechanism.[307] These consultation procedures are consistent with those outlined in the USMCA. Additionally, similar to the USMCA, the CPTPP allows the complaining party to bring disputes related to labour rights and the environment to the WTO DSB.[308] However, once a panel has been established under the CPTPP's rules, the complaining party forfeits the option to refer the dispute to the WTO DSB.[309]

The CPTPP differs from the USMCA in two notable ways. First, Article 28.3 (2) of the CPTPP explicitly prohibits parties from initiating non-violation nullification or impairment claims concerning intellectual property rights. This provision is quite rare, as most contemporary FTA dispute settlement rules typically cover such disputes, thereby addressing gaps in the WTO's jurisdiction. Under Article 64 (2) of the TRIPS Agreement, WTO members have been temporarily barred from bringing non-violation nullification or impairment complaints before the WTO DSB, and this moratorium remains in effect.[310] Nevertheless, the CPTPP does not entirely rule out

[304] See Article 6.8 (3) of the CPTPP.

[305] See Article 28.3 of the CPTPP.

[306] See Article 28.4 of the CPTPP.

[307] See Article 19.15 of the CPTPP.

[308] See Articles 19.15 (12) and 20.23 (1) of the CPTPP.

[309] See Article 28.4 (2) of the CPTPP.

[310] See WTO General Council (2019).

the possibility of addressing these disputes through its dispute settlement mechanism in the future. Article 28.3 (2) of the CPTPP states that:

> No later than six months after the effective date that members of the WTO have the right to initiate non-violation nullification or impairment complaints under Article 64 of the TRIPS Agreement, the parties shall consider whether to amend paragraph 1 (c) to include Chapter 18 (Intellectual Property).

Second, the CPTPP includes Chapter 23, 'Development,' which underscores the parties' commitment to fostering an open trade and investment environment aimed at enhancing welfare, reducing poverty, raising living standards, and creating new employment opportunities to support development.[311] This chapter is distinctive because although the UN 2030 Agenda for Sustainable Development incorporates economic development alongside social rights and environmental concerns as core dimensions of sustainable development, no other FTAs contain a dedicated development chapter. The CPTPP's inclusion of such a chapter is a notable innovation.

However, the chapter's impact is considerably diminished by two key provisions. First, Article 23.8 establishes that other chapters take precedence over Chapter 23, 'Development.' Second, the CPTPP includes a clause that excludes Chapter 23 from the dispute settlement mechanism outlined in Chapter 28, 'Dispute Settlement,' meaning that disputes arising under Chapter 23 cannot be resolved through this mechanism.[312]

It is worth noting that the WTO Agreements do not include a chapter equivalent to Chapter 23 of the CPTPP. Consequently, CPTPP parties will have no means to resolve disputes arising from this chapter. This situation is quite problematic, as it reveals that the CPTPP's development chapter is essentially non-binding.

9.3 Conclusion

The EU has a long-standing tradition of shaping its trade policy around sustainable development principles.[313] The 2017 European Development Policy, which includes the PEDP, is grounded in normative values and the interests of its citizens.[314] These normative values encompass democracy, the rule of law, human rights, equality, solidarity, the principles of the United Nations Charter, and international law.[315] The interests of EU citizens involve their well-being both within Europe and beyond its borders. The EU is dedicated to incorporating the PEDP into its domestic laws and foreign policies to address a range of non-trade concerns, including human rights,

[311] Article 23.1 (1) of the CPTPP.

[312] Article 23.9 of the CPTPP.

[313] See Ziegler (1996), pp. 220–242; See also Reid (2016), pp. 353–361.

[314] See Sect. 9.1.2.1.

[315] See Sect. 9.1.2.1.1.

economic sanctions, labour practices, agriculture, fisheries, and trade in goods and services.[316]

Additionally, the EU has adopted three strategies to achieve the SDGs: embracing higher international standards, taking a comprehensive approach to sustainable development, and employing principled pragmatism.[317] These strategies guide EU policymakers in refining the EU's international trade policy in alignment with the PEDP. WTO members can refer to these EU strategies when formulating the common policy for the sustainable development club.

The EU enacted the Green Deal and launched its domestic emission reduction mechanisms based on the PEDP.[318] The Green Deal mandates revising legislation in areas such as industrial production, energy, mobility, chemical products, ecosystems and biodiversity, and agriculture.[319] By developing specific policies and laws in these areas, the EU aims to meet the emission reduction targets committed to in the Paris Agreement and mitigate biodiversity damage caused by trade.[320] The domestic emission reduction mechanisms, including the ETS and the CBAM, aim to achieve a climate-neutral EU by 2050, ensuring net-zero carbon dioxide emissions within EU member states.

Additionally, the EU has integrated various sustainability provisions into its FTAs to implement its development policies. These provisions protect labour rights, safeguard the environment, and address climate change issues.[321] However, gaps remain in EU trade policy. Developments in the EU's environmental and human rights laws will influence how trade regulators can enhance these policies.[322] Although the EU has significantly improved its domestic legislation on sustainable development in recent years, it still needs to enact more comprehensive sustainable development rules and ensure their enforcement.[323]

Furthermore, the EU often compromises sustainability provisions in trade negotiations to gain economic benefits.[324] This compromise is reflected in the dispute settlement mechanisms of EU FTAs, which often lack binding effect.[325] As WTO members negotiate the TSDA, they could incorporate these EU rules and policies into the agreement and make necessary improvements.

Similar to the European development policy, the sustainability provisions and dispute settlement mechanisms in FTAs can serve as valuable references for drafting

[316] See Sect. 9.1.2.1.2.

[317] See Sect. 9.1.2.2.

[318] See Sects. 9.1.2.3 and 9.1.2.4.

[319] European Commission (2019b), p. 3.

[320] See Ibid., pp. 20–21.

[321] See Sect. 9.1.3.

[322] See Sect. 9.1.4.1.

[323] Ibid.

[324] See Sect. 9.1.4.2.

[325] Ibid.

the TSDA.[326] Although differences in sustainability rules across various FTAs result in inconsistencies in countries' governance models,[327] WTO members generally share similar governance frameworks.[328] FTA sustainability provisions can be categorised into sustainability standards provisions, adjustment provisions, and support provisions.[329] These provisions reflect the common elements of national legislation on sustainable trade rules, suggesting that WTO members can use them to harmonise their trade rules and draft the TSDA.[330]

It is worth noting that FTAs incorporate these sustainability provisions at varying levels.[331] Section 9.2.2 provides a detailed introduction to the various sustainability provisions included in FTAs and concludes that WTO members should include as many of these provisions as possible in the TSDA.

WTO members must also improve their FTA dispute settlement mechanisms to enforce TSDCs. Section 9.2.3 elaborates on the dispute settlement mechanisms of today's most prominent FTAs and highlights several key findings.

First, the RCEP's dispute settlement rules have weakened the enforcement of sustainability provisions, significantly undermining the capacity of FTA dispute settlement mechanisms.[332]

Secondly, the CETA's dispute settlement mechanism has exclusive jurisdiction over labour rights and environmental provisions. However, it does not maximise the use of the FTA to ensure the enforcement of sustainability provisions. Additionally, CETA panels' decisions in environmental disputes are more legally binding than those in labour disputes.[333]

Third, the USMCA's dispute settlement mechanism does not provide exclusive jurisdiction over sustainable development disputes. It only requires that USMCA parties complete consultation procedures before referring labour rights-related and environmental disputes to the WTO's DSB. While these consultation rules somewhat enhance the implementation of the USMCA's human rights and environmental provisions, they are not as effective as CETA rules.[334]

Fourth, the CPTPP's dispute settlement mechanism applicable to sustainable development-related disputes is nearly identical to the USMCA's. The CPTPP differs from the USMCA in two ways: Article 28.3 (2) of the CPTPP stipulates that parties do not have the right to initiate non-violation nullification or impairment complaints with respect to intellectual property rights. Additionally, the CPTPP contains a unique Chapter 23, 'Development,' which affirms the parties'

[326] See Sect. 9.2.1.

[327] Ibid.

[328] Ibid.

[329] See Sect. 9.2.2.

[330] Ibid.

[331] Ibid.

[332] See Sect. 9.2.3.1.

[333] See Sect. 9.2.3.2.

[334] See Sect. 9.2.3.3.

commitment to promoting and strengthening an open trade and investment environment aimed at improving welfare, reducing poverty, raising living standards, and creating new employment opportunities in support of development. However, this development chapter is entirely non-binding.[335]

In view of these findings, WTO members should incorporate a unified dispute settlement mechanism into the TSDA to enhance the enforcement of sustainability provisions. This mechanism should ensure that FTAs' dispute settlement systems have exclusive jurisdiction over sustainable development disputes until WTO reforms are completed.

Jurisprudence

WTO Panel Report, EC-Asbestos, WT/DS135/R, 18 September 2000
WTO Appellate Body Report, EC-Asbestos, WT/DS135/AB/R, adopted on 5 April 2001
WTO Panel Report, United States-Continued Suspension of Obligations in the EC-Hormones Dispute, WT/DS320/R, adopted on 14 November 2008
WTO Panel Report, India-Agricultural Products, WT/DS430/R, adopted on 19 June 2015
WTO Panel Report, United States-Safeguard Measure on Imports of Crystalline Silicon Photovoltaic Products, WT/DS562/R, circulated 2 September 2021
CJEU's Advisory Opinion 2/15, ECLI:EU:C:2017:376, 16 May 2017
Arbitral Panel Established Pursuant to Chapter Twenty of the CAFTA-DR, Report on Guatemala-Issues Relating to the Obligations Under Article 16.2.1 (a) of the CAFTA-DR, 14 June 2017
Panel of Experts Proceeding Constituted Under Article 13.15 of the EU-Korea Free Trade Agreement, Report of the Panel of Experts, 20 January 2021

References

Bibliography

Andrade PG (2019) The role of the European Parliament in the adoption of non-legally binding agreements with third countries. In: Vara JS, Sánchez-Tabernero SR (eds) The democratisation of EU international relations through EU law. Routledge, New York, pp 115–131
Anke C-P et al (2020) Coal phase-outs and carbon prices: interactions between EU emission trading and national carbon mitigation policies. Energy Policy 144:111647–111657
Antons C, Hilty RM (eds) (2015) Intellectual property and free trade agreements in the Asia-Pacific region. Springer, Berlin and Heidelberg

[335] See Sect. 9.2.3.4.

Bacchus J (2021) Legal issues with the European Carbon Border Adjustment Mechanism. CATO Briefing Paper. 9 August 2021. Available via https://www.cato.org/sites/cato.org/files/2021-08/briefing-paper-125.pdf. Accessed 21 July 2024

Barfield CE (2001) Free trade, sovereignty, democracy: the future of the World Trade Organization. Chic J Int Law 2(2):403–415

Bercero IG (2006) Dispute settlement in European Union free trade agreements: lessons learned? In: Bartels L, Ortino F (eds) Regional trade agreements and the WTO legal system. OUP, New York, pp 383–406

Brink C et al (2016) Carbon pricing in the EU: evaluation of different EU ETS reform options. Energy Policy 97:603–617

Bronckers M, Gruni G (2021) Retooling the sustainability standards in EU free trade agreements. J Int Econ Law 24(1):25–51

Cali M (2020) The significance of the regional economic partnership agreement. Brookings. 20 November 2020. Available via https://www.brookings.edu/blog/future-development/2020/11/20/the-significance-of-the-regional-economic-partnership-agreement/. Accessed 21 July 2024

Cottier T, Caplazi A (2011) Labour standards and world trade law: interfacing legitimate concerns. Institute of European and International Economic Law. Available via https://www.humanrights.ch/cms/upload/pdf/000303_cottier_caplazi.pdf. Accessed 21 July 2024

De Sadeleer N (2014) EU environmental law and the internal market. OUP, New York

De Schutter O (2015) Trade in the service of sustainable development: linking trade to labour rights and environmental standards. Hart, Oxford and Portland, Oregon

Douglas-Scott S (2011) The European Union and human rights after the Treaty of Lisbon. Hum Rights Law Rev 11(4):645–682

Durán GM, Morgera E (2012) Environmental integration in EU's external relations: beyond multilateral dimensions. Hart, Oxford and Portland, Oregon

Fermeglia M (2021) Recent developments in EU environmental policy and legislation. J Eur Environ Plan Law 18(3):313–323

Fleurke F (2016) EU climate law and human rights: new prospects for judicial environmental activism? In: Delimatsis P (ed) Research handbook on climate change and trade law. Edward Elgar, Cheltenham and Northampton, pp 375–393

Fukunaga Y (2021) The SPS Chapter under the TPP agreement and its implications. In: Huerta-Goldman JA, Gantz DA (eds) The comprehensive and progressive trans-pacific partnership: analysis and commentary. CUP, Cambridge and New York, pp 253–276

Furculita C (2020) FTA dispute settlement mechanisms: alternative fora for trade disputes-the case of CETA and EUJEPA. In: Weiß W, Furculita C (eds) European yearbook of international economic law-special issue: global politics and EU trade policy facing the challenges to a multilateral approach. Springer, Cham, pp 89–111

Giovanella F (2017) Copyright and information privacy: conflicting rights in balance. Edward Elgar, Cheltenham, Northampton

Gouritin A (2016) EU environmental law, international environmental law, and human rights law. Brill and Nijhoff, Leiden

Grübler J et al (2021) Wanted! Free trade agreements in the service of environmental and climate protection. The Vienna Institute for International Economic Studies. January 2021. Available via https://wiiw.ac.at/wanted-free-trade-agreements-in-the-service-of-environmental-and-climate-protection-dlp-5589.pdf. Accessed 21 July 2024

Hájek M et al (2019) Analysis of carbon tax efficiency in energy industries of selected EU countries. Energy Policy 134:110955–110965

Han JS (2021) The EU-Korea labour dispute: a critical analysis of the EU's approach. Eur Foreign Aff Rev 26(4):531–552

Harrison J et al (2019) Labour standards provisions in EU free trade agreements: reflections on the European Commission's reform agenda. World Trade Rev 18(4):635–657

Hoggard KC (1993) Is this land really our land?: Impacts of free trade agreements on U.-S. environmental protection. Environ Law 23(2):635–681
Horn H, Weiler JHH (2003) EC-Asbestos: European communities-measures affecting asbestos and asbestos-containing products. World Trade Rev 2(S1):14–40
Hsieh PL (2022) The roadmap to the ASEAN-EU FTA in the post-pandemic era. Leg Issues Econ Integr 49(2):125–148
Hunnicutt T, Holland S (2021) White House Says Biden Believes Nord Stream 2 Bad Deal for Europe. Reuters. London. 16 February 2021. Available via https://www.reuters.com/article/us-nordstream-usa-biden-idUSKBN2AG29A. Accessed 21 July 2024
Jackson JH (1996) Appraising the launch and functioning of the WTO. German Yearb Int Law 39: 20–41
Johnson P (2021) Intellectual property, free trade agreements and the United Kingdom: the continuing influence of European Union law. Edward Elgar, Cheltenham and Northampton
Jurua M (2017) Access to drugs at risk: securing access to medicines for least developed countries. Africa Dev 42(1):101–120
Kaufmann C (2007) Globalisation and labour rights: the conflict between core labour rights and international economic law. Hart, Oxford
Keith KJ (2015) John Dugard Lecture-2015 The International Rule of Law. Leiden J Int Law 28: 403–417
Kennedy M (2011) Why are WTO panels taking longer? And what can be done about it? J World Trade 45(1):221–253
Kilian B, Elgström O (2010) Still a Green Leader? The European Union's role in international climate negotiations. Coop Confl 45(3):255–273
Kim M (2012) Disguised protectionism and linkages to the GATT/WTO. World Polit 64(3): 426–475
Knox JH (2015) Human rights, environmental protection, and the sustainable development goals. Wash Int Law J 24(3):517–536
Kwak K, Marceau G (2006) Overlaps and conflicts of jurisdiction between the World Trade Organization and regional trade agreements. In: Bartels L, Ortino F (eds) Regional trade agreements and the WTO legal system. OUP, New York, pp 465–524
Lydgate E (2019) Environmental standards and regulation. In: West J, Vergels E (eds) Ensuring a post-Brexit level playing field. European Policy Centre, pp 33–46
Lydgate E (2021) The Carbon Border Adjustment Trilemma. UK Trade Policy Observatory Briefing Paper. May 2021. Available via http://sro.sussex.ac.uk/id/eprint/102016/1/BP56.pdf. Accessed 21 July 2024
Macleod I et al (1996) The external relations of the European communities. OUP, Oxford and New York
Malhotra A, Babu RR (2020) United States-Mexico-Canada Agreement (USMCA) (formerly the North American Free Trade Agreement (NAFTA)). Yearb Int Environ Law 31(1):283–287
Marcoux J-M, Bjorklund AK (2022) Duty evasion in free trade agreements: norm emergence and implications. J World Trade 56(2):215–238
Moïse E, Rubínová S (2021) Sustainability impact assessments of free trade agreements. OECD Trade Policy Paper. Available via https://www.oecd-ilibrary.org/docserver/65b1a07e-en.pdf?expires=1651878253&id=id&accname=guest&checksum=F0A9BE2720D3E3F0BFDED3F85449207D. Accessed 21 July 2024
Murry J (2001) Transnational labour regulation: the ILO and EC compared. Kluwer Law International, The Hague
O'Toole C (2021) TPP Trade and Environment Chapter. In: Huerta-Goldman JA, Gantz DA (eds) The comprehensive and progressive trans-pacific partnership: analysis and commentary. CUP, Cambridge and New York, pp 633–649
Panagariya A (2019) Free trade and prosperity. OUP, New York

Pauwelyn J (2020) The real rot in the system: delays are making WTO Dispute Settlement Irrelevant, especially during a pandemic. 27 March 2020. Available via https://www.linkedin.com/pulse/real-rot-system-delays-making-wto-dispute-settlement-during-pauwelyn/. Accessed 21 July 2024

Peel J (2010) Science and risk regulation in international law. CUP, New York

Petersmann E-U (2019) How should WTO Members react to their WTO crises? World Trade Rev 18(3):503–525

Piermartini R, Buedetta M (2009) A mapping of regional rules on technical barriers to trade. In: Estevadeordal A et al (eds) Regional rules in the global trading system. CUP, Cambridge, pp 250–315

Pusceddu P (2018) Assessing access to medicines in preferential trade agreements: from the trans-pacific partnership to the comprehensive and progressive agreement for trans-pacific partnership. Int Rev Intellect Prop Compet Law 49:1048–1079

Ramcharan BG (2015) The law, policy and politics of the UN Human Rights Council. Brill and Nijhoff, Leiden and Boston

Reid E (2016) EU climate law and the WTO. In: Delimatsis P (ed) Research handbook on climate change and trade law. Edward Elgar, Cheltenham and Northampton, pp 352–374

Roessler F (2017) Dispute settlement in the WTO from a deliberately designed to a spontaneously grown order. In: Elsig M, Hoekman B (eds) Assessing the World Trade Organization. CUP, Cambridge, pp 99–119

Romanchyshyna I (2020) Tackling technical barriers to trade in EU 'New Generation' FTAs: an example of open or conflicting regionalism? In: Weiß W, Furculita C (eds) European yearbook of international economic law-special issue: global politics and EU trade policy facing the challenges to a multilateral approach. Springer, Cham, pp 41–64

Rudloff B, Simons J (2004) Comparing EU free trade agreements: sanitary and phytosanitary standards. European Centre for Development Policy Management. July 2004. Available via https://ecdpm.org/application/files/2016/5547/2863/IB-6B-Comparing-EU-Free-Trade-Agreements-Sanitary-Phytosanitary-2004.pdf. Accessed 21 July 2024

Segger M-CC (2006) Sustainable development in regional trade agreements. In: Bartels L, Ortino F (eds) Regional trade agreements and the WTO legal system. OUP, New York, pp 313–340

Shear MD, Davenport C (2021) Biden to Cancel Keystone XL Pipeline in Inauguration Day Executive Order. The New York Times. New York. 18 January 2021. Available via https://www.nytimes.com/2021/01/18/us/politics/biden-cancel-keystone-xl-pipeline.html. Accessed 21 July 2024

Trachtman JP (2017) WTO trade and environment jurisprudence: avoiding environmental catastrophe. Harv Int Law J 58(2):273–309

van't Wout D (2022) The enforceability of the trade and sustainable development chapters of the European Union's free trade agreements. Asian Europe J 20:81–98

Vara JS (2019) The European Parliament in the conclusion of international agreements Post-Lisbon: entrenched between values and prerogatives. In: Vara JS, Sánchez-Tabernero SR (eds) The democratisation of EU international relations through EU law. Routledge, New York, pp 63–81

Vidal-Quadras M (2019) Analysis of EU Regulation 2019/933 on the SPC Manufacturing Waiver Exception. IIC-Int Rev Intellect Prop Compet Law 50:971–1005

Zachariadis T et al (eds) (2020) Economic instruments for a low-carbon future. Edward Elgar, Cheltenham and Northampton

Ziegler AR (1996) Trade and environmental law in the European community. Oxford Clarendon Press, Oxford

Ziegler AR (2022) Common Commercial Policy (CCP). In: Laurence G (ed) Encyclopedia of EU law. OUP, Oxford. May 2022. Available via https://opil.ouplaw.com/display/10.1093/law-oeeul/law-oeeul-e143?rskey=kymXRp&result=17&prd=OEEUL. Accessed 21 July 2024

Documents

Bellmann C, Tipping AV (2015) The Role of Trade and Trade Policy in Advancing the 2030 Development Agenda. International Development Policy. Available via https://journals.openedition.org/poldev/2149#tocto1n1. Accessed 21 July 2024

Euromines (2020) Public Consultation Inception Impact Assessment: Amendment of the EU Emissions Trading System (ETS). November 2020. Available via http://www.euromines.org/files/euromines_statement_regarding_the_ets_amendment_0.pdf. Accessed 21 July 2024

European Commission (2007) Furthering Human Rights and Democracy Across the Globe. Available via https://www.ab.gov.tr/files/ardb/evt/1_avrupa_birligi/1_6_raporlar/1_3_diger/commision_report_furthering_human_rights_and_democracy_across_the_globe.pdf. Accessed 21 July 2024

European Commission (2010a) Communication from the Commission to the Council, the European Parliament, the Europe Economic and Social Committee and the Committee of the Regions: The EU Role in Global Health. Brussels, 31.3.2010. COM (2010) 128 final. Available via https://eur-lex.europa.eu/LexUriServ/LexUriServ.do?uri=COM:2010:0128:FIN:EN:PDF. Accessed 21 July 2024

European Commission (2010b) Trade as a Driver of Prosperity. European Commission Staff Working Document. COM (2010) 612, SEC (2010) 1268

European Commission (2011) Communication from the Commission to the European Parliament, the Council, the European Economic and Social Committee and the Committee of the Regions: Increasing the impact of EU Development Policy: An Agenda for Change. COM (2011) 637 final. Available via https://eur-lex.europa.eu/legal-content/EN/TXT/?uri=celex%3A52011DC0637. Accessed 21 July 2024

European Commission (2017) The New European Consensus on Development: 'Our World, Our Dignity, Our Future. 26 June 2017). Available via https://international-partnerships.ec.europa.eu/policies/european-development-policy/european-consensus-development_en. Accessed 21 July 2024

European Commission (2018a) Proposal for a Regulation of the European Parliament and of the Council: Establishing the Neighbourhood, Development and International Cooperation Instrument. Brussels, 14.6.2018. COM (2018) 460 final, 2018/0243 (COD). Available via https://eur-lex.europa.eu/resource.html?uri=cellar:d2c24540-6fb9-11e8-9483-01aa75ed71a1.0002.02/DOC_1&format=PDF. Accessed 21 July 2024

European Commission (2018b) Sustainable Garment Value Chains. Available via https://www.switch-asia.eu/site/assets/files/2466/eu_garment_industry.pdf. Accessed 21 July 2024

European Commission (2018c) The EU Special Incentive Arrangement for Sustainable Development and Good Governance ('GSP+') Assessment of Sri Lanka Covering the Period 2016-2017. European Commission Joint Staff Working Document, Brussels. 19.1.2018. SWD (2018) 31 final. Available via https://eur-lex.europa.eu/legal-content/EN/TXT/?uri=CELEX:52018SC0031. Accessed 21 July 2024

European Commission (2019a) Communication 'United in delivering the Energy Union and Climate Action-Setting the Foundations for a Successful Clean Energy Transition'. Brussels, 18.6.2019. COM (2019) 285 final. Available via https://eur-lex.europa.eu/legal-content/EN/TXT/PDF/?uri=CELEX:52019DC0285&from=EN. Accessed 21 July 2024

European Commission (2019b) The European Green Deal. Brussels, 11.12. 2019. COM (2019) 640 final. Available via https://eur-lex.europa.eu/resource.html?uri=cellar:b828d165-1c22-11ea-8c1f-01aa75ed71a1.0002.02/DOC_1&format=PDF. Accessed 21 July 2024

European Commission (2020a) Commission Staff Working Document Evaluation of the Regulation (EC) No 469/2009 of the European Parliament and of the Council Concerning the Creation of a Supplementary Protection Certificate for Plant Protection Products. Brussels, 25.11.2020. SWD (2020) 292 final. Available via https://eur-lex.europa.eu/legal-content/EN/TXT/PDF/?uri=CELEX:52020SC0292&rid=7. Accessed 21 July 2024

European Commission (2020b) Communication from the Commission to the European Parliament, the Council, the European Economic and Social Committee and the Committee of the Regions. Brussels, 19.10.2020. COM (2020) 690 final

European Commission (2021a) Proposal for a Directive of the European Parliament and of the Council: Amending Directive 2003/87/EC Establishing a System for Greenhouse Gas Emission Allowance Trading within the Union, Decision (EU) 2015/1814 Concerning the Establishment and Operation of a Market Stability Reserve for the Union Greenhouse Gas Emission Trading Scheme and Regulation (EU) 2015/757. Brussels, 14.7.2021. COM (2021) 551 final. Available via https://ec.europa.eu/info/sites/default/files/revision-eu-ets_with-annex_en_0.pdf. Accessed 21 July 2024

European Commission (2021b) Proposal for a Regulation of the European Parliament and of the Council Establishing a Carbon Adjustment Mechanism. COM (2021) 564 final, 2021/0214 (COD). 14 July 2021. Available via https://eur-lex.europa.eu/resource.html?uri=cellar:a95a4441-e558-11eb-a1a5-01aa75ed71a1.0001.02/DOC_1&format=PDF. Accessed 21 July 2024

European Commission (2022) Proposal for a Directive of the European Parliament and of the Council on Corporate Sustainability Due Diligence and Amending Directive (EU) 2019/1937. Brussels, 23.2.2022. 2022/0051(COD). Available via https://eur-lex.europa.eu/resource.html?uri=cellar:bc4dcea4-9584-11ec-b4e4-01aa75ed71a1.0001.02/DOC_1&format=PDF. Accessed 21 July 2024

European Commission (2024) Commission Implementing Regulation (EU) 2023/1773 of 17 August 2023 laying down the rules for the application of Regulation (EU) 2023/956 of the European Parliament and of the Council as regards reporting obligations for the purposes of the carbon border adjustment mechanism during the transitional period. L 228/94. 15 September 2023. Available via https://eur-lex.europa.eu/legal-content/EN/TXT/PDF/?uri=CELEX:32023R1773 . Accessed 29 July 2024

European External Action Service (2017) Shared Vision, Common Action: A Stronger Europe-A Global Strategy for the European Union's Foreign and Security Policy. 9 November 2017. Available via file:///C:/Users/petit/Downloads/OF0116825ENN.en.pdf. Accessed 21 July 2024

European Parliament (2022) Deal Confirms Zero-Emissions Target for New Cars and Vans in 2035. Brussels, 27 October 2022. Available via https://www.europarl.europa.eu/news/en/press-room/20221024IPR45734/deal-confirms-zero-emissions-target-for-new-cars-and-vans-in-2035. Accessed 21 July 2024

Feiring B et al (2016) Human Rights in Follow-up and Review of the 2030 Agenda for Sustainable Development. The Danish Institute for Human Rights. February 2016. Available via https://www.humanrights.dk/sites/humanrights.dk/files/media/dokumenter/sdg/dihr-fur_paper_final_draft_29_02_16.pdf. Accessed 21 July 2024

Global Affairs Canada (2020) 2020-21 Departmental Plan. FR2-24E-PDF

Hart NM (2021) Congressional Research Service, 'USMCA: Legal Enforcement of the Labour and Environmental Provisions. CRS Report. 14 May 2021. Available via https://sgp.fas.org/crs/row/R46793.pdf. Accessed 21 July 2024

OECD (1996) Trade, Employment and Labour Standards, A Study of Core Workers' Rights and International Trade. 9 September 1996. Available via https://www.oecd-ilibrary.org/docserver/9789264104884-en.pdf?expires=1649837269&id=id&accname=ocid195770&checksum=3906F98FB7732DC30409B1FDDE90DAAA. Accessed 21 July 2024

OECD (2005) Environmental Requirements and Market Access. OECD Trade Policy Studies. Available via https://www.oecd-ilibrary.org/docserver/9789264013742-en.pdf?expires=1651947849&id=id&accname=ocid195770&checksum=03FA9A6F0857A7259927094F511E82C0. Accessed 21 July 2024

Rattray S (2019) Human Rights and the SDGs-Two Sides of the Same Coin. UNDP. 5 July 2019. Available via https://www.undp.org/blog/human-rights-and-sdgs-two-sides-same-coin. Accessed 21 July 2024

The European Parliament and the Council (2003) Establishing a System for Greenhouse Gas Emission Allowance Trading within the Union and Amending Council Directive 96/61/EC. Directive 2003/87/EC. 13 October 2003
Titievskaia J (2021) Sustainability Provisions in EU Free Trade Agreements: Reviewing of the European Commission Action Plan. European Parliament Briefing. November 2021. Available via https://www.europarl.europa.eu/RegData/etudes/BRIE/2021/698799/EPRS_BRI(2021)698799_EN.pdf. Accessed 21 July 2024
UNGA (2015) Transforming Our World: the 2030 Agenda for Sustainable Development. UNGA Res/70/1. 21 October 2015
Weko S et al (2020) The Global Impacts of an EU Carbon Border Adjustment Mechanism. IASS Policy Brief. Available via https://publications.iass-potsdam.de/rest/items/item_6000630_8/component/file_6000869/content. Accessed 21 July 2024
WTO General Council (2019) TRIPS Non-violation and Situation Complaints Moratorium. WT/L/1080. 11 December 2019. Available via https://docs.wto.org/dol2fe/Pages/SS/directdoc.aspx?filename=q:/WT/L/1080.pdf&Open=True. Accessed 21 July 2024
Zamfir L, Dobreva A (2019) EU Supports for Democracy and Peace in the World. European Parliament Briefing. June 2019. Available via https://www.europarl.europa.eu/RegData/etudes/BRIE/2018/628271/EPRS_BRI(2018)628271_EN.pdf. Accessed 21 July 2024

Chapter 10
Suggestions for the Next Steps

Abstract WTO members are currently leveraging plurilateral negotiations to advance reforms of WTO rules. Among these efforts, the TESSD has been established to facilitate plurilateral discussions on environmental sustainability. However, TESSD's role in promoting sustainable development within WTO negotiations is somewhat constrained. This mechanism specifically focuses on advancing plurilateral negotiations related to environmental issues and does not address broader concerns such as economic development, social rights, or FTAs.

To achieve comprehensive reform of international trade rules, WTO members must establish an effective mechanism for coordinating their actions both within and beyond the WTO framework. This approach should encompass all aspects of sustainable development. To address this need, I propose the creation of a sustainable development club. This club would enhance cooperation among like-minded countries and institutionalise their collective efforts (See Sect. 7.3.3.). Through this mechanism, member countries could align their sustainable development and trade policies, develop harmonised legal norms and dispute resolution processes, and bolster their efforts to reform WTO rules.

I propose that WTO members committed to sustainable development undertake a three-step process to reform the WTO's multilateral trade rules.

First, these members should sign the TSDA outside the WTO framework to establish a comprehensive sustainable trade regime. The TSDA should encompass the sustainable development values pursued by the parties, strategies to achieve these values, legal norms, dispute resolution mechanisms, and a periodic review process. Ideally, the parties would include the majority, if not all, members of the RCEP, EU FTAs, USMCA, and the CPTPP. By coordinating their development and trade policies through this sustainable development club, these major FTA parties can work together to reform WTO rules in alignment with their shared policies.

Second, after signing the TSDA, the parties should issue a joint ministerial statement at the WTO Ministerial Conference to expand the TESSD's scope to include all trade-related sustainable development issues. This step would enable the sustainable development club to influence the reform of WTO rules across all dimensions of sustainable development.

X. Zhao, *Integrating the UN SDGs into WTO Law*, EYIEL Monographs - Studies in European and International Economic Law 41,
https://doi.org/10.1007/978-3-031-73876-0_10

Finally, the TSDA parties would negotiate various plurilateral sustainable development agreements within the sustainable development club and invite other WTO members to join these negotiations. I will elaborate on these proposals in the following sections.

10.1 Trade and Sustainable Development Agreement on Establishing the Sustainable Development Club

The TSDA aims to institutionalise WTO members' efforts to reform international trade rules by establishing unified development and trade policies through agreements, and by strengthening solidarity and cooperation among members.[1] To achieve these objectives, the TSDA should incorporate specific rules, policies, and commitments. This section will outline the essential elements that need to be included in the TSDA.

10.1.1 Sustainable Development Values

The values of sustainable development embody the economic, social, and environmental standards that humanity aspires to achieve. They are crucial for guiding the formulation and implementation of sustainable development policies and laws. Therefore, the contracting parties should explicitly incorporate these values into the preamble of the TSDA.

To achieve sustainable development, the parties must uphold the economic, social, and environmental values outlined in the UN 2030 Agenda for Sustainable

[1]To some extent, WTO members' political unwillingness will undermine to some extent the efficiency of the sustainable development club. The weaponization of economic interdependent or China's totalitarian denial of human rights could shrink common policies under the sustainable development club. For example, the club may achieve more consensus on trade and climate issues, such as the compliance of CBAMs with WTO law. Nonetheless, the sustainable development club remains the most promising way for WTO members to break through the existing impasse in multilateral negotiations. It is a plurilateral mechanism that requires members to promote the full realisation of the UN SDGs. Achieving one aspect of the SDGs in isolation is impractical. The universal nature of the UN SDGs makes it clear that economic, environmental and social interests are indivisible. Therefore, WTO members cannot limit the alliance's work to one aspect of sustainable development, such as climate policies. Moreover, the important role of the plurilateral mechanism, 'the sustainable development club,' is to exert pressure on large economies that reject sustainable development by forming a robust sustainable development alliance. Thus, the absence of certain economies does not detract from the important role of the sustainable development club in achieving sustainable development. On the contrary, it illustrates the significance of the club's existence.

Development and other relevant international conventions.[2] Sustainable development is inherently a bottom-up, citizen-centred initiative.[3] Therefore, the contracting parties should commit to realising the comprehensive economic, social, and environmental rights that humanity requires rather than pursuing their own national interests.

Moreover, sustainable development is universal and entails two key principles.[4] First, the values of sustainable development are interconnected and must be pursued holistically to be effective.[5] Second, the realisation of individual rights is crucial to achieving sustainable development. As the UN 2030 Agenda emphasises,[6] no one should be left behind in this process.[7] Consequently, the contracting parties should ensure that their efforts do not focus on just one dimension of sustainable development or come at the expense of particular groups or individuals.

Furthermore, it is essential to recognise that the rule of law is a fundamental prerequisite for achieving sustainable development.[8] To ensure the effective implementation of sustainable development rules, the contracting parties must acknowledge the importance of the rule of law in realising these values.

First, they should embed sustainable development as a clear legal norm within their national constitutions and other relevant laws. Secondly, they must support an international legal order centred around the UN dedicated to advancing sustainable development. This involves ensuring global security and peace,[9] adhering to all international conventions related to sustainable development, and collaboratively advancing international legislative measures in this area. The contracting parties should specify that these legislative measures encompass human rights, environmental protection, and economic regulations, including trade and investment.

Additionally, the sustainable development club must focus on reforming multilateral trade rules and ultimately advancing the constitutionalisation of the WTO. This effort is part of a broader process of global constitutionalisation.[10] As previously discussed, the WTO's constitutionalisation must be decentralised,[11] requiring collaborative efforts among participating countries to uphold the WTO's constitutional values through multilevel governance.

---

[2]These international conventions include, particularly, multilateral environmental conventions and the core international human rights treaties. To know more details, please see, The United Nations Human Rights Office of the High Commissioner (2014); see also Chasek and Wagner (2012).

[3]See, For example, Swarnakar et al. (2017).

[4]See Long (2015), p. 203.

[5]See Fonseca et al. (2020), pp. 3361–3363.

[6]Please see The United Nations Human Rights Office of the High Commissioner (2023).

[7]See UNGA (2015), preamble and paras. 4 and 72.

[8]See Michel (2020), p. 5.

[9]See UNGA (2015), para. 35.

[10]To know more details about global constitutionalisation, please see, for example, O'Donoghue (2014).

[11]See Sect. 6.2.2.

The sustainable development club is not intended to be a large institution. Instead, it functions as a small coordinating body to align members' actions. The contracting parties should recognise their obligation to coordinate their development and trade policies and laws under the TSDA, promoting sustainable development through the implementation of harmonised norms. They should also commit to the multilevel governance of sustainable development, which includes advocating for reforms in national laws, FTAs, multilateral trade rules (such as WTO rules), and other relevant international agreements.

To reform WTO rules, the contracting parties should establish a sustainable development club within the WTO framework and work towards inviting additional WTO members to join. This effort will facilitate modifications to WTO Agreements in accordance with TSDA principles.

10.1.2 Strategies to Achieve the Trade-Related UN SDGs

Integrating sustainability into international trade rules necessitates well-defined strategies. The TSDA contracting parties should outline their strategies for reforming international trade rules to enhance the transparency and predictability of TSDA policies and regulations, thereby enabling the public to monitor the sustainable development club's actions effectively.

The EU's approach to sustainable trade offers valuable insights and should be considered a model for the TSDA. I recommend that the contracting parties incorporate the EU's three key sustainable development strategies into the TSDA framework and align their international trade rule reforms accordingly. These strategies include adopting higher sustainable development standards, employing a comprehensive approach, and applying principled pragmatism.[12]

The parties should adhere to international sustainability standards while striving to exceed these benchmarks with their own sustainable development measures. Currently, although many FTAs address this obligation, the provisions vary significantly across different agreements.[13] Some FTAs entirely lack protection level clauses,[14] while others impose varying obligations for different aspects of sustainable development. For instance, the USMCA and CPTPP use non-regression clauses for environmental standards but apply non-derogation clauses for labour rights standards. These discrepancies have resulted in uneven promotion of sustainable development standards among countries.

To establish a unified development and trade policy, the parties must agree to apply non-regression clauses uniformly across all sustainability standards. This means they should commit to continuously upgrading their sustainability standards

[12] See Sect. 9.2.2.3.

[13] See Sect. 9.2.2.3.

[14] Ibid.

by amending relevant laws and regulations.[15] Additionally, this commitment entails promoting scientific research related to sustainable development and ensuring that scientific criteria guide development and trade policies.[16]

The contracting parties should also embrace a comprehensive approach to advancing the reform of international trade rules. Members of the sustainable development club must acknowledge the interconnectedness of the SDGs and address sustainable development issues in a holistic manner. As previously mentioned, this approach necessitates broad partnerships.[17] The contracting parties should commit to fostering global cooperation through various means, including diplomatic efforts. It is anticipated that the sustainable development club will have access to greater diplomatic resources than the EU, particularly in facilitating North-South and South-South cooperation.[18]

Finally, the contracting parties must commit to a principled pragmatism approach to drive the reform of international trade rules. This approach, as previously discussed, requires states to balance pragmatism with principled stances in fostering international cooperation.[19] In the current turbulent international environment, the sustainable development club must uphold the international rule of law, maintain global peace, and protect fundamental human rights. Only by doing so can the club ensure that a comprehensive range of sustainability values is embedded in international trade rules. Specifically, the contracting parties should first uphold universal international constitutional norms that encompass human rights, the rule of law, and sustainable development. Second, they must pledge not to undermine these norms in their actions.

10.1.3 Legal Norms

The coherence of legal norms is essential for achieving effective multi-level governance.[20] Therefore, it is crucial for TSDA parties to reach a consensus on sustainability rules. Existing FTAs offer a solid foundation for integrating these rules. A review of sustainability provisions in current FTAs reveals a range of approaches, particularly concerning climate change, ecological biodiversity, waste management, chemicals, and oceans.[21] However, in areas such as SPS measures, technical regulations, labour rights, environmental protection, and intellectual property, the norms

[15] See Lydgate (2019), pp. 37–39.

[16] See Spangenberg (2011), p. 257.

[17] See Sect. 9.2.2.3.2.

[18] The sustainable development club members will include the CPTPP signatories, many of them are developing countries capable of promoting North-South and South-South cooperation.

[19] See Sect. 9.2.2.3.3.

[20] See Lamy (2012).

[21] See Sect. 9.2.2.1.

are largely consistent across FTAs.[22] The parties should consolidate these sustainability provisions into a unified rulebook. They must endorse this rulebook and promptly update their FTAs to align with it. Additionally, the parties need to address any deficiencies in their existing FTA sustainability provisions. They should commit to enhancing standards related to labour rights,[23] economic development, human rights, and environmental protection, and revise their FTA provisions, accordingly, based on the consolidated rulebook.

10.1.4 Dispute Settlement Mechanisms

Until the reform of international trade rules is fully realised, the sustainable development club members must actively contribute to sustainable development by implementing the sustainability rules outlined in the TSDA. An FTA dispute settlement mechanism with exclusive jurisdiction can significantly enhance the enforcement of FTA sustainability provisions by preventing parties from seeking favourable rulings through WTO dispute settlement procedures.[24]

Currently, only a few FTAs include dispute settlement mechanisms with exclusive jurisdiction over sustainability disputes.[25] Even when such clauses are present, they often do not cover all disputes related to sustainability provisions.[26] To ensure the effective enforcement of sustainability provisions, TSDA parties should revise their FTAs to strengthen the jurisdictional provisions applicable to sustainability disputes. They should commit to using the FTA dispute settlement mechanisms for resolving disputes arising from their FTA sustainability provisions, except in cases where WTO Agreements provide identical legal norms.[27]

10.1.5 Review and Monitoring

Review and monitoring mechanisms are crucial for maintaining an effective sustainable development club, as they ensure transparency in its operations. Numerous studies highlight the role of transparency in good governance.[28] The TSDA parties must establish a regular monitoring mechanism to track the club's progress and

---

[22] Ibid.

[23] Some scholars opine that the existing FTAs' labour rights provisions cannot be referred to as human rights clauses. See Gammage (2018), pp. 12–13. See also Micara (2019), p. 1452.

[24] See Henckels (2008), p. 581.

[25] See Sect. 9.2.3.

[26] Ibid.

[27] See Henckels (2008), p. 597.

[28] See Weiss and Steiner (2006), pp. 1553–1571; See also Ala'i (2018), p. 2.

develop annual work plans. Many contemporary international agreements, including FTAs, incorporate such provisions.[29]

Additionally, TSDA parties should recognise the importance of public participation in reviewing and monitoring the club's activities. Grassroots public engagement is essential for assessing and tracking progress toward the UN SDGs.[30] To facilitate this, TSDA parties must ensure that citizens are actively involved in the club's policymaking, annual reviews, and oversight of FTA dispute settlement mechanisms. Specifically, parties should explicitly welcome the oversight of citizen groups, such as civil society organisations and NGOs, to enhance accountability and inclusivity.[31]

10.2 Joint Ministerial Statement on the Sustainable Development Club

Joint ministerial statements have become a key mechanism for WTO members to reform WTO rules.[32] The TSDA parties should use a joint ministerial statement at the Ministerial Conference to announce the establishment of a sustainable development club. This club will serve as the primary body within the WTO for TSDA parties to advance plurilateral sustainability negotiations. Ideally, the TSDA parties could transform the existing TESSD into this sustainable development club. If this approach is adopted, the TSDA parties should revise their consensus on the TESSD accordingly and reflect these changes in the joint ministerial statement.

In their joint ministerial statement, TSDA parties must unequivocally commit to ensuring that the sustainable development club they establish adheres to TSDA rules and actively promotes the reform of WTO rules. This commitment entails supporting the WTO's constitutionalisation through a multilevel governance approach to international trade. In other words, TSDA members should translate their agreed-upon development and trade policies under the TSDA framework into plurilateral WTO Agreements.

Additionally, TSDA parties should commit in the joint ministerial statement to adopt the TSDA as the charter for the sustainable development club. They should pledge to promote plurilateral negotiations within the WTO based on TSDA rules and commitments, seek to expand club membership, and ultimately drive the reform of multilateral trade rules under the WTO framework. Furthermore, the joint ministerial statement should stipulate that other WTO members must also sign the TSDA as a prerequisite for joining the sustainable development club.

---

[29] See Remáč (2019).

[30] See, for example, Stelwagen et al. (2021), p. 958.

[31] To know citizen groups' contribution to monitoring, see Khan et al. (2003), p. 909; See also Tsampi (2021), p. 102.

[32] See Kelsey (2020), p. 2.

10.3 Developing WTO Plurilateral Agreements

After establishing a sustainable development club within the WTO under TSDA rules, the TSDA parties should initiate WTO plurilateral negotiations to address gaps in existing international trade rules, whether related to FTAs or WTO regulations. Based on my research, I recommend that TSDA parties focus on two key areas for these negotiations: sustainable development commitments and dispute settlement rules that assess the policy space for national regulation.[33] The plurilateral negotiations should yield outcomes that become WTO rules that are adhered to by TSDA parties. Concurrently, these rules should serve as the basis for amending their FTAs, ensuring consistency across international sustainable trade regulations. This approach will harmonise sustainability rules in all TSDA parties' FTAs and re-establish the WTO as a leading platform for advancing sustainable trade negotiations.

10.3.1 Plurilateral Negotiations on Sustainable Development Commitments

10.3.1.1 Trade Liberalisation Commitments

As previously discussed, some WTO members are actively seeking to further liberalise international trade, focusing on four key areas: trade in services, electronic commerce, investment facilitation, and MSMEs.[34] These negotiations are noteworthy and are expected to significantly boost global economic growth,[35] enhancing the contribution of WTO rules to the economic dimension of sustainable development. However, these negotiations also have limitations, particularly concerning services trade, MSMEs, and e-commerce. Since WTO members address only a limited range of investment issues within the WTO framework, TSDC members may face constraints in expanding market access for foreign investors in the future.[36] In this context, they may focus on improving regulations related to services and e-commerce management.

In the context of services trade and MSMEs, current negotiations primarily aim to reduce trade barriers, such as unreasonable or discriminatory approval procedures.[37] TSDA parties should continue to advocate for further trade liberalisation in services and push for WTO members to open additional service sectors to foreign traders. Regarding e-commerce, the negotiation outcome does not address data flows. It is

[33] See Chaps. 2 and 3.

[34] See Sect. 7.1.

[35] See OECD (2021a). See OECD (2017). See OECD (2021b). See also Sauvant (2021).

[36] See Bernasconi-Osterwalder and Bonnitcha (2020).

[37] See Sects. 7.1.1 and 7.1.4.

essential to recognise that a truly free international e-commerce environment relies on an open Internet, unrestricted data flow, and zero customs.[38] TSDA parties must be prepared to advance these discussions in the future to address these fundamental requirements.

10.3.1.2 Development Commitments

Development commitments aim to use trade to eradicate poverty and hunger in a targeted manner. Currently, such commitments do not exist within the WTO framework. The closest existing commitment is to reduce economic inequality between developing and developed countries.[39] However, this commitment is not specifically designed to achieve the SDGs of ending hunger and poverty.[40] TSDA contracting parties should establish specific WTO commitments targeting these two SDGs. They could draw inspiration from Chapter 23 of the CPTPP, explicitly incorporating poverty and hunger eradication into WTO plurilateral agreements.[41] It is crucial that the WTO dispute settlement system provides equal protection for these commitments alongside other sustainable development commitments. Notably, Chapter 23 of the CPTPP does not safeguard these commitments through its FTA dispute settlement mechanism.[42]

10.3.1.3 Subsidies and Green Trade Remedy Measures

Subsidies and remedial measures are closely linked to sustainable development.[43] Research has shown that many subsidies can be detrimental to sustainable development.[44] For instance, agricultural export subsidies can create unfair competition among agricultural producers and jeopardise other countries' food security.[45] Similarly, fisheries subsidies promote overfishing and undermine marine resource conservation,[46] while fossil fuel subsidies exacerbate air pollution and climate change.[47]

TSDA parties should advocate for eliminating subsidies that harm sustainable development through plurilateral agreements. Fortunately, WTO members have already made significant strides in this area. At the WTO's 10th Ministerial

---

[38] See Andrenelli and González (2019); See also González and Ferencz (2018).

[39] See Sect. 2.3.2.

[40] Ibid.

[41] See Sect. 9.2.3.2.

[42] Ibid.

[43] See OECD (2006).

[44] See Kicia and Rosenstock (2015), p. 83.

[45] See The FAO Commodities and Trade Division (2001).

[46] See Vivas-Eugui et al. (2022).

[47] See Merrill et al. (2015).

Conference in 2015,[48] members unanimously decided to eliminate agricultural export subsidies. Additionally, at the 2022 Ministerial Conference, WTO members agreed to remove harmful fisheries subsidies.[49]

However, WTO members still need to advance negotiations to eliminate fossil fuel subsidies.[50] This issue presents additional challenges, as only forty-seven WTO members are currently willing to participate in these negotiations. This starkly contrasts the negotiations on agricultural export subsidies and fisheries subsidies, which involved all WTO members. TSDA parties must persist in advocating for and pushing forward negotiations aimed at eliminating fossil fuel subsidies.

In addition to eliminating fossil fuel subsidies, TSDA parties need to initiate negotiations within the WTO to support green subsidies and green trade remedy measures, such as green anti-dumping and countervailing duties. Green anti-dumping and countervailing duties aim to limit imports of products detrimental to sustainable development. Green countervailing duties target subsidised goods that harm sustainable development, while anti-dumping duties are used to prevent dumped green products from unfairly competing with domestic environmental industries. These measures ensure the healthy development of green industries and promote innovation.

It is important to note that governments must use green subsidies in conjunction with anti-dumping duties to develop domestic green industries. Numerous studies have confirmed the positive impact of green subsidies on combating climate change.[51] To effectively slow global warming, governments should implement green incentive policies, such as feed-in tariff (FIT) programmes, often referred to as renewable energy subsidies,[52] alongside contingent protection measures.[53] However, WTO panels have prohibited FIT programmes due to their domestic content requirements. Therefore, WTO members should negotiate new norms related to green subsidies and contingent protection measures in the future.[54]

In addition, international cooperation is essential for effectively implementing green anti-dumping and countervailing duties. TSDA parties need to enhance the enforcement of these duties through plurilateral negotiations to prevent circumvention, where countries facing such duties export their products through third countries. Currently, Chapter 10 of the USMCA includes excellent provisions for

---

[48] See WTO (2015).

[49] See WTO (2022).

[50] These environmental harmful subsidies include direct budgetary grants, tax reductions or exemptions, concessionary loans at reduced interest rates, state guarantees below costs, lack of full cost pricing, indirect support by for example financing infrastructure, and reducing liability for environmental disasters. See Kicia and Rosenstock (2015), p. 83.

[51] Kent and Jha (2014), p. 249; Weber (2015), p. 161.

[52] Espa and Holzer (2018), p. 435; Borlini and Montanaro (2018), p. 86.

[53] Bougette and Charlier (2018), p. 182.

[54] See Zhao (2023).

preventing duty evasion.[55] TSDA parties should consider adopting these USMCA rules to strengthen their measures.

10.3.1.4 Climate, the Environment, and Ecosystem Biodiversity

So far, WTO members have not initiated plurilateral negotiations on climate, the environment, or ecosystem biodiversity. Regarding climate issues, TSDA parties should work to fulfil their emission reduction commitments under the Paris Agreement through trade measures. They should advocate for agreements that include the elimination of tariffs on goods that contribute to emission reduction and the establishment of trade rules that promote the dissemination of such products and technologies.[56] For example, they can support the conclusion of a carbon neutrality agreement proposed by Japan[57] and resume the suspended environmental goods agreement negotiations.[58]

In addition to climate issues, the WTO Agreements currently lack comprehensive commitments on the environment and biodiversity. TSDA parties should champion plurilateral negotiations to address these gaps. Many existing FTAs already include numerous provisions related to environmental protection and biodiversity conservation.[59] TSDA parties could draft their plurilateral agreements by consolidating these FTA provisions while seeking to improve existing standards.[60]

10.3.1.5 Labour Rights

As previously mentioned, the WTO Agreements lack commitments on labour rights, leaving the WTO dispute settlement mechanism without a legal basis to address these issues. Recent trade disputes stemming from FTAs have highlighted the necessity of incorporating labour provisions into multilateral trade rules.[61] TSDA parties should work to include their consolidated labour rights commitments in WTO rules through plurilateral agreements.

---

[55] See Sect. 9.2.2.3.3.

[56] Many scholars have stressed the importance of removing barriers to trade in environmental products. See Howse (2009); Selivanova (2011); Rubini (2012), pp. 525–579; Holzer et al. (2017), pp. 356–389; Gudas (2016).

[57] See WTO (2021).

[58] To know the benefits of the Environmental Goods Agreement, please see Dent (2018), pp. 746–747.

[59] See Sect. 9.2.2.1.2.

[60] To know some ideas for improving climate clauses, please see Frey (2015), pp. 273–282.

[61] See Sect. 9.2.2.1.3.

10.3.2 The Sustainability Test for Assessing the Policy Space for National Regulations

In addition to new sustainable development commitments, TSDA parties need to advance plurilateral negotiations on implementing a sustainability test.[62] As previously mentioned, existing WTO dispute settlement rules significantly restrict WTO members' ability to enact and enforce domestic sustainable development measures.[63] These negotiations should focus on applying the sustainability test among TSDA parties to assess their policy space for domestic regulation of sustainable development. This will ensure that parties comprehensively achieve the UN SDGs.

10.4 Conclusion

WTO members committed to sustainable development should adopt a three-step process based on club theory to reform WTO rules. First, they must sign the TSDA to coordinate their trade and development policies. The TSDA should encompass the sustainable development values pursued by the contracting parties, strategies to achieve these values, legal norms, dispute settlement rules, and a periodic review mechanism. My suggestions are as follows:

(1) The contracting parties should include sustainable development values in the TSDA's preamble. This will ensure the fulfilment of economic, social, and environmental standards for all, uphold the international rule of law, harmonise national policies, promote multilevel governance of sustainable development, guide the development of TSDA rules, and expand membership in the sustainable development club.[64]
(2) The contracting parties should embed the EU's three sustainable development strategies within the body of the TSDA and use these strategies to drive the reform of international trade rules.[65]
(3) The contracting parties should codify the sustainability provisions from existing FTAs into a consolidated TSDA framework. This will facilitate the revision of their FTAs and elevate sustainability standards.[66]
(4) The contracting parties should update the jurisdiction provisions in their FTAs to ensure that disputes arising from sustainability provisions are resolved through

[62] See Chaps. 4 and 5.

[63] See Chap. 3.

[64] See Sect. 10.1.1.

[65] See Sect. 10.1.2.

[66] See Sect. 10.1.3.

the FTA dispute settlement mechanisms unless WTO Agreements provide identical legal norms.[67]

(5) The contracting parties must establish a regular monitoring mechanism to review the progress of the sustainable development club's activities and develop annual work plans. They should also recognise the importance of public participation in reviewing and monitoring its work and accept the oversight of citizen groups.[68]

After concluding the TSDA, the contracting parties should announce the establishment of a sustainable development club at the Ministerial Conference. In their joint ministerial statement, they must commit to advancing plurilateral negotiations within the WTO based on TSDA rules and principles, expanding the membership of the club, and ultimately driving the reform of multilateral trade rules under the WTO framework.[69] Subsequently, the TSDA parties should initiate WTO plurilateral negotiations focused on sustainable development commitments and dispute settlement mechanisms to address existing gaps in international trade rules.[70]

References

Bibliography

Ala'i P (2018) From the periphery to the center? The evolving WTO jurisprudence on transparency and good governance. J Int Econ Law 11(4):1–24

Andrenelli A, González JL (2019) Electronic transmissions and international trade-shedding new light on the moratorium debate. OECD Trade Policy Papers No.233. Available via https://www.oecd-ilibrary.org/docserver/57b50a4b-en.pdf?expires=1651498610&id=id&accname=guest&checksum=ACDE2528618918326B2D09AEFD493BA7. Accessed 21 July 2024

Bernasconi-Osterwalder N, Bonnitcha J (2020) Investment negotiations at the WTO and the IIA regime: anticipating unintended interactions. IISD Analysis. 5 October 2020. Available via https://www.iisd.org/itn/en/2020/10/05/investment-negotiations-at-the-wto-and-the-iia-regime-anticipating-unintended-interactions-nathalie-bernasconi-osterwalder-jonathan-bonnitcha/. Accessed 21 July 2024

Borlini LS, Montanaro F (2018) Climate change and trade: challenges and lingering questions on the relationship between renewable energy subsidies and WTO disciplines. China-EU Law J 6: 81–101

Bougette P, Charlier C (2018) Antidumping and feed-in tariffs as good buddies? Modeling the EU-China Solar Panel Dispute. Energy J 39(6):171–188

Chasek PS, Wagner LM (2012) The roads from Rio: lessons learned from twenty years of multilateral environmental negotiations. Routledge, New York and London

Dent CM (2018) Clean energy trade governance: reconciling trade liberalism and climate interventionism? New Polit Econ 23(6):728–747

---

[67] See Sect. 10.1.4.

[68] See Sect. 10.1.5.

[69] See Sect. 10.2.

[70] See Sects. 10.3.1 and 10.3.2.

Espa I, Holzer K (2018) Negotiating 21st century rules on energy: what is at stake for the European Union, the United States and the BRICS? J World Invest Trade 19(3):415–443
Fonseca LM et al (2020) Mapping the Sustainable Development Goals Relationships (2020). Sustainability 12(8):3359–3373
Frey C (2015) Mega-regional trade agreements and post-2015 climate protection: bridging the gap. J Eur Environ Plan Law 12(3–4):264–285
Gammage C (2018) A critique of the extraterritorial obligations of the EU in relation to human rights clauses and social norms in EU free trade agreements. Europe World: Law Rev 2(1):1–20
González JL, Ferencz J (2018) Digital trade and market openness. OECD Trade Policy Papers No.217. Available via https://www.oecd-ilibrary.org/docserver/1bd89c9a-en.pdf?expires=1651498942&id=id&accname=guest&checksum=B1EE8AF5D55249265604DB78D2AD21AF. Accessed 21 July 2024
Gudas K (2016) The external dimension of cross-border electricity systems planning. In: Heffron RJ, Little GFM (eds) Delivering energy law and policy in the EU and the US. Edinburgh University Press, Edinburgh, pp 191–196
Henckels C (2008) Overcoming jurisdictional isolationism at the WTO-FTA Nexus: a potential approach for the WTO. J Int Econ Law 19(3):571–959
Holzer K et al (2017) Promoting green electricity through differentiated electricity tax schemes. In: Cottier T, Espa I (eds) International trade in sustainable electricity: regulatory challenges in international law. CUP, Cambridge, pp 356–389
Kelsey J (2020) The illegitimacy of joint statement initiatives and their systemic implications for the WTO. Int J Econ Law 25(1):2–24
Kent A, Jha V (2014) Keeping up with the changing climate: the WTO's evolutive approach in response to the trade and climate conundrum. J World Invest Trade 15:254–271
Khan K et al (2003) Monitoring and evaluation: civil society organisations' competitive edge in effective poverty alleviation. Pak Dev Rev 42(2):909–924
Kicia M, Rosenstock M (2015) Reforming fossil fuel subsidies: will it make a difference? In: Kreiser L et al (eds) Environmental pricing. Edward Elgar, Cheltenham and Northampton, pp 83–99
Lamy P (2012) Global governance: from theory to practice. In: Petersmann E-U (ed) Multilevel governance of interdependent public goods: theories, rules and institutions for the central policy challenge in the 21st century. EUI Working Papers, Italy, pp 25–30
Long G (2015) The idea of universality in the sustainable development goals. Ethics Int Aff 29(2): 203–222
Lydgate E (2019) Environmental standards and regulation. In: West J, Vergels E (eds) Ensuring a post-Brexit level playing field. European Policy Centre, pp 33–46
Merrill L et al (2015) Fossil-fuel subsidies and climate change: options for policy-makers within their intended nationally determined contributions. Nordic Working Papers. Available via http://norden.diva-portal.org/smash/get/diva2:786861/FULLTEXT02.pdf. Accessed 21 July 2024
Micara A (2019) Human rights protection in new generation's free trade agreements of the European Union. Int J Hum Rights 23(9):1447–1469
O'Donoghue A (2014) Constitutionalism in global constitutionalisation. CUP, Cambridge
Rubini L (2012) Ain't Wastin' time no more: subsidies for renewable energy, the SCM agreement, policy space and law reform. J Int Econ Law 15(2):525–579
Sauvant KP (2021) A multilateral investment facilitation agreement can help advance development. East Asia Forum. 2 May 2021. Available via https://www.eastasiaforum.org/2021/05/02/a-multilateral-investment-facilitation-agreement-will-help-advance-development/. Accessed 21 July 2024
Selivanova Y (ed) (2011) Regulation of energy in international trade law: WTO, NAFTA and Energy Charter. Kluwer Law International, Alphen aan den Rijn
Spangenberg JH (2011) Sustainability science: a review, an analysis and some empirical lessons. Environ Conserv 38(3):257–287

Stelwagen RE et al (2021) A bottom-up approach to model the environmental impact of the last-mile in an urban food-system. Sustain Prod Consum 26:958–970
Swarnakar P et al (eds) (2017) 'Bottom-up' approaches in governance and adaptation for sustainable development: case studies from India and Bangladesh. SAGE, New Delhi
Tsampi A (2021) The role of civil society in monitoring the executive in the case-law of the European Court of Human Rights: recasting the rule of law. Utrecht Law Rev 17(2):102–115
Vivas-Eugui D et al (2022) Fisheries subsidies deal: why we need it and how to implement it. The UNTAD News Blog. 28 March 2022. Available via https://unctad.org/news/blog-fisheries-subsidies-deal-why-we-need-it-and-how-implement-it. Accessed 21 July 2024
Weber RH (2015) Renewable energy: subsidies and taxes as competition distortion. In: Kreiser L et al (eds) Environmental pricing. Edward Elgar, Cheltenham and Northampton, pp 161–176
Weiss F, Steiner S (2006) Transparency as an element of good governance in the practice of the EU and the WTO: overview and comparison. Fordham Int Law J 30(5):1545–1586
Zhao X (2023) The WTO Panel Report on US-safeguard measure on PV products: a decisive victory for the fight against climate change? In: Bäumler J et al (eds) European Yearbook of International Economic Law 2022. Springer, Cham, pp 175–204

Documents

Howse R (2009) World Trade Law and Renewable Energy: The Case of Non-Tariff Barriers. UNCTAD. Available via https://unctad.org/system/files/official-document/ditcted20085_en.pdf. Accessed 21 July 2024
Michel J (2020) The Rule of Law and Sustainable Development. A Report of the CSIS Program on Prosperity and Development. June 2020. Available via https://csis-website-prod.s3.amazonaws.com/s3fs-public/publication/200626_Michel_RuleOfLaw_Web.pdf. Accessed 21 July 2024
OECD (2006) Subsidy Reform and Sustainable Development: Economic, Environmental and Social Aspects. 19 April 2006. Available via https://www.oecd-ilibrary.org/docserver/9789264025653-en.pdf?expires=1651517360&id=id&accname=ocid195770&checksum=926590E79BBE8B849B0591FE9858C583. Accessed 21 July 2024
OECD (2017) Services Trade Policies and the Global Economy. 8 June 2017. Available via https://www.oecd.org/publications/services-trade-policies-and-the-global-economy-9789264275232-en.htm. Accessed 21 July 2024
OECD (2021a) Joint Initiative on Services Domestic Regulations: Reducing Services Trade Costs in the G20 and Beyond. OECD Trade Policy Brief. October 2021. Available via https://issuu.com/oecd.publishing/docs/reducing_services_trade_costs_in_the_g20_and_beyon?fr=sZTJhMzI5ODk2NDE. Accessed 21 July 2024
OECD (2021b) Trade in the Time of Parcels. OECD Trade Policy Paper. May 2021. Available via https://www.oecd-ilibrary.org/docserver/0faac348-en.pdf?expires=1651497433&id=id&accname=guest&checksum=2FDE2F85F02EECE77FD6363B878558CC. Accessed 21 July 2024
Remáč M (2019) International Agreements-Review and Monitoring Clauses: A Rolling Check-List. European Parliamentary Research Service, PE 631.768. October 2019. Available via https://www.europarl.europa.eu/RegData/etudes/STUD/2019/631768/EPRS_STU(2019)631768_EN.pdf. Accessed 21 July 2024
The FAO Commodities and Trade Division (2001) Incorporating Food Security Concerns in a Revised Agreement on Agriculture. 20 July 2001. Available via https://www.fao.org/3/y3733e/y3733e03.htm#TopOfPage. Accessed 21 July 2024
The United Nations Human Rights Office of the High Commissioner (2014) The Core International Human Rights Treaties. Available via https://www.ohchr.org/sites/default/files/documents/publications/coretreatiesen.pdf. Accessed 21 July 2024

The United Nations Human Rights Office of the High Commissioner (2023) Sustainable Development Goals Related Human Rights. Available via https://www.ohchr.org/sites/default/files/Documents/Issues/MDGs/Post2015/SDG_HR_Table.pdf. Accessed 21 July 2024

UNGA (2015) Transforming Our World: the 2030 Agenda for Sustainable Development. UNGA Res/70/1. 21 October 2015

WTO (2015) Nairobi Ministerial Declaration. WT/MIN(15)/DEC. 19 December 2015. Available via https://docs.wto.org/dol2fe/Pages/FE_Search/FE_S_S009-DP.aspx?language=E&CatalogueIdList=225910,225713&CurrentCatalogueIdIndex=0&FullTextHash=&HasEnglishRecord=True&HasFrenchRecord=True&HasSpanishRecord=True. Accessed 21 July 2024

WTO (2021) Trade and Environmental Sustainability Discussions (TESSD): Proposal on Contributions to Achieving Global Carbon Neutrality at the WTO. INF/TE/SSD/W/10. 23 March 2021

WTO (2022) Agreement on Fisheries Subsidies, Ministerial Decision of 17 June 2022. WT/ MIN (22) /33 WT/ L/ 1144. 22 June 2022

Chapter 11
Conclusion

Abstract This monograph examines the limitations of current WTO rules in supporting the SDGs. While WTO agreements touch on various sustainability issues, they fall short of providing a comprehensive framework for sustainable development. The monograph contends that existing rules excessively limit members' regulatory policy space and proposes replacing the necessity test with a sustainability test. This new test would prioritise sustainable development values over trade concerns, ensuring that domestic regulations align with the SDGs and balance different sustainable development interests. Additionally, this monograph advocates for incorporating trade-related sustainable development commitments and the sustainability test into the WTO framework. It also offers recommendations for formalising these efforts through the establishment of a sustainable development club within the WTO.

Sustainable development is the zeitgeist of our era,[1] reflecting the principles of good governance and addressing fundamental human needs across economic, social, and environmental dimensions. The UN 2030 Agenda for Sustainable Development outlines 17 SDGs and 169 targets that member states are expected to achieve by 2030.[2] It is essential for humanity to strive earnestly to meet these goals for the well-being of both present and future generations. Although progress to date has been less than ideal, many countries have begun to implement policies that advance sustainable development. These positive changes are gradually guiding humanity toward a more sustainable future.

Among the numerous efforts to advance sustainable development, reforming international trade rules is especially significant. International trade regulations profoundly influence the economic aspects of sustainable development and are

[1] See Blewitt (2008), p. ix.

[2] See UNGA (2015) Transforming Our World: the 2030 Agenda for Sustainable Development. UNGA Res/70/1. 21 October 2015, Preamble.

X. Zhao, *Integrating the UN SDGs into WTO Law*, EYIEL Monographs - Studies in European and International Economic Law 41,
https://doi.org/10.1007/978-3-031-73876-0_11

intricately linked to its social and environmental dimensions.[3] Despite varying opinions on how best to implement trade reforms, there is broad consensus on the critical role that sustainable trade rules play in achieving overall sustainable development goals.

It is encouraging to see that WTO members are acknowledging the need to reshape global trade rules to meet the SDGs outlined in the UN 2030 Agenda. Since the WTO's Buenos Aires Ministerial Conference in 2017,[4] many members have actively pursued reform through plurilateral negotiations. This monograph examines the challenges posed by current WTO rules concerning sustainable development, suggests amendments to address these issues, and outlines how these proposed changes can be integrated into WTO law through plurilateral negotiations.

It is evident that current WTO rules pose challenges to achieving the UN SDGs. While WTO rules permit members to implement domestic regulatory measures that may restrict international trade, these measures must align with legitimate purposes recognised in the WTO Agreements.[5] In the context of sustainable development, these legitimate purposes are reflected in WTO members' commitments. International trade intersects with the economic, social, and environmental dimensions of sustainable development.[6] To effectively support these SDGs, international trade rules must incorporate specific commitments related to sustainable development, thereby upholding individuals' economic, social, and environmental rights.[7]

The WTO Agreements encompass commitments to trade liberalisation, reducing economic disparities between developed and developing nations, industry protection, public health, environmental sustainability, animal welfare, and labour rights.[8] However, these commitments fall short of aligning international trade with the broader requirements of sustainable development.[9] Consequently, the WTO Agreements do not provide a robust legal framework for members to pursue regulatory objectives related to sustainable development. As a result, members must rely on their own regulatory autonomy to introduce and implement measures that address sustainability issues.[10]

Unfortunately, WTO rules excessively constrain members' policy space for domestic regulation. Article 11 of the DSU mandates that WTO panels assess the legality of a member's domestic regulatory measures using an objective standard of review.[11] This means that panels must independently evaluate the extent of a

[3] See Sect. 2.1.

[4] See Chap. 7.

[5] See Sect. 3.3; See also Trachtman (2017), pp. 276–288.

[6] See Sect. 2.1.

[7] See Ibid.

[8] See Sect. 2.3.

[9] See Ibid.

[10] See the conclusion of Chap. 2.

[11] See WTO Appellate Body Report, Canada-Aircraft, WT/DS70/AB/R, adopted on 20 August 1999, para. 187.

member's regulatory autonomy in each case. The objective standard requires panels to develop their own analysis of the disputed issues,[12] consider all evidence, and base their decisions on that evidence.[13] Panels have the discretion to draw reasonable inferences from circumstantial evidence,[14] rather than relying solely on direct evidence, which grants them considerable latitude to question the purpose of a member's regulatory measures.[15]

As a result, panels often judge the intent behind domestic regulatory measures and determine whether they are excessively trade-restrictive in relation to their stated objectives.[16] Additionally, the stringent requirements for scientific evidence in panels' assessments make it even more challenging for members to justify the legality of their SPS measures.[17] Consequently, WTO law can impede the implementation of the UN SDGs by restricting members' ability to adopt and enforce regulations aimed at sustainable development.

I proposed that WTO law should adopt a sustainability test to replace the current necessity test, in order to better achieve SDGs.[18] Unlike the necessity test, which focuses on whether a regulatory measure is essential and minimally trade-restrictive, the sustainability test prioritises sustainable development values over trade values.[19] This means it does not primarily focus on the trade-restrictive effects of a member's domestic regulations.

However, the sustainability test ensures the protection of economic benefits linked to sustainable development, including efforts to eradicate poverty and hunger.[20] It evaluates whether a member's regulatory measures align with the UN SDGs and effectively balance various elements of sustainable development within those measures.[21] The test should encompass all trade-related SDGs and aim to harmonise different sustainable development interests.[22] In essence, the sustainability test mandates that legitimate domestic regulatory measures must achieve one SDG without undermining others.

To conduct the sustainability test effectively, WTO panels must consider all relevant sustainability elements when balancing various sustainable development values. The specific elements that panels need to weigh will vary depending on the

[12] See Sect. 3.2.2.

[13] See WTO Appellate Body Reports, Australia-Tobacco Plain Packaging (Honduras), WT/DS435/AB/R, adopted on 29 June 2020, para. 6.210.

[14] See WTO Appellate Body Report, Canada-Aircraft, WT/DS70/AB/R, adopted on 20 August 1999, para. 198.

[15] See the conclusion of Chap. 3.

[16] See Trachtman (2017), pp. 290–300.

[17] See the conclusion of Chap. 3.

[18] See Sect. 4.1.

[19] See Sect. 4.3.

[20] See Ibid.

[21] See Sect. 4.1.

[22] See Sect. 4.3.

conflicts between different SDGs. I examined several scenarios that panels may encounter as they assess and balance these sustainable development values.

These scenarios involve poverty eradication and inequality reduction, food security and sustainable agriculture, human rights and life, climate change, the Ocean and clean energy, and intellectual property rights.[23] For each scenario, I outlined the specific sustainability elements that panels must consider when evaluating the legitimacy of domestic regulatory measures. If panels were to apply the sustainability test, they would balance conflicting sustainable development interests without compromising any of them, thereby ensuring that members' domestic regulations support all dimensions of sustainable development in a balanced and effective manner.

It is important to note that the existing WTO Agreements do not currently enable WTO panels to conduct a sustainability test. I argued that WTO members should incorporate TSDCs and the proportionality principle required for applying the sustainability test into WTO law. This would provide a legal basis for WTO panels to implement the test effectively.[24] By doing so, international trade regulators, including national governments and international organisations, will apply consistent sustainable trade rules. This multilevel governance approach will help ensure that international trade aligns with the requirements of sustainable development.

It is argued that sustainable development commitments and the proportionality principle necessary for effective multilevel governance are akin to international constitutional norms in the context of international constitutional theory.[25] Incorporating these principles into WTO rules would represent the constitutionalisation of the WTO.[26] Thus, it is essential for WTO members to undertake this process of constitutionalisation to embed sustainability within WTO rules.

Despite significant scepticism surrounding the WTO's constitutionalisation,[27] some members are actively engaged in reforming WTO rules through plurilateral negotiations that encompass the full spectrum of sustainable development—addressing economic, social, and environmental dimensions. Specifically, negotiations focusing on the economic aspects of sustainable development include services domestic regulation, e-commerce, investment facilitation, and MSMEs.[28] These initiatives aim to further liberalise international trade, thereby fostering global economic growth. It is important to recognise that economic prosperity serves as the foundation for achieving the SDGs. There is no doubt that the outcomes of these negotiations will significantly contribute to the eradication of poverty and hunger.

Negotiations on the social dimension of sustainable development are currently focused on gender equality. WTO members engaged in these discussions are

[23] See Chap. 5.

[24] See Jackson (2001), p. 78.

[25] See Sect. 6.1.1.

[26] See Jackson (2001), p. 78.

[27] See Sect. 6.2.

[28] See Sect. 7.1.

working to reform trade rules to eliminate barriers to women's entrepreneurship and participation in international trade.[29] Regarding the environmental dimension, plurilateral negotiations are addressing key issues such as plastic pollution and fossil fuel subsidies.[30] While many of these negotiations are still in their early stages, and their immediate impact may be limited, they are noteworthy. They reflect a growing commitment among certain WTO members to advance the WTO's constitutionalisation by integrating sustainability into its rules. In this context, the trade and environment discussions have effectively established a nascent coalition within the WTO dedicated to furthering this constitutionalisation process.

From the experience of WTO members' plurilateral negotiations, it is evident that club theory can be a valuable tool for advancing the reform of WTO rules. I propose that establishing a sustainable development club within the WTO represents a pragmatic approach to constitutionalising the organisation.[31] By forming such a club, WTO members could incorporate constitutional norms, such as TSDCs and the principle of proportionality, into WTO rules.

Critics argue that plurilateral negotiations undermine the WTO's multilateral framework, raising questions about their legitimacy.[32] Despite these concerns,[33] I contend that the establishment of a sustainable development club is a necessary step for WTO reform.[34] Furthermore, involving public participation can help mitigate the potential drawbacks of plurilateral strategies on the multilateral trading system. In conclusion, while the formation of a sustainable development club will undoubtedly encounter challenges, its potential benefits make it a promising avenue for integrating sustainability into WTO rules.[35]

It can be argued that current development and trade policies, such as those of the EU, along with the sustainable development provisions in several major FTAs, offer valuable insights for shaping the trade rules and policies of a sustainable development club.[36] The EU's development and trade policies, in particular, provide important lessons on how countries can achieve sustainable trade. They elucidate the core values of sustainable development policies, strategies for reforming international trade rules, and methods for advancing sustainability through FTAs.[37] Similarly, mega FTAs provide valuable guidance on integrating national development and trade policies into international trade rules, harmonising disparate trade regulations, and enhancing the implementation of sustainable trade practices.[38]

[29] See Sect. 7.2.

[30] See Sect. 7.3.

[31] See Sect. 8.1.

[32] For example, see Kelsey (2022), pp. 2–24.

[33] See Sect. 8.2.1.

[34] See Sect. 8.2.2.

[35] See Sect. 8.4.

[36] See Sect. 9.1.2.

[37] See Sect. 9.1.3.

[38] See Sect. 9.2.

However, this monograph also identifies limitations in the existing sustainable trade rules, highlighting areas for further improvement.[39]

The book concludes with recommendations on how WTO members can establish a sustainable development club and outlines the follow-up actions required. It suggests that WTO members, such as those involved in TESSD, should first sign the TSDA outside the WTO framework to formally commit to creating this club within the WTO. Subsequently, they should formalise this commitment through a joint ministerial statement.[40]

Additionally, the book provides recommendations on the content of the TSDA. It highlights that the TSDA should institutionalise WTO members' efforts to reform international trade rules, foster common development and trade policies through agreements, and enhance solidarity and cooperation.[41] To achieve this, the TSDA should define sustainable development values, strategies, legal norms, a dispute settlement mechanism, and a review and monitoring process.[42] Following the establishment of the TSDA, its parties should integrate these commitments into the WTO framework and advocate for necessary reforms to WTO rules.[43]

The book offers specific recommendations for integrating sustainability into WTO rules, addressing existing shortcomings. It proposes that TSDA parties promote plurilateral negotiations within the WTO to cover both TSDCs and the application of a sustainability test.[44] The former should include areas such as trade liberalisation, development commitments, subsidies, green trade remedy measures, climate change, environmental protection, ecosystem biodiversity, and labour rights.[45]

If WTO members follow the recommendations in this monograph to reform WTO rules, one would expect to see the formal establishment of a WTO sustainable development club at a Ministerial Conference, accompanied by the initiation of plurilateral negotiations to integrate the UN SDGs into WTO rules. While the reform process will be gradual and may not achieve all objectives by 2030, it will mark the beginning of constitutionalising the WTO and setting the stage for comprehensive WTO rule reforms by 2030. Although this timeline might fall short of ideal, it would still be a significant and welcome advancement. The pursuit of sustainable development will extend beyond 2030, as the UN and governments are likely to continue launching initiatives for the benefit of current and future generations. By creating a sustainable development club and constitutionalising the WTO, members can proactively establish sustainable multilateral trade rules at the earliest opportunity.

[39] See Sect. 9.1.4 and the conclusion of Chap. 9.

[40] See Sect. 10.1.

[41] See Ibid.

[42] See Sects. 10.1.1–10.1.5.

[43] See Sect. 10.2.

[44] See Sect. 10.3.

[45] See Sect. 10.3.1.

Jurisprudence

WTO Appellate Body Report, Canada-Aircraft, WT/DS70/AB/R, adopted on 20 August 1999

WTO Appellate Body Reports, Australia-Tobacco Plain Packaging (Honduras), WT/DS435/AB/R, adopted on 29 June 2020

References

Bibliography

Blewitt J (2008) Understanding sustainable development. Routledge, Abingdon, Oxon and New York

Jackson JH (2001) The WTO 'constitution' and proposed reforms: seven 'mantras' revisited. J Int Econ Law 4(1):67–78

Kelsey J (2022) The illegitimacy of joint statement initiatives and their systemic implications for the WTO. J Int Econ Law 25(1):2–24

Trachtman JP (2017) WTO trade and environment jurisprudence: avoiding environmental catastrophe. Harv Int Law J 58(2):273–309

Documents

UNGA (2015) Transforming Our World: the 2030 Agenda for Sustainable Development. UNGA Res/70/1. 21 October 2015

WTO Appellate Body Report, Canada-Aircraft, WT/DS70/AB/R, adopted on 20 August 19

The manufacturer's authorised representative in the EU is Springer Nature Customer Service Centre GmbH, Europaplatz 3, 69115 Heidelberg, Germany. If you have any concerns regarding our products, please contact ProductSafety@springernature.com

Printed and bound by CPI Group (UK) Ltd, Croydon, CR0 4YY
30/09/2026
02225599-0008